W9-AAF-066

ICE IN THE OCEAN

ICE IN THE OCEAN

Peter Wadhams

Scott Polar Research Institute
University of Cambridge, UK

GORDON AND BREACH SCIENCE PUBLISHERS
Australia • Canada • France • Germany • India • Japan
Luxembourg • Malaysia • The Netherlands • Russia
Singapore • Switzerland

Amsteldijk 166
1st Floor
1079 LH Amsterdam
The Netherlands

British Library Cataloguing in Publication Data

A catalogue record for this book is available from the British Library.

ISBN 90-5699-296-1

To my wife

Maria Pia

whose warmth melts all kinds of ice

"Polar ice cannot be studied as other branches of science, philosophy, medicine or law are studied. The study of sea ice belongs among the most exhausting disciplines which have to be studied on the spot, *in loco nascendi et vitae,* and which require strong men, absolutely sound in mind and body, courageous, willing and fit to renounce all comfort, thoroughly prepared both in theory for the work and ready to face all the hardships that may come like a bolt from the blue and in the most unfavourable moments. The sea-ice has to be studied far away, in the north or south, in white deserts of ice, where there is nothing, nothing else, no shelter, no help; and where, on those vast plains in which the chasms of the sea keep tearing open and insurmountable obstacles in the shape of mountains of ice keep piling up, the Lord only is with man."

J. Zukriegel
Cryologia Maris
The Geographical Institute of the Charles IV
University, Prague, 1935

CONTENTS

PREFACE

The ice-covered seas which surround the poles of our planet occupy about seven per cent of the total area of the world ocean. They are of enormous importance for global climate and human activity, and their study forms an important branch of the science of oceanography, a branch which we might call 'solid state oceanography'. There is even the exciting possibility that something akin to sea ice exists elsewhere in our solar system, on Jupiter's satellite Europa. Yet for the beginning student in oceanography, or for the general scientist or layman interested in the nature of the frozen seas, there has been no simple introductory text on the subject of sea ice although there are several excellent advanced monographs or multi-author works.

The present book offers an introduction to our modern knowledge of sea ice and icebergs and the role that they play in the ocean system. It arose from a course of lectures for final-year geography students that I have given for many years at Cambridge University, but I have gone further into the physics of sea ice processes without making the mathematics too forbidding. It shows an unashamed bias towards phenomena rather than models.

I am most grateful to the National Institute of Polar Research, Tokyo, for giving me the opportunity to write much of this book during two periods as a Visiting Professor, in particular to the Director-General, Dr Takeo Hirasawa, and to my hosts Dr Nobuo Ono and Dr Takashi Yamanouchi. Further thanks are due to a host of people who have helped me during a lifetime spent in the dangerous, exhilarating and thankless task of seeking knowlege of the polar seas by working in them, on them and under them. Firstly my wife Maria Pia, for being Italian and for her active assistance on several field programmes. Then those who helped me get started and who have supported me steadily over the years: the US Office of Naval Research, with its Arctic Program managers Ron McGregor, Leonard Johnson and Tom Curtin; the Royal Navy submarine fleet; the Commission of the European Communities; Walter Munk; Gordon Robin; and Bosko Loncarevic. Then, friends and colleagues who are no longer with us but who were an inspiration at different times: Adrian Gill, George Deacon, Bill Campbell, Malcolm Mellor and Thomas Viehoff. Finally, those numberless friends and colleagues throughout the world who make up the present polar research community, a special kind of people united in the quest for understanding of the polar regions and in love of these most remote parts of our beautiful world. They are too many to name.

For illustrations I am grateful to Natasha Egorova and Anne Howe for original drawings, and to numerous colleagues for permission to reproduce figures.

1. THE FROZEN OCEANS

You are in a world of white. The air temperature hovers at –45 degrees Centigrade. You are clad in a voluminous parka, wind pants and mukluks. Every inch of your face is covered, by goggles and a face mask. The parka hood is drawn tightly around your face in a kind of tube so that the wolverine trimming allows the air to be warmed before you breathe it in. The sun burns low on the horizon through a scintillating suspension of ice crystals in the air, called "diamond dust". All around you lies a pure white landscape of snow, like a snow-covered field, with its gentle hummocks, valleys and occasional rocky lumps softened by the snowy mantle. The field is bordered by what looks like a white stone wall, created by some civilisation long gone, and now fallen into decay. Great white rock monoliths rear into the sky, while others lie at random at their feet and in some places there is a semblance of a regularly laid pattern of blocks. It is a pressure ridge. The snow piled up on its windward side has been carved by the wind into a serrated pattern called sastrugi. The low sun casts a long irregular shadow from the ridge onto the snow. Suddenly a part of the shadow moves and a pale yellow mass separates itself from the random mass of ridge blocks and moves slowly towards you. It is a polar bear. You reach for your red signal flare to frighten him off. Beneath your feet are three metres of sea ice. Beneath that are five thousand metres of water. You are walking on the Arctic Ocean.

1.1. A WORLD OF WHITE

The white world of the polar oceans covers 7% of our planet, an area greater than Europe and North America combined. Few people venture into it, especially in winter. Only recently has it become a possible destination for tourists in icebreakers. But scientists have worked on the polar oceans for many decades, and for a very good reason. The ice-covered seas represent the cold end of the enormous ocean heat engine which enables our planet to have temperatures suitable for human life over most of its surface. Solar radiation received at equatorial latitudes is absorbed by the ocean and transported north and south at a rate of about 2 PW (2×10^{15} W) towards the poles, where it is lost through the sea ice to the atmosphere. The rate of loss is determined by the extent, thickness and consistency of the ice cover, so this thin and variable skin, constantly in motion, serves a vital regulatory function for the climate of planet Earth.

The sea ice cover has other and more subtle effects. As it grows it rejects salt into the ocean, and it is in the polar seas that surface water, made more dense in this way, can sink and so carry oxygen and dissolved carbon dioxide down into the dark abyss. These convection regions, on the Arctic Ocean continental shelves, in the Greenland Sea and Antarctic Ocean, themselves drive a three-dimensional global pattern of deep and shallow currents known as the thermohaline circulation or the "Great Ocean Conveyor Belt" (Broecker, 1995). The oxygen carried down in this way makes life possible in the deep ocean all over the world. The sea ice with its snow cover is also very bright, reflecting 80–90% of the solar radiation falling on it whereas open water reflects only about 10%, and so a shrinking of the sea ice cover due to climate warming would lead to more radiation being absorbed by the earth and to the rate of warming being increased — a positive feedback mechanism which, according to global climate models, enhances the rate of global change.

All of these global effects are due to a material which itself experiences a huge annual cycle of growth and decay, certainly in the Antarctic where almost all of the ice disappears in summer while in the winter its area grows to 20 million square kilometres. The seasonal variation of sea ice extent in both hemispheres, together with the variation of snow cover extent on land in the Northern Hemisphere, represents the biggest and most noticeable seasonal change in the appearance of the Earth as seen from space. For all of these reasons, geophysicists regard sea ice as one of the most important components of the planetary surface and the key to understanding many of the basic questions about the energy balance of the earth.

In this book we introduce some of the properties of the crystalline solid which has created the strange white marine landscapes of the far north and south. However, in the long run there is no substitute for going there yourself. Once you have seen it, you will always want to go back.

1.2. THE PHYSICAL STRUCTURE OF THE POLAR OCEANS

1.2.1. The Topography

Arctic

The Arctic Ocean (fig. 1.1) is like the Mediterranean Sea in that it is an almost closed basin with only one deep passage through which water can be easily exchanged with the

Figure 1.1. The bathymetry of the Arctic Ocean.

rest of the world ocean, the so-called Fram Strait (still an unofficial name, although everybody uses it) between Svalbard and Greenland. A much narrower passage of lesser depth, the Nares Strait between Greenland and Ellesmere Island, allows some connection with Baffin Bay, while there are many shallow links with the Atlantic and Pacific, notably through the Barents Sea, the Canadian Arctic Archipelago and Bering Strait.

The Arctic Ocean has a complex bathymetry. It is divided into two deep basins, the Eurasian Basin and the Canada Basin, by a narrow straight ridge called the Lomonosov Ridge which crosses the whole Arctic from Siberia to Greenland. The basins exceed 4300 m in depth, while typical depths along the ridge crest are about 1000 m; a shallow peak at only 610 m was discovered as recently as 1994. The Lomonosov Ridge is not a true mid-ocean ridge in the sense of a spreading centre where new crust is being created; instead it is thought to be a narrow strip of old continental crust which was split off the edge of Siberia when the Eurasian Basin opened up some 100 million years ago. The true spreading centre, the Arctic Mid-Ocean Ridge, lies along the centreline of the Eurasian

OCEAN DEPTHS

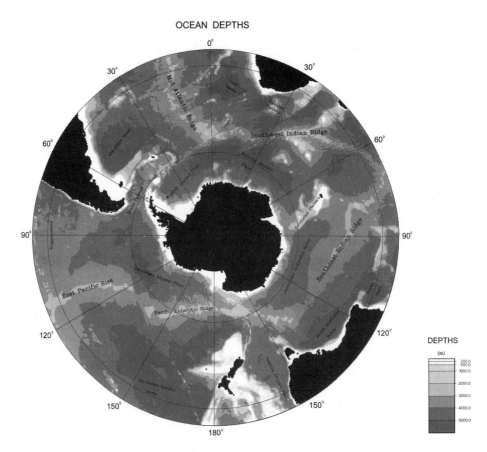

Figure 1.2. The bathymetry of the Antarctic Ocean.

Basin and is much deeper than the Lomonosov Ridge. In the Canada Basin there is a very intricate bottom topography, with the Alpha Ridge (actually a convoluted plateau region) probably representing a very ancient spreading centre from a much earlier period of geological history.

An important characteristic of the Arctic Ocean is the extent of the continental shelves bordering it. About one-third of the ocean area is taken up by shelf seas, of typical depth 100 m or less, and with the widest expanse fringing the north of Russia. The East Siberian Shelf is the widest continental shelf in the world, with water less than 50 m deep extending out 600 km from shore. Groups of islands divide the Russian shelves into a number of separate seas — the Chukchi, East Siberian, Laptev, Kara and Barents. The result is that the Arctic Ocean as a whole has the smallest mean depth (1800 m) of any ocean. The outer edges of the shelves are cut by a series of deep canyons or troughs, which, as we shall see in section 1.2.4, are important in providing conduits for newly-produced dense water in winter to run off the shelves into the deep basin. Fig. 1.1 shows the locations of the most important canyons — Barrow Canyon off Alaska, Herald Canyon in the Chukchi Sea, and the St. Anna and Voronin Troughs in the Kara Sea.

Antarctic

The Antarctic Ocean (fig. 1.2), by contrast with the Arctic, is an intimate part of the world ocean system. In fact, the polar stereographic projection shows that it is really the mother of all oceans, with the Indian, Atlantic and Pacific Oceans being merely northerly off-shoots from this great circumpolar water mass. The zone which carries a sea ice cover in winter (south of about 55°S) contains the southerly parts of the main mid-ocean ridge systems, in the Pacific and Atlantic-Indian Ocean sectors, as well as the Scotia Arc, which connects the Andes to the islands and mountains of the Antarctic Peninsula via a loop of seamounts and islands which include South Georgia, the South Sandwich and South Orkney Islands, and the submerged Burdwood Bank.

The continental shelves around Antarctica are very narrow, except in the Weddell and Ross Sea embayments where wide continental shelves exist under the ice shelves. Along a more typical coastline there is a very narrow continental shelf followed by a deepening to a so-called **shelf basin** and then a sill of 200–300 m depth, this being a terminal moraine deposited when the Antarctic ice sheet was slightly further advanced than it is today. This sill structure may be important in holding shelf waters close to the coast to enable them to mix with intruding warmer deep water, this being a mechanism for Antarctic bottom water production (see section 1.2.4).

1.2.2. The Water Masses

Arctic

When we take a salinity and temperature profile in the Arctic Ocean (fig. 1.3) we see a three layer system. The uppermost layer is called **polar surface water.** It is up to 200 m thick, is at or near the freezing point, and has a very low surface salinity of 30–32 psu, compared to 35 psu as an average for the world ocean (one psu, or practical salinity unit, is equivalent to one part per thousand of dissolved salt). The low salinity occurs because of the very large influx of fresh water into the Arctic Ocean from large river systems such as the Ob', Lena, Yenisei and Mackenzie which drain huge areas of the Asian and North American continents. Fig. 1.4 shows the annual average fresh water fluxes from these rivers, and also shows the watersheds of rivers draining into the Arctic; it is astonishing to note that the Arctic drainage basin includes about half of the Asian continent. Fig. 1.5 shows how surface salinity varies over the Arctic, demonstrating that the lowest values are found near the mouths of these great rivers. The river discharge mixes with sea water over the shallow continental shelves and then spreads out over the central Arctic Ocean as a surface layer, except under special circumstances in early winter associated with sea ice formation (see section 1.2.4).

It can be seen from fig. 1.3 that the depth at which the temperature rapidly increases, called the **thermocline**, does not necessarily coincide with the depth at which the salinity rapidly increases, the **halocline**. Part of the Arctic Ocean, mainly the Beaufort Sea region, is called the **cold halocline** zone, where the deeper part of the surface water layer, in the region of 150–200 m depth, remains cold but already has a salinity which is rising towards its deeper water value (e.g. profile 2). In the Eurasian Basin and Greenland Sea the temperature and salinity rise more or less together (e.g. profile 5). The reason for this is that in the Beaufort Sea the surface water characteristics are affected by an inflow from

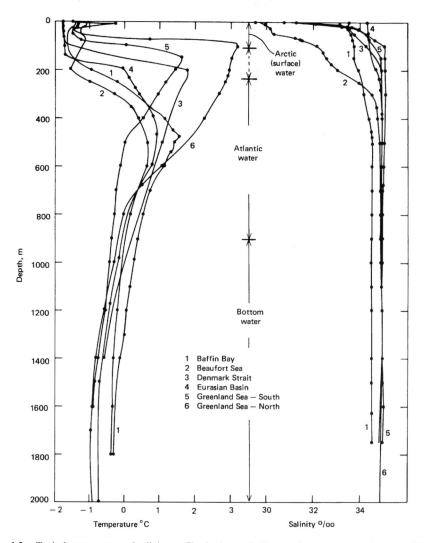

Figure 1.3. Typical temperature and salinity profiles in the Arctic Ocean (after Aagaard and Carmack, 1989). The locations of the numbered profiles are shown in fig. 1.1.

Bering Strait of cold, high salinity water, which enters the Arctic Basin at a depth of 150–200 m as it comes off the Chukchi and Alaskan shelves. The influence of this water extends almost to the Lomonosov Ridge. However, in recent years there is evidence that the cold halocline has retreated far back into the Beaufort Sea (Steele and Boyd, 1998), indicating a diminution of the range of influence of Bering Sea water as opposed to water of Atlantic origin. This is one of several major changes which are going on in the marine climate of the Arctic.

Below the polar surface water lies the **Atlantic water** layer, which is remarkably warm (about 3°C) and saline (about 35 psu). It has entered the Arctic partly through Fram Strait

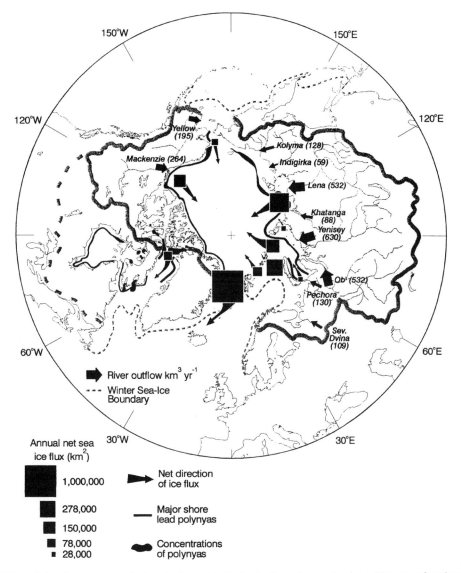

Figure 1.4. Annual average fresh water fluxes into the Arctic Ocean from major rivers. The areas of northern hemisphere land surface draining into the Arctic area also shown (after Macdonald, in press).

(figures 1.7, 1.9), where a warm current, the North Atlantic Current (a continuation of the Gulf Stream) runs up the west side of Spitsbergen as the West Spitsbergen Current and then sinks as it encounters the less dense polar surface water and spreads around the Arctic at mid-depths. This current follows the shelf break and skirts the western edge of the Barents Sea as it moves north. However, some of the water moves in across the Barents Sea, and then enters the Arctic Ocean further to the east via the St. Anna and Voronin troughs, also contributing to the Atlantic water layer.

Figure 1.5. Variation of surface salinity over the Arctic Ocean. Values given in psu.

From about 900 m depth to the bottom, the water temperature has sunk below 0°C again and continues to slowly decrease with increasing depth. This large water mass is called the **Arctic Ocean deep water**. It is again of Atlantic origin, but suffers much modification in its very sluggish circulation around the basin (Rudels, 1995). Because of the sill depth of the Lomonosov Ridge, the deep water of the Eurasian Basin is slightly different in its characteristics from that of the Canada Basin, and also from the deep water of the Greenland Sea, although it is capable of mixing with the latter via the deep part of Fram Strait.

Antarctic

Antarctic water masses are more complex because of the unconfined nature of the Southern Ocean. Fig. 1.6 shows a typical slice stretching from the Antarctic continent northwards. Again there is a cold **Antarctic surface water** mass, with a low salinity

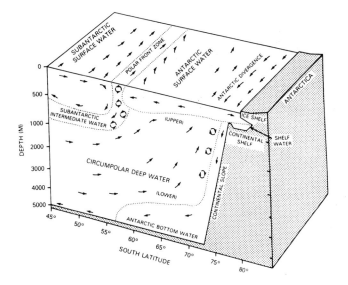

Figure 1.6. Water masses of the Antarctic Ocean (after Gordon and Goldberg, 1970).

maintained by the melt of icebergs and ice shelves and by the excess of precipitation over evaporation in the Southern Ocean. However, the salinity is higher than in the Arctic because the fresh water inflow is much less, and so the boundary between the surface water and the underlying water masses — the so-called **pycnocline** (density discontinuity) — is less steep than in the Arctic and so has less effect in preventing vertical mixing and heat transfer. The surface water lies at the freezing point in winter, when it carries an ice cover, but warms during the ice-free summer months.

The meridional extent of the Antarctic surface water is not limited by the presence of land masses, as it is in the Arctic, but instead it reaches a natural boundary with the warmer surface water masses which dominate lower-latitude regions. This water mass boundary is known as the **Antarctic Convergence** or Antarctic Polar Front. Its mean position is shown on the current chart of fig. 1.10 as a bold line. It lies at about 50°S around most of Antarctica, but lies nearer to 60°S in the New Zealand — Pacific sector then moves diagonally north-eastward through Drake Passage.

Below the surface water is the **Circumpolar Deep Water**, again a warm and more saline water mass originating at lower latitudes. Finally, there is a dense water mass near the seabed known as **Bottom Water**, which is produced by a complex mixing process on shelves between polar surface water and circumpolar deep water, with an additional contribution due to cooling under ice shelves and the addition of salt from sea ice formation. This is discussed in section 1.2.4.

1.2.3. The Current Systems

Arctic

As a long-term average, the motion of ice and surface water in an ice-covered ocean are the same. The surface current system in the Arctic (fig. 1.7) therefore applies both to ice

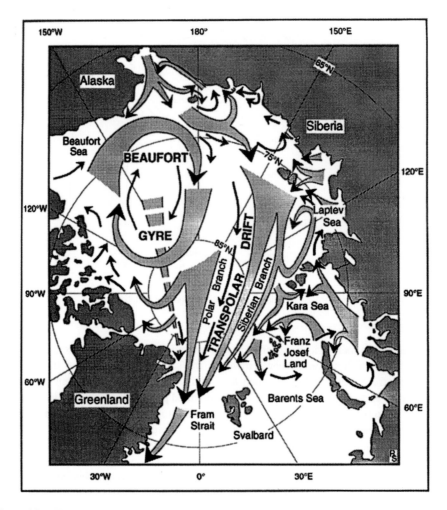

Figure 1.7. The long-term average surface circulation of ice and water in the Arctic Ocean.

and water. It is largely wind-driven and consists of an anticyclonic (clockwise) gyre in the Canada Basin, the **Beaufort Gyre**, and a motion of translation, the **Trans Polar Drift Stream,** in the Eurasian Basin. Ice in the Beaufort Gyre requires 7–10 years for a complete circuit. The Trans Polar Drift Stream collects ice and water from the Eurasian shelves and transports it across the Pole and down towards Fram Strait, requiring about three years for this drift. The Drift Stream is renamed the **East Greenland Current** after it passes through Fram Strait and enters the Greenland Sea. This is the route for most of the ice which leaves the Arctic Basin and so it is through Fram Strait that most water and heat exchange occur between the Arctic Ocean and the rest of the world ocean. Much of this heat exchange occurs in the form of latent heat transported northward as the ice moves southward. The wind system which drives this flow arises because of the presence of the polar high over the centre of the Beaufort Sea, with a ridge extending over Greenland, although again this appears to have changed in recent years (see section 8.3.5).

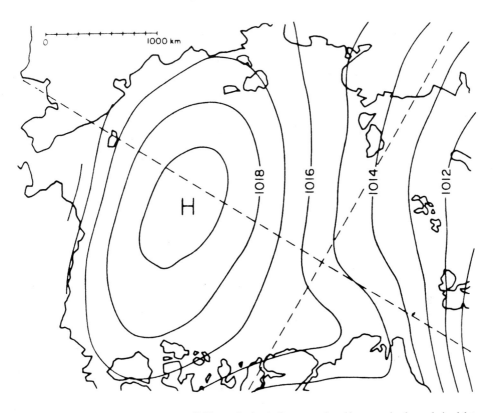

Figure 1.8. Long-term average pressure field over the Arctic Ocean, produced by averaging buoy-derived data for five years.

Figure 1.8 shows the long-term average pressure field over the Arctic Ocean. This contour map was generated by averaging over five years the air pressures measured by the array of drifting satellite-tracked buoys which have been deployed over the Arctic Ocean by the Arctic Ocean Buoy Program (now known as the International Arctic Buoy Programme, administered in Oslo), which has done so much to tell us about the motion of ice in the Arctic. The map shows a high-pressure centre over the Beaufort Sea, at about 80°N, 140°W, with a ridge of high pressure extending over the high, intensely cold ice sheet of northern Greenland. Annual or monthly average pressures can differ greatly from this simple pattern.

A rule of thumb for ice drift in unconstrained conditions, developed independently by Nansen and Zubov and thus known as the *Nansen Rule* or the *Zubov Law*, is that ice moves parallel to the isobars of a surface pressure field. Thus the pressure distribution of fig. 1.8 should give rise to a clockwise circulation of ice in the Beaufort Sea, and a current on the Eurasian side of the Arctic which moves ice from the seas north of Russia across the North Pole and down towards the entrance to the Greenland Sea. This is just what is observed. An extra, but minor, contribution to this circulation comes from the effect of the Earth's rotation on the surface water of the Arctic Basin. The water forms a low-density surface lens which tries to slump outward (i.e. southward) under centrifugal force, but is turned to the right by Coriolis force to give a clockwise rotation, which is then split by the presence of Greenland into a clockwise gyre and a motion of translation, enhancing the pattern of fig. 1.7 (Wadhams *et al.*, 1979).

The great current system of the Trans Polar Drift Stream and East Greenland Current is the chief means by which ice and cold water are exported from the Arctic Ocean. Fram Strait is therefore the Arctic's great two-way water highway, because the main warm water input into the Arctic occurs up the eastern side of the Strait. The famous whaling captain and Arctic scientist William Scoresby was the first to postulate the existence of a transpolar current because of his observations of great masses of old ice passing into the Greenland Sea from the Arctic Basin, but it was Fridtjof Nansen (1861–1930) who put the hypothesis to the test by freezing his specially constructed ship "Fram" into the ice off the New Siberian Islands in 1893, hoping to drift across the Pole. The ship missed the Pole, but in 1896 emerged in what is now known as Fram Strait. Nansen was inspired to carry out this drift by the discovery off south Greenland of wreckage from the exploration ship "Jeanette", which had been crushed in the ice off Wrangel Island north of Siberia.

As fig. 1.7 shows, other current systems in the Arctic do not carry much water to or from the Basin because they occur in shallow water. There is a net influx into the Basin through Bering Strait, and in the Canadian Archipelago a net eastward flux of water from the Beaufort Sea towards Baffin Bay. This is mainly driven by the pressure head between the Pacific Ocean (which stands higher) and the Atlantic. In Baffin Bay itself a northward current, the **West Greenland Current,** carries cold water and icebergs up the western side of Greenland, translating into the southward flowing **Baffin Island Current** down the east coast of Baffin Island, with an addition of polar water out of the Arctic Basin brought by a southward current through the very narrow but deep Nares Strait. The Baffin Island Current in turn passes its ice or iceberg burden on to the cold **Labrador Current** which carries it down as far as the Grand Banks of Newfoundland, where the cold water has a sharp front with the warm water of the Gulf Stream.

The intermediate and deep currents in the Arctic Basin are much more difficult to measure, and appear to be quite different in sense from the surface currents. Fig. 1.9 shows some present ideas about their nature, which are described in greater detail by Aagaard and Carmack (1994) and Rudels (1995). Briefly, the warm Atlantic water moves north up the coast of Norway, then part of it continues along the shelf break to Svalbard as the West Spitsbergen Current and part moves eastward over the shallow Barents Sea and eventually enters the Arctic Basin from the Kara Sea at the St. Anna Trough. The Svalbard branch sinks NW of Svalbard, and part of the water turns west and recirculates back through Fram Strait as a lower part of the East Greenland Current. The rest turns eastward and follows the northern edge of the Russian shelf, joined by the Barents Sea branch, until the water reaches the beginning of the Lomonosov Ridge at the edge of the Laptev Sea. Here much of the water turns northward and follows the nearside (i.e. European side) of the Lomonosov Ridge, ending north of Greenland where it joins the lower part of the Trans Polar Drift — East Greenland Current system. The rest of the water crosses the Lomonosov Ridge and continues to follow the shelf break until it is joined by water of Bering Strait origin; the combined water masses continue eastward around the edge of the Beaufort Sea shelf, joining the other water mass north of Greenland. The mixing of water masses in the Beaufort Sea causes subsurface eddies to form in this region, shown in the diagram. Since deep currents are difficult to measure, this picture is still very

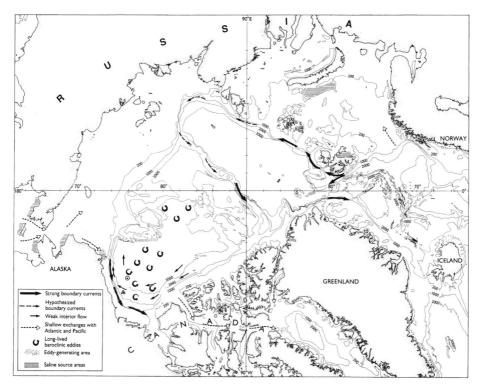

Figure 1.9. Intermediate and deep currents in the Arctic Basin (after Aagaard and Carmack, 1994).

speculative. The two broad aspects of the picture are clear, however: a net flow from the Pacific to the Atlantic, which is in any case required because the Pacific Ocean stands higher than the Atlantic; and the eventual recirculation back into the Atlantic of Atlantic water entering from the North Atlantic Current — subject, however, to many modifications and vicissitudes en route around the Arctic Basin. The tendency of the water to hug shelf breaks, keeping the shelves on its right hand, is of course a consequence of Coriolis force.

Antarctic

The surface circulation in the Antarctic, again largely wind-driven (fig. 1.10), consists of an enormous eastward flowing current, the **Antarctic Circumpolar Current**, or West Wind Drift, which is one of the greatest current systems in the world, transporting some 100–150 million cu m of water per second (100–150 Sverdrups). Near the coast there is a westward flowing current, the **Antarctic Coastal Current,** transporting some 8 Sv, which governs the inital drift direction of icebergs which calve from the continent. Finally there is evidence of gyral circulations in the Weddell and Ross Seas. Again, these motions are largely driven by the wind field.

Figure 1.11 shows an instantaneous pattern of surface pressure over the Antarctic Ocean and a longer-term average. The immediate pattern is familiar to anyone who sails

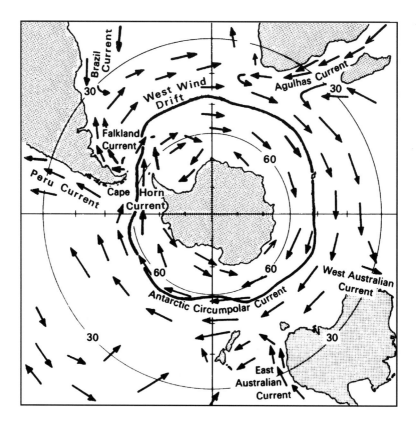

Figure 1.10. Surface current patterns in the Antarctic Ocean.

in the Southern Ocean; a sequence of five or six low-pressure storm centres are constantly found over the ocean, so that a ship will experience a new storm coming from the west every few days. The storm centres form and collapse, and when the pressure fields are averaged over the long term, a more stable pattern emerges of a prevailing westerly geostrophic wind over most of the Antarctic Ocean, but with remaining low-pressure centres offshore north of the Antarctic coastline (fig. 1.11b). Application of the Nansen rule shows that the resulting pattern of wind-driven motion is the familiar Antarctic Circumpolar Current over most of the Southern Ocean, with the coastal easterly winds due to the offshore low pressure centres being responsible for the Antarctic Coastal Current. A complementary, and more local, way of imagining the Antarctic Coastal Current is that winds blow outwards from the high-pressure centre over the Antarctic continent, are turned to the left by Coriolis force, and are further accelerated as they move down the slope of the domed Antarctic ice sheet, producing locally strong easterly winds (**katabatic winds**) near the coast.

The location in the ocean at which the influence of the Antarctic Coastal Current gives way to that of the Circumpolar Current is called the **Antarctic Divergence**. It is located about half-way between the coast and the Antarctic Polar Front. It is rightly called a

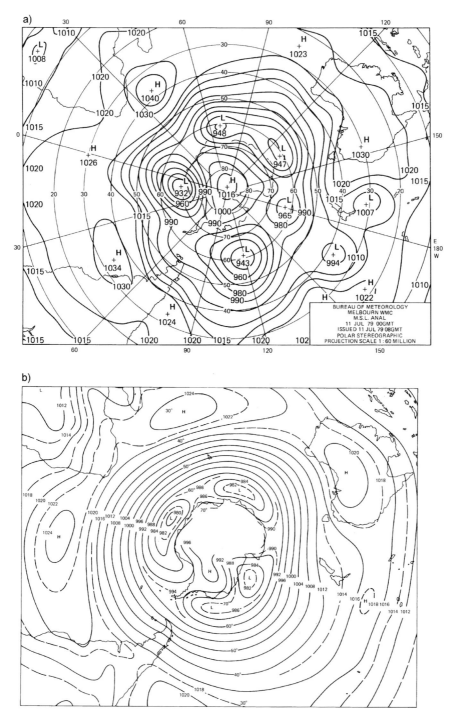

Figure 1.11. An instantaneous and a long-term average surface pressure field for the Antarctic (after Deacon, 1984).

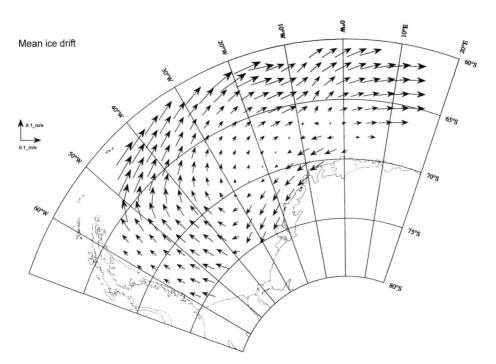

Figure 1.12. Long-term average ice motion in the Weddell Sea, from the buoys of the International Programme for Antarctic Buoys (after Kottmeier *et al.*, 1997).

divergence because the theory of wind-driven surface currents developed by Ekman shows that a westward current in the Southern Hemisphere implies a net *southward* transport of water (since the current direction turns with increasing depth), while an eastward current implies a net *northward* transport. The boundary between two such currents therefore has a horizonal outflow of water in both directions, implying that upwelling must be occurring to preserve continuity. In fact, it is at the Antarctic Divergence that we sometimes see lower ice concentrations in winter, suggesting that warmer water is reaching the surface from below, and this has been suggested as one mechanism for the existence of the so-called Weddell Polynya, which we consider in section 1.3.

In the Ross Sea and Weddell Sea there are partial gyres, caused by the Coastal Current coming up against a zonal barrier and being diverted out to sea so that it joins the Circumpolar Current. In the case of the Weddell Sea there have been sufficient drifting ice buoys deployed by the International Programme for Antarctic Buoys (IPAB) to establish that the gyre is not closed at its eastern end, so that it is unlikely that ice can continually recirculate in the way that it does in the Beaufort Gyre. Fig. 1.12 shows the long-term average ice (and hence surface water) motion from all relevant IPAB buoys.

The deep flow in the Antarctic is shown schematically in fig. 1.13. Near the sea bed the Antarctic Bottom Water flows northward from its formation regions, along the deeper parts of the ocean basins, seeking passages between ocean ridge systems. As a result, it

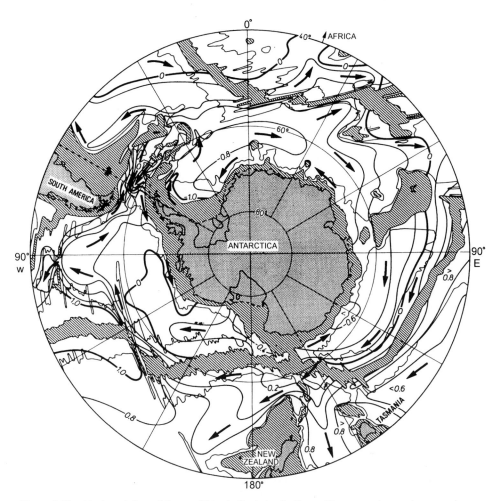

Figure 1.13. Northward flow of Bottom Water in the Antarctic Ocean. Deep ocean temperatures are shown, and arrows give direction of flow.

is detectable at some longitudes as far north as the Equator. The formation of this water type is discussed in the next section.

1.2.4. Water Mass Transformations

The polar oceans are not only a heat sink for our planet, but also play an important role in maintaining the vertical circulation of the ocean, that is in transferring surface water, with all its useful components such as dissolved oxygen, into the deep ocean, so that elsewhere deep water with other useful components such as nutrients can rise to the surface. This vertical circulation is called the global **thermohaline circulation** because, unlike the wind-driven surface circulation, it is driven by heat and salt differences between different parts of the ocean. When surface water sinks it is said to **ventilate** the deep ocean, i.e. it adds oxygen to the deep water, although it also carries down dissolved carbon dioxide, providing a sink for this climatically active gas. We will mention here some processes by which ice-covered oceans contribute to the global thermohaline circulation, by virtue of the salinity changes induced by freezing and melting.

Greenland Sea convection

There are very few locations where ventilation occurs through open ocean convective activity, not associated with coastal processes. In the Northern Hemisphere they are the Labrador Sea, the western Mediterranean and the Greenland Sea, all occurring in winter. In the first two cases intense atmospheric cooling increases the surface water density to the point where overturning and sinking can occur. In the third, sea ice is involved. A full description of Greenland Sea convection is given in section 6.7, but we introduce the topic here.

The convection occurs in the centre of the Greenland Sea gyre, at about 74–75°N, 0–5°W. This region is bounded to the west by the southward-flowing East Greenland Current, and to the east by the warm northward-flowing Norwegian Atlantic Current. Its boundary to the south is a cold current which diverts from the East Greenland Current at about 72–73°N because of bottom topography and wind stress. This is called the Jan Mayen Polar Current, and in winter it develops its own local ice cover of **frazil** and **pancake** ice (described further in section 2.3.2), forming a tongue-shaped feature called **Odden**. These types of ice can grow very quickly, with most of the brine content of the freezing sea water rejected back into the ocean. The salinity increase caused by brine rejection triggers overturning of the surface water and the formation of convective plumes which carry surface water down through the pycnocline into the intermediate and deep layers. Of course, over a whole year ice formation and ice melt balance out so that the net overall salt flux is zero. However, the ice formation and melt regions are geographically separated — the ice growth occurs on the western side of Odden while the ice formed is moved eastward by the wind to melt at the eastern, outer edge of the ice feature — so there is a net positive salt flux in a zone which is found to be the most fertile source of deep water. Evidence from recent hydrographic and tracer studies has shown that convection in this region has become weaker and shallower in recent years, corresponding to a decline in ice formation within Odden. It is an active research area to determine whether this is a trend deriving from global warming, or a cyclic effect associated with a particular pattern of wind field over the Greenland Sea. In section 8.3.5 we consider these climate change implications together with other Arctic changes.

Shelf-slope convection in the Arctic

At the end of summer the sea ice cover has retreated from much of the area of the Siberian shelves (see fig. 1.16 maps for August and September), leaving an extensive, though interannually very variable, area of open water against the coast. This permits icebreaker-assisted marine transport in summer along the so-called Northern Sea Route. As autumn begins and the air temperature drops significantly below zero, the water loses the heat that it has absorbed during the summer and freezing starts. However, it is seldom the case that calm conditions prevail throughout the freeze-up period so that a smooth first-year ice sheet can form. More usual is a sequence of storms, or windy periods. In shallow water a storm creates waves and turbulence which reach right down to the bottom, which helps transfer oceanic heat to the surface and so enhances the rate of cooling. The turbulence at seabed level also suspends sediment particles into the water column, from which they can become incorporated in young ice as it grows at the surface, a process called **suspension freezing**. This process is discussed in section 8.2.2 since it provides a route for pollutants, carried downstream in Siberian rivers and deposited at their mouths, to become incorporated in sea ice and transported around the Arctic.

The most important role of the early winter storms, however, is to force much of the earliest freezing on the shelves to occur in the form of frazil ice, like the case of Odden described above. Frazil ice is a suspension of ice crystals in water. These are especially good at scavenging sediment particles out of the water column, either as nuclei for the ice crystals or adhering to the crystals. The frazil suspension also leaves the sea surface exposed to rapid heat loss, rather than forming an insulating sheet as is the case with young ice forming in calm water. This permits a continuing high freezing rate, especially if the frazil is itself transported northward out of the zone of formation, a mechanism which occurs in coastal polynyas (see below) – in this sense the freezing shelf seas can be thought of as very wide coastal polynyas. The rapid freezing rate causes salt to be rejected back into the water column so that the shelf waters quickly acquire new water mass characteristics, a much higher salinity associated with a freezing temperature throughout. Such a water mass can become much denser than the surface water in the Arctic Ocean beyond the shelf break, and therefore finds ways to move down off the shelf and inject itself into the Arctic Ocean at some depth commensurate with its density. The canyons at the edge of the shelves, e.g. the St. Anna and Voronin Troughs discussed above, may provide easy routes for this transport. The whole process is called **shelf-slope convection** (Rudels, 1995) and is responsible for many of the characteristics of intermediate and deeper waters of the Eurasian Basin.

Antarctic Bottom Water formation

Antarctic Bottom Water is an important component of the Southern Ocean system, responsible for about 34 Sv of northward flow, and providing cold water at depth to all the oceans of the southern hemisphere. Its formation involves a mixing of water types, helped by surface interaction. An example of how this happens on a local scale is the case of bottom water formed off Adélie Land, at about 132°E, which was investigated by Gordon and Tchernia (1972) and later Rintoul (1998). Fig. 1.14 shows temperature and salinity sections across the shelf and slope of the region in summer. There is a narrow shelf with a characteristic shelf basin, as mentioned in section 1.2.1, with the deepest part of the shelf, the Dibble Depression at 700 m, being separated from the ocean by a 200 m sill. The water on the shelf is very cold, near freezing throughout, and has a relatively high salinity, although lower than the water outside the shelf at the same depth. Its properties were formed in the previous winter by sea ice production and brine rejection, which would have been enhanced if katabatic winds had kept open a winter polynya against the coast (see next section). The temperature section shows warm circumpolar deep water approaching from the left (with a core at 400–800 m depth) and mixing with

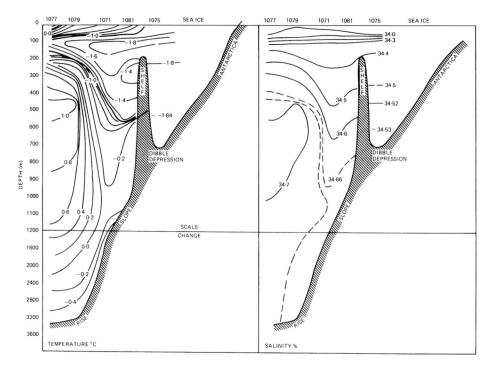

Figure 1.14. Temperature and salinity sections across the Antarctic shelf and slope at 132°E in summer, showing mixing leading to bottom water formation (after Gordon and Tchernia, 1972).

the shelf water just outside the sill, producing cold, saline water which is sinking close up against the slope.

The fact that a mixing process can produce this effect was first noted by Fofonoff (1956). He showed how the non-linear density properties of sea water lead to the possibility of a warm high-salinity water mass mixing with a cold low-salinity water mass to produce a third water mass which is more dense than either of its constituents, a seemingly counter-intuitive result. This process is called **cabelling**, and is clearly what is occurring in fig. 1.14. Further additions of salt, from active sea ice formation, can only enhance the mixing process.

It has been found that the most prolific source of Antarctic Bottom Water is the huge Weddell Sea shelf, where possibly 50% of the total volume is produced (Gill, 1973; Reid *et al.*, 1977). Second to this is the Ross Sea shelf (Jacobs *et al.*, 1970, 1985), with some sites in the sector from 30°E to 170°E also being possible sources (Baines and Condie, 1998). The source areas, and the directions of spread of the bottom water, can be identified by the "signature" of water temperature at depths below 4000 m (fig. 1.13), where the coldest temperature in any sector identifies a source region.

Water mass modification in coastal polynyas

Antarctic Bottom Water formation, and the necessary preconditioning process of water

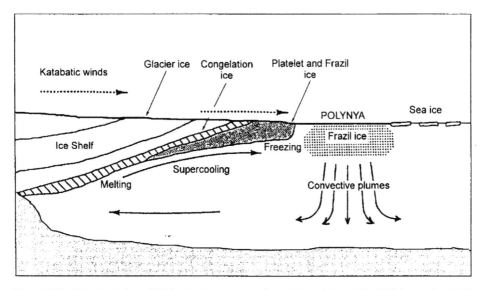

Figure 1.15. The circulation which is thought to occur under an Antarctic ice shelf which is associated with a coastal wind-driven polynya.

mass modification on the shelves and shelf basins, are helped by the presence along the Antarctic coastline of **coastal polynyas** at certain well-defined sites. These are areas of open water, typically only a few tens of km in diameter, kept free of a continuous ice cover in winter by powerful katabatic winds which funnel down off the Antarctic ice sheet between mountain peaks and which blow newly-forming ice out to sea. As in the case of Siberian shelves, the sea surface is thus kept open for new ice production. These regions are thus "ice factories" for the Antarctic. Coastal polynyas in both polar regions are discussed in further detail in section 2.10, and their ice dynamics are analysed in section 4.4.

In terms of water mass modification, the main role of Antarctic coastal polynyas is to enhance the rate of salinity increase of shelf water during winter, which promotes the formation of very dense mixtures by subsequent cabelling. When the coastal polynya occurs at the edge of an ice shelf, however, there is also an interaction with the ice shelf itself. The circulation which is surmised to occur is shown schematically in fig. 1.15. Intense cooling in the open water seaward of the ice shelf front causes ice production, mainly as frazil ice, which is blown out to sea by katabatic winds to keep the polynya open, ensuring continued unhampered ice production. The salt rejected into the surface water causes convective plumes to sink towards the sea floor and from there the water flows inwards under the ice shelf. Where it first encounters the ice shelf, near its grounding line, the water causes melting of the ice shelf bottom. This dilutes the circulating water, which flows outwards and upwards along the bottom of the ice shelf, reaching a lesser depth where the pressure-induced depression of freezing point is less so that freezing can occur. It is thought therefore that a layer of sea ice forms on the underside of the ice shelf near its front, but this has yet to be confirmed by observations.

1.3. THE DISTRIBUTION OF ICE IN THE OCEANS

1.3.1. The Arctic

We now introduce the sea ice itself, the subject of this book. The best way of surveying its extent and variability is by the use of satellite imagery, and the most useful imagery on the large scale has been passive microwave, which identifies types of surface through their natural microwave emissions, a function of surface temperature and emissivity. Figure 1.16 shows ice extent and concentration maps for the Arctic for each month, averaged over the period 1979–1987, derived from the multifrequency SMMR sensor (scanning multichannel microwave radiometer) aboard the Nimbus 7 satellite (Gloersen *et al.*, 1992). This instrument, described in section 1.4, gives ice concentration and, through comparison of emissions at different frequencies, the percentage of the ice cover that is multi-year ice, i.e. ice which has survived at least one summer of melt. The ice concentrations are estimated to be accurate to ±7%.

It can be seen that at the time of maximum advance, in February and March (fig. 1.16a), the ice cover fills the entire Arctic Ocean. The Siberian shelf seas are also ice-covered to the coast, although the warm inflow from the Norwegian Atlantic Current keeps the western part of the Barents Sea open. There is also a bight of open water to the west of Svalbard, kept open by the warm West Spitsbergen Current and formerly known as Whalers' Bay because it allowed sailing whalers to reach high latitudes. It is here that open sea is found closest to the Pole in winter — beyond 81° in some years. The east coast of Greenland has a sea ice cover along its entire length (although in mild winters the ice fails to reach Cape Farewell); this is the ice transported out of Fram Strait by the Trans Polar Drift Stream and advected southward in the East Greenland Current, the strongest part of the current (and so the fastest ice drift) being concentrated at the shelf break. The Odden ice tongue at 72–75°N can be seen in these averaged maps as a distinct bulge in the ice edge rather than a tongue, and visible from January until April with an ice concentration of 20–50%.

Moving round Cape Farewell there is a thin band of ice off West Greenland (called the "Storis"), which soon meets the dense ice cover of Baffin Bay and Davis Strait. The whole of the Canadian Arctic Archipelago, Hudson Bay and Hudson Strait are ice-covered, and the Davis Strait ice edge leads smoothly down until it joins the ice stream of the Labrador Current, carrying ice southward towards Newfoundland. The southernmost ice limit of this drift stream is usually the north coast of Newfoundland, where the ice is separated by the bulk of the island from an independently-formed ice cover filling the Gulf of St. Lawrence, with the ice-filled St. Lawrence River and Great Lakes behind. Further to the west a complete ice cover extends across the Arctic coasts of NW Canada and Alaska and fills the Bering Sea, at somewhat lower concentration, as far as the shelf break. Sea ice also fills the Sea of Okhotsk and the northern end of the Sea of Japan, with the north coast of Hokkaido experiencing the lowest-latitude sea ice (44°) in the northern hemisphere.

In April the ice begins to retreat from its low-latitude extremities. By May the Gulf of St Lawrence is clear, as is most of the Sea of Okhotsk and some of the Bering Sea. The Odden ice tongue has disappeared and the ice edge is retreating up the west coast of Greenland. By June the Pacific south of Bering Strait is ice-free, with the ice concentration reducing in Hudson Bay and several Arctic coastal locations. August and September (fig. 1.16b) are the months of greatest retreat, constituting the brief Arctic summer. During these months the Barents and Kara Seas are ice free as far as the shelf break, with the Arctic pack retreating to (and occasionally beyond) northern Svalbard and Franz Josef Land. The Laptev and to some extent the East Siberian seas are ice-free, allowing ship passage across the north of Siberia except for the effect of the Taimir Peninsula which ends in the Vilkitsky Strait south of Severnaya Zemlya. This allows marine transport through a Northern Sea Route across the top of Russia, but with a need for icebreaker escort through the central ice-choked region. In East Greenland the ice has retreated northwards to about 72–73° (a latitude which varies from year to year), while the whole system of Baffin Bay, Hudson Bay and Labrador is ice free. Occasionally a small mass of ice, called the "Middle Ice", remains at the northern end of Baffin Bay. In the Canadian Arctic Archipelago the winter fast ice which filled the channels usually breaks up and partly melts or moves out, but in some years ice remains to clog vital channels, and the Northwest Passage is not such a dependably navigable seaway as the Northern Sea Route. There is usually a slot of open water across the north of Alaska, but again in some years the main Arctic ice edge moves south to touch the Alaskan coast, making navigation very difficult for anything but a full icebreaker.

By October new ice has formed in many of the areas which were open in summer, especially around the Arctic Ocean coasts, and in November – January there is steady advance everywhere towards the winter peak. The Sea of Okhotsk acquires its first ice cover in December, and the Odden starts to appear; Baffin Bay and Hudson Bay are already fully ice-covered.

This cycle varies in detail from year to year, and there is evidence from a continuation of the passive microwave record to the present day that a steady decrease of the overall ice extent in the Arctic has been taking place (see section 8.3.5) for every season of the year. However, the annual cycle of Arctic sea ice extent appears to follow a remarkably similar shape from year to year, at least during the SMMR period (fig. 1.17a — "extent" here is defined as the total area of sea within the 15% ice concentration contour). This made it possible to derive an averaged seasonal cycle (fig. 1.17b), which gave a maximum extent of 15.7×10^6 km^2 in late March, and a minimum of 9.3×10^6 km^2 in early September. For sea ice area, derived as extent multiplied by concentration, the figures were 13.9 and 6.2×10^6 km^2 in winter and summer.

Since the emissivities of first-year and multi-year ice differ, and vary with frequency, it is possible to use combinations of frequencies in SMMR to derive algorithms which give multi-year ice fraction for the winter months (October – April). In summer the presence of surface water eliminates the differences in emissivity and the method fails. Results show that in the Arctic multi-year is found in the highest concentrations within the central Arctic Ocean, in the area controlled by the Beaufort Gyre. This is not surprising, since the area is permanently ice-covered, and floes circulate on closed paths which take 7–10 years for a complete circuit. It was found (Gloersen et al., 1992) that multi-year fractions of 50–60% are typical for the Gyre region, rising to 80% in the very centre. Multiyear fractions of 30–40% are found in the part of the Trans Polar Drift Stream fringing the Beaufort Gyre, while in the rest of this current and in peripheral areas of the Arctic the multi-year fraction is 20% or less.

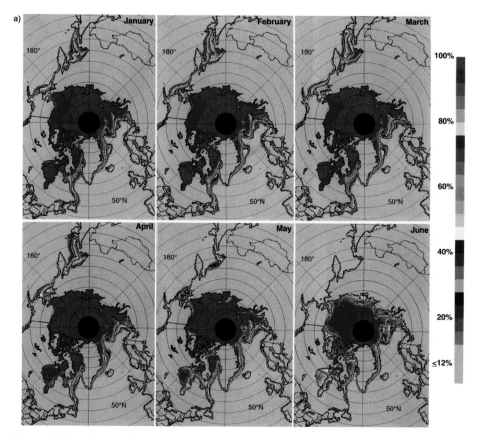

Figure 1.16(a). Northern Hemisphere sea ice extent during January–June, from passive microwave imagery (after Gloersen *et al.*, 1992).

1.3.2. The Antarctic

The sea ice cover in the Antarctic is one of the most climatically important features of the southern hemisphere. Its enormous seasonal variation in extent greatly outstrips that of Arctic sea ice, and makes it second only to northern hemisphere snow extent as a varying cryospheric feature on our planet's surface. Figure 1.18 shows monthly averaged sea ice extent and concentration maps for the Antarctic, derived in the same way as fig. 1.16 from SMMR passive microwave data (Gloersen *et al.*, 1992), and covering the same period 1979–1987.

With the seasons reversed, the maximum ice extent occurs in August and September. It can be seen that the maximum ice cover is circumpolar in extent. Moving clockwise, the ice limit reaches 55°S in the Indian Ocean sector at about 15°E, but lies at about 60°S around most of the rest of East Antarctica, then slips even further south to 65°S off the Ross Sea. The edge moves slightly north again to 62°S at 150°W, then again shifts southward to 66°S off the Amundsen Sea before moving north again to engulf the South

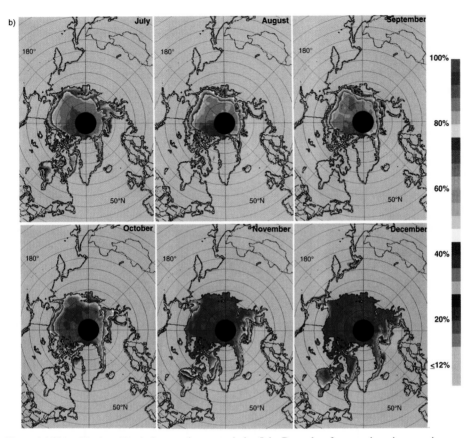

Figure 1.16(b).   Northern Hemisphere sea ice extent during July–December, from passive microwave imagery (after Gloersen *et al.*, 1992).

Shetland and South Orkney Islands off the Antarctic Peninsula and complete the circle. The zonal variation in latitude of this winter maximum therefore amounts to some 11°, which is not negligible. It has been shown (Zwally *et al.*, 1983) that the winter advance of the ice edge follows closely the advance of the 271.2°C isotherm in surface air temperature (freezing point of seawater) and almost coincides with this isotherm at the time of maximum advance. The ice limit is therefore mainly determined thermodynamically, with the gross zonal variations in the winter ice limit matching zonal variations in the freezing isotherm (due to the distribution of continents in the southern hemisphere). Smaller-scale variations in the maximum ice limit may be related to deflections in the Antarctic Circumpolar Current as it crosses submarine ridges. We note that within the ice limit the ice concentration is generally less than the almost 100% concentration found in the Arctic Ocean in winter. Even in the areas of greatest concentration, the central Weddell and Ross seas, it is only in the range 92–96%, while there is a broad marginal ice zone facing the open Southern Ocean over which the concentration steadily diminishes

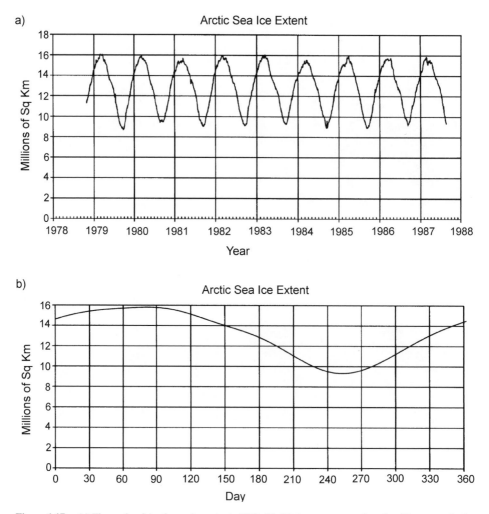

Figure 1.17. (a) The cycle of Arctic sea ice extent, 1979–87. (b) Average seasonal cycle of ice extent in the Arctic (after Gloersen *et al.*, 1992).

over an outer band of width 200–300 km. We shall see in chapter 2 that this is a zone over which the advancing winter ice edge is composed of pancake ice.

Ice retreat begins in October and is rapid in November and December. Again the retreat is circumpolar but has interesting regional features. In the sector off Enderby Land at 0–20°E a large gulf opens up in December to join a coastal region of reduced ice concentration which opens in November. This is a much attenuated version of a winter polynya which was detected in the middle of the pack ice in this sector during 1974–6 (Zwally *et al.*, 1976, 1983; Zwally and Gloersen, 1977) but which has not been seen as a full open water feature since that date. It was known as the Weddell Polynya

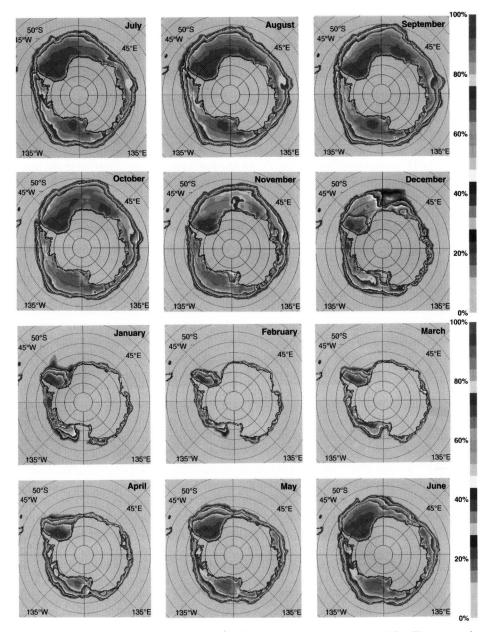

Figure 1.18. Southern Hemisphere sea ice extent, from passive microwave imagery (after Gloersen *et al.*, 1992).

and lay over the Maud Rise, a plateau of reduced water depth. The area was investigated in winter 1986 by the Winter Weddell Sea Project cruise of the German icebreaker FS "Polarstern", and it was found that the region is already part of the Antarctic Divergence, where upwelling of warmer water can occur, and that additional circulating currents and the doming of isopycnals (layers of constant water density) over the Rise could allow enough heat to reach the surface to keep the region ice-free in winter (Bagriantsev *et al.*, 1989; Gordon and Huber, 1990). Since this has not occurred since 1976, the region is presumably balanced on the edge of instability as far as its winter ice cover is concerned. The 1986 winter cover was of high concentration but was very thin (Wadhams *et al.*, 1987). The December distribution also shows an open water region appearing in the Ross Sea, the so-called Ross Sea Polynya, with ice still present to the north. In November and December a series of small coastal polynyas can be seen to be actively opening along the East Antarctica coast.

By January further retreat has occurred. The Ross Sea is now completely open, east Antarctica has only a narrow fringe of ice around it, and large ice expanses are confined to the eastern Ross Sea, the Amundsen-Bellingshausen Sea sector ($60°$–$140°$W) and the western half of the Weddell Sea. The month of furthest retreat is February. Ice remains in these three regions, but most of the East Antarctic coastline is almost ice free as is the tip of the Antarctic Peninsula. This is the season when supply ships can reach Antarctic bases, when tourist ships visit Antarctica, and when most oceanographic research cruises are carried out. It can be seen that the ice concentration in the centre of the western Weddell Sea massif is still 92–96%. This is the region which bears the most resemblance to the central Arctic Ocean; it is the only part of the Antarctic to contain significant amounts of multi-year ice, it is very difficult to navigate, and consequently even its bathymetry is not as well known as that of other parts of the Antarctic Ocean.

By March the very short Antarctic summer is over and ice advance begins. The first advances take place within the Ross and Weddell seas, then circumpolar advance begins in April. During May and June the Weddell Sea ice swells out to the north-east, while around the whole of Antarctica the ice edge continues to advance until the August peak.

Figure 1.19 is the Antarctic equivalent of fig. 1.17. The annual cycle of ice extent can be seen to have a much higher amplitude than in the Arctic, and the year-to-year variability of the peaks and troughs is also somewhat greater. During the 8.8-year record, the February average extent varies from 3.4 to 4.3×10^6 km^2, while the September average extent varies from 15.5 to 19.1×10^6 km^2, both covering ranges of ± 10–12%. The overall average cycle (fig. 1.19b) shows a retreat which is steeper than the advance. The mean minimum extent, at the end of February, is 3.6×10^6 km^2, while the mean maximum extent, in the middle of September, is 18.8×10^6 km^2. Because of low average ice concentrations within the pack, the corresponding minimum and maximum ice areas are 2.1 and 15.0×10^6 km^2.

Thus in the winter we see that the ice extent in the Antarctic exceeds that of the Arctic winter, while the summer minima are very much lower. This implies that the combined Arctic and Antarctic sea ice extent should be greatest during the Arctic summer. Fig. 1.20 shows that in fact the peak occurs in October after a plateau during the summer and that the global minimum occurs in late February. The range is approximately 19 to 29×10^6 km^2, with a high interannual variability for both maxima and minima.

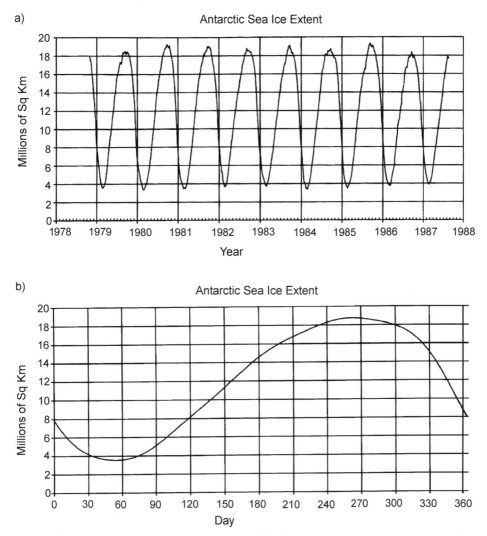

Figure 1.19. (a) The cycle of Antarctic sea ice extent, 1979–87. (b) Average seasonal cycle of ice extent in the Antarctic (after Gloersen *et al.*, 1992).

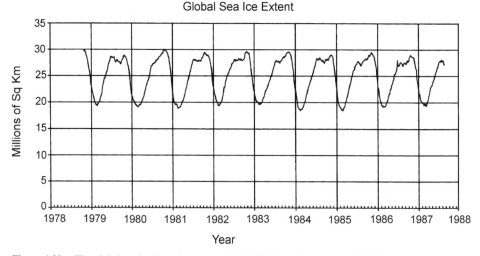

Figure 1.20. The global cycle of sea ice extent, 1979–87 (after Gloersen *et al.*, 1992).

1.3.3. Icebergs

Another type of floating ice which varies in distribution both seasonally and interannually is ice of glacial origin, i.e. icebergs and ice islands. In the Antarctic several thousand icebergs per year calve from the ice shelves, aprons of floating ice which surround about a third of the continent and which are the final product of ice creeping seaward from the continental ice sheet. Typically they begin by drifting westward in the Antarctic Coastal Current, often running aground for long periods, then eventually move into the Antarctic Circumpolar Current. As we shall show in chapter 7, they develop a northward component of movement across the axis of this current because of the strong effect of Coriolis force, and this combined with their high thermal capacity enables them to reach lower latitudes than sea ice. They are most common south of the Antarctic Convergence but can be seen at much lower latitudes. They gradually warm up, become eroded by wave action, and finally break up into fragments and melt.

Arctic icebergs calve from the narrow, steeper mountain glaciers that drain the Greenland ice sheet and other mountainous islands and coasts of the Arctic. They often capsize a number of times in their early lifespans, and they are usually rougher, more rugged and irregular than Antarctic icebergs. Their trajectory takes them down the coast of East Greenland, up the West Greenland coast and then southward down the coasts of Baffin Island and Labrador. At all times they are in a coastal current with land on their right hand, so their response to Coriolis force keeps them within the bounds of this current "conveyor belt", with frequent groundings. They do not therefore stray so far from the sea ice limits as Antarctic bergs, although the end of the conveyor deposits them on the Grand Banks, often beyond the sea ice edge. Some bergs also move southward off Cape Farewell and enter the ice-free part of the Labrador Sea. Iceberg trajectories are discussed in more detail in chapter 7.

Finally, Arctic ice islands are thin tabular bergs which calve from a small number of ice shelves in the Arctic — mainly from two, Ward Hunt Ice Shelf on Ellesmere Island and the Flade Isblink on NE Greenland. Their trajectories correspond to sea ice within the Beaufort Gyre and East Greenland Current respectively, and they are chiefly useful as platforms for drifting laboratories. Again we discuss them in chapter 7.

1.4. MAPPING THE FROZEN OCEANS

The preceding section has shown the value of satellite mapping of sea ice. Until the advent of the first Earth satellites, however, the only way of mapping sea ice extent and concentration was from ship reports, later augmented by aerial surveys carried out by government agencies. In the era of ship reports the ice limits that were most reliably reported were those where human activity was most frequent. In the Arctic this meant the East Greenland ice edge, regularly visited in the 19th Century by British whalers (Scoresby, 1815, 1820) and Scandinavian sealers. Sealers visited the ice edge region in spring and summer, and Norwegian ice maps for the April-September period date back to 1853 (Vinje et al., 1996). A continuous monthly set of ice charts for the European Arctic has been produced by the Danish Meteorological Institute since 1893. In the Antarctic the first chart of circumpolar ice limits, by Mackintosh and Herdman (1940), was based on reports from whaling ships and oceanographic ships of the British Discovery Committee, while a recent and more controversial study (De La Mare, 1997) claimed to have discovered a significant amelioration of ice conditions in the Antarctic during the 1960s from changes in the latitudes attained by whaling vessels. Examples of airborne mapping are the work of Soviet aircraft along the Northern Sea Route since the 1930s; of the Canadian Ice Patrol maintained by the Atmospheric Environment Service, Downsview, using initially human observers and later synthetic aperture radar (SAR); and the work of the International Ice Patrol in using aircraft to map icebergs in the North Atlantic shipping lanes. Another important contribution to knowledge was the series of BIRDS EYE research flights across the Arctic Basin run by the US Navy from 1962 to 1971 (USNOO, 1963–71), in which aerial photography and laser profiles were obtained.

Inevitably, ships and aircraft could only give a partial view of the global distribution of ice. All this changed when the first Earth satellites were launched. From the very start it was seen that ice and snow are such distinctive features of the Earth's surface seen from space that mapping of them was an easy and early satellite application. Initially the sensors used were low-resolution visual band sensors on weather satellites, i.e. the equivalent of photography. The NOAA (National Oceanic and Atmospheric Administration) series of satellites began in the early 1960s and still continues, now carrying the AVHRR sensor (Advanced Very High Resolution Radiometer) with a 1 km resolution. Similar sensors at higher resolution (500 m) fly aboard the DMSP satellites (Defense Meteorological Satellite Program), while very high resolution photography has been available since the early 1970s in the ERTS (later LANDSAT) series of satellites, augmented since 1990 by the French SPOT satellite (Massom, 1991).

The difficulty with visual sensors is that much of the polar seas, especially the ice edge regions, are swathed in cloud for most of the time, so that the ice surface cannot be seen. The solution was to penetrate cloud and darkness with microwave sensors, either passive

— relying on the natural microwave emissions from the Earth's surface due to its temperature — or active, i.e. the backscatter from the surface of radar signals emitted by the satellite. The classic pair of sensor types currently in use for ice mapping are multifrequency **passive microwave**, in which the variability of emissivity of different types of ice surface enable both ice concentration and distribution of ice type (e.g. fraction of multi-year ice) to be mapped; and **synthetic aperture radar** (SAR), which sends out a fan-shaped beam and turns the backscattered energy into a geometrically corrected image of the ice surface. Other sensors which have found use in the polar regions are the **scatterometer** and the **radar altimeter**.

Passive microwave radiometers detect the natural emission of radiation from the earth's surface, which has a peak in the microwave band. At any frequency the emission is a function of surface temperature and emissivity of the surface, and is affected only slightly by moisture in the intervening atmosphere. The radiometer responds to a particular frequency and polarisation of emitted radiation, and measures its intensity over a particular area, or **pixel**, of the viewed surface at a known viewing angle. An almost continuous record of passive microwave data from the polar regions has been obtained from NASA spacecraft since 1972. The first sensor, ESMR (Electrically Scanning Microwave Radiometer), operated at a single frequency, 19.35 GHz, and scanned the surface across the satellite track. From the brightness temperature it was possible to estimate ice concentration, and data from 1973–76 were used to compile atlases of Antarctic (Zwally et al., 1983) and Arctic (Parkinson *et al.*, 1987) ice conditions.

The next advance, in 1978, was the Nimbus-7 satellite carrying SMMR (scanning multichannel microwave radiometer), which used five frequencies (6.63, 10.69, 18, 21 and 37 GHz) and both horizontal and vertical polarisation. The lower frequencies were used to estimate sea surface temperature, wind speed and atmospheric water content, while sea ice concentration was derived from the 18 and 37 GHz channels. The different variation in emissivity of first-year and multi-year ice with frequency also allowed multi-year ice fraction in winter to be estimated from the data, rather than just ice concentration, although the best algorithm to use, and the best way of validating it, have been continuing research problems. The radiometer scanned conically, giving a constant incidence angle of 50° and a pixel size of approximately 30 km at 37 GHz. The data have been used to compile a newer and more complete atlas of both Arctic and Antarctic ice conditions from 1978-1987 (Gloersen *et al.*, 1992). The present sensor, launched in June 1987 on a DMSP satellite, is the SSM/I (Special Sensor Microwave/Imager), also conically scanning at 53°, and carrying channels at 19.35, 22.235, 37 and 85.5 GHz. The new high frequency measures rainfall over land but is also of value in discriminating thin sea ice from open water. Future sensors will be launched as part of NASA's EOS programme.

Because of their wide swath width, passive microwave radiometers can obtain an almost complete global coverage, at least in the polar regions where orbits converge, in 1–3 days. This is not yet true of synthetic aperture radar (SAR). SAR is an active sensor which transmits a beam across a swath to the side of the spacecraft and discriminates individual resolution cells according to their range and the doppler shift in frequency of the reflected radiation caused by spacecraft motion. The resolution can be very high — a few tens of metres — but this implies a high data rate. The first satellite-borne SAR was launched on NASA's Seasat in 1978, but the satellite failed after only 3 months. The next SAR satellite was the European Space Agency's ERS-1 in 1991, with a C-band (5.3 GHz) SAR, and a swath width of some 100 km. The main problems with SAR are the high power requirements and very high data acquisition rate, so that the ERS-1 satellite could not collect SAR data throughout its orbit, and could also only acquire data if it could simultaneously transmit it to a ground station. Ground stations were built to cover the whole Arctic, but complete coverage of the Antarctic ice cover has not been achieved. ERS-1 was followed by the similar ERS-2 in 1995, while Japan launched the L-band (1.275 GHz) JERS-1 in 1992. In 1995 Canada launched Radarsat, an innovative satellite with a wide swath width option (450 km at 100 m resolution) and the possibility of recording data on a tape recorder on board, for later transmission to a ground station. The next ESA SAR satellite, Envisat, will also have a wider swath (400 km) and on-board recording, while NASA plans an advanced SAR for its EOS programme.

Interpretation of imagery from passive and active microwave sensors is not straightforward, and is discussed extensively in multi-author works such as Carsey (1992) and Tsatsoulis and Kwok (1998). In the Antarctic, for instance, problems with passive microwave algorithms due to ice surface flooding and to the unusual ice types found near the advancing ice edge, and problems due to disagreement between algorithms, cause real uncertainty in the delineation of total ice area (i.e. extent x concentration) and multi-year ice distribution. It is important that these quantities be reliably monitored in order to give us warning of any trend indicating that global warming may be having an impact on the area of sea ice. Furthermore, it would be very desirable for passive microwave interpretations to be reliable enough to take over from advanced very high resolution radiometer (AVHRR) visible imagery as the prime means of generating ice charts and ice extent interpretations. More direct validation measurements are needed before this can happen.

The future of sea ice monitoring is an exciting one, because the advent of SAR sensors with wider swaths and on-board tape recorders offers the possibility of real global repeat mapping rather than the narrow swaths, long repeat periods, and dependence on ground stations of ERS-1 and ERS-2. Techniques have been developed to map ice motion from the superimposition of successive images, and this works both for passive microwave at low resolution and for SAR at high resolution (Kwok and Rothrock, 1999).

Information may also be obtained from other types of sensor. For instance, an instrument which is aboard ERS-2 and which operates continuously is the radar altimeter. This shows ice edge position and also the depth of penetration of ocean wave energy, hence the likely width of the pancake ice zone. However, the relationship between altimeter return and true ice surface topography in heavy pack ice (from which ice thickness may possibly be inferred) has not yet been demonstrated, although there may be some possibility of obtaining valid information. A future sensor which might yield better estimates of topography is the satellite-borne laser altimeter. Finally we should mention the scatterometer, which obtains coarse-resolution measurements of backscatter from the ocean surface, intended to be used to measure wind speed and direction from the naure of the wave spectra produced. It has been found (Ezraty and Cavanie, 1999) that scatterometer data over sea ice can also be processed to yield coarse-resolution sea ice imagery, useful for delineating the position of the ice edge.

1.5. SEA ICE AND THE HISTORY OF EXPLORATION

We end this introductory chapter by a brief look at how the presence of sea ice in the Arctic and Antarctic has been a factor hampering the exploration of these regions. By the late 19th Century most of the world's land masses had been explored and mapped, from the Himalayas to the source of the Nile. Not so the polar regions. On the ceiling of the museum in my laboratory in Cambridge, the Scott Polar Research Institute, there are two "domes of discovery", painted in 1934 by Mcdonald Gill. The idea was to paint coloured maps of the two polar regions, surrounded by the names of the great explorers of each pole and showing small pictures of their ships (these were the days when universities could afford such extravagances). The Arctic map at first sight looks correct. But then we notice things missing. Where are Prince Charles Island and Air Force Island, two quite substantial islands to the west of Baffin Island? The answer is that they were not discovered until 1949, when the Royal Canadian Air Force carried out a complete aerial photographic survey of the Canadian Arctic. They are not even in the remote High Arctic, but further south where ships have been trying to sail for hundreds of years from the time of Frobisher. And where is Stefansson Island? It is still joined to Victoria Island; nobody had realised that there was a strait between them. The Antarctic is even worse. The Antarctic Peninsula and the coastline of the Ross Sea are mapped out, and a few other stretches of coast, but for most of the circumpolar Antarctic the artist has drawn a faint dotted line which he has tried to hide among masses of sea ice. The coastline was simply not known. And this was less than 70 years ago. The reason was that the sea ice offered a major barrier to exploration.

People have lived in the Arctic, of course, for tens of thousands of years, developing cultures which were designed to cope with the presence of sea ice. Most advanced of all were the Inuit, or Eskimos, who invented the technology of kayaks, umiaks, igloos, parkas and dog sledges. For southern European man, however, living in the warm lands that were the cradle of civilisation, the Arctic was a closed book, a frightening place in which lived a mythical race called the Hyperboreans. The first recorded encounter with sea ice by classical man was the voyage of Pytheas of Massilia (about 325 BC), a Greek explorer who sailed north from the Pillars of Hercules and reached Ultima Thule, a place where there was a "curdled sea", a "sea-lung" which sounds very much like frazil or pancake ice. This could have been the Odden ice tongue region near Jan Mayen. Later the Vikings challenged the sea ice to settle in south and west Greenland, but it seems that their colony died out in the 15th Century because the worsening climate (the "Little Ice Age" replacing the "Little Climatic Optimum" — in reality a change of only about 0.5–1°C in mean temperatures) reduced their ability to grow hay for their animals whilst the advancing sea ice cut them off from supply voyages from Scandinavia. In Iceland too the population fluctuated enormously during the 1000 years of settlement, depending on whether the winter sea ice in the East Greenland Current tended to engulf the island or, as today, stay in the centre of Denmark Strait.

The role of sea ice in exploration is nowhere more clearly seen than in British exploration of the Arctic. In 1817 the great Arctic whaling captain William Scoresby noted that the ice in the Greenland Sea had undergone a significant retreat, and the ice edge now lay further north than at any time in the memory of the whalers. The Admiralty responded by sending two expeditions in 1818, one to seek to reach the North Pole via what is now known as Fram Strait, and the other into Baffin Bay, to try to extend discoveries made two centuries earlier by Baffin and Bylot, and perhaps find a Northwest Passage. In neither case was Scoresby, by far the most experienced Arctic sailor-scientist in Britain or the world at that time, allowed to play a role; instead the Navy sent officers who had no previous experience of the Arctic.

As Scoresby had predicted, the expedition towards the North Pole by Buchan was a complete failure, since the southward ice drift in the Trans Polar Drift Stream pushed the explorers back almost as fast as they tried to move north. Further, they were burdened with heavy boats, which they intended to launch into the "open polar sea", a popular concept of that period which Scoresby recognised as ludicrous. The Northwest Passage expedition under John Ross, was more successful, exploring the entrances to Smith, Jones and Lancaster Sounds, although Ross mistakenly thought that Lancaster Sound was a bay, spotting mountains (probably a mirage of sea ice) just inside. There followed a sequence of expeditions which should have taught lessons about the variability of ice conditions. Parry in 1819 sailed into Lancaster Sound and, finding an unusually open ice year, was able to penetrate in one summer almost through the Northwest Passage, as far as Melville Island. Subsequent expeditions, by Parry himself (1822–4), Ross again (1829–34) and, disastrously, by Franklin (1845), could make comparatively little progress during the brief navigation seasons between long ten-month spells in winter quarters. In every case (except Ross, a private expedition) the Admiralty sent two large sailing ships, with upwards of a hundred men, to explore shallow and intricate channels which are blocked by ice for most of the year.

The tragedy of the loss of Franklin and his 128 men was made more certain by the failure on the part of anyone in an official position, including former Arctic naval explorers, to understand the variability of sea ice. Franklin sailed into Peel Sound in 1846 after wintering in Beechey Island, and was trapped for the following winter in heavy, multi-year ice in Victoria Strait NW of King William Island. The ice did not break up in the summer of 1847, and Crozier (after Franklin's death) abandoned the ships in April 1848 and tried to walk southward, resulting in the death of all. His assumption must have been that if the ice did not break up in 1847 it would not do so in 1848, and yet there is evidence that it did. The search for the lost expedition was bedevilled by the same narrow view. In 1849–50 sledge journeys from search expeditions were made into the northern part of Peel Sound. The heavy ice conditions encountered there convinced the officers concerned that Franklin could not have come this way, so the search was diverted elsewhere, and in fact much of the Canadian Arctic was explored during the next few years as a result of this erroneous view. It never occurred to anyone that the ice severity could vary from year to year.

Even after the lands surrounding the Arctic Ocean had been explored, the central basin was slow to give up its secrets. Only Nansen in the "Fram", deliberately freezing himself into the ice to drift across the Arctic Basin in 1893–6, was able to explore the oceanography of the Arctic. Even in the 1920s it was still thought possible that large undiscovered islands existed in parts of the Arctic Ocean that had never been seen from the surface or the air, and as late as 1931 one of the justifications put forward by Sir Hubert Wilkins for a submarine crossing of the Arctic Ocean was to explore the large areas of the Beaufort Sea that were still blanks on the map and might contain land. It was, in fact, the nuclear submarine which freed Arctic Ocean exploration from the restrictions of ice conditions, and it is mainly due to submarines that we now have reasonably adequate bathymetric charts of the Arctic Ocean. But our knowledge of Arctic Ocean structure, chemistry, biology and sediments is still weak compared to other oceans, because such detailed work is done much better from surface ships, and in polar regions such ships cannot operate freely. Drifting stations on ice floes and ice islands, from which most scientific work has been done, have the disadvantage that they cannot be steered to any desired location. Today there are powerful icebreakers that are capable of crossing the Arctic Ocean, and a systematic cruise involving such a crossing was carried out in 1994, in summer, by two such icebreakers, the USCG "Polar Sea" and the Canadian "Louis S. St. Laurent" (Wheeler, 1997). However, such voyages are difficult and expensive and can only be done safely in summer.

In the Antarctic exploration began much later and identical problems were encountered, although the virtual disappearance of the ice in summer made the phase of geographical exploration easier. The most important case of sea ice interfering with exploration was of course the famous expedition of Sir Ernest Shackleton in the "Endurance" where the ship, on attempting to land a party in the eastern Weddell Sea for a transantarctic crossing, became trapped in the ice and drifted around the Weddell Sea (1914–6) until she was crushed in the NW Weddell Sea, leaving the crew to travel by ice drift and boat to Elephant Island followed by Shackleton's heroic boat journey to South Georgia to get help. The trajectory of "Endurance" closely matches the ice drift vectors shown by the buoy data of fig. 1.12. Another earlier case was that of the "Belgica" expedition of 1898, a Belgian expedition led by Adrien de Gerlache on which later polar explorers gained their first experience — Amundsen as Second Mate and Frederick Cook (later to claim to be the first at the North Pole) as surgeon. "Belgica" became trapped for a year in the ice of the Amundsen-Bellingshausen Sea, becoming the first ship in history to pass the winter in the Antarctic as she drifted westwards in the coastal current in a region where the ice dynamics even nowadays are not fully understood.

2. FORMATION, GROWTH AND DECAY OF SEA ICE

The icebreaker has spent the night feeling her way towards the Antarctic ice edge through a Southern Ocean gale. Keeping her head to the sea she has shipped great sheets of icy spray over the foredeck, which have frozen to every crane and fitting and have welded the hatchcovers shut with a coating of hundreds of tons of smooth, clinging ice. The icing threatens the very stability of the ship, so under floodlights the crew set to work with hammers, axes and steam hoses to remove it. You and the other scientists go out to help them. Now, as the weak midwinter sun rises in late morning, you look up and see that the ship has entered the ice. But it is not the familiar Arctic ice of heavy floes. It is a heaving, breathing mass of small cakes, floating in a milky fluid and stretching to the horizon. The cakes are constantly colliding and the milky fluid, which is a suspension of tiny ice crystals, slops over their edges and builds up a rim to each cake, making it look just like a white pancake. You go up in the helicopter to map the sizes of the pancakes with an aerial camera. The whole ocean is a mass of interlocking pancakes, like the scales of an immense white fish, and under this constraining skin the sea moves like a living creature, with great lines of swell sweeping majestically forward into the distance. Suddenly you notice a movement. In the vast perfect expanse of white a single imperfection appears. A pancake has mysteriously removed itself from its place and slid on top of its neighbour, leaving a black circular hole of sea. But the hole is not empty. Slowly a long black snout with a pale grooved underside slides up out of the hole and a little puff of vapour rises from it. It is a minke whale, smallest of the fast-swimming rorquals. He has made use of the fact that the pancakes are loose to create a breathing hole. Despite being an air-breathing mammal he is still absolutely free to wander the pancake ice zone of the outer Antarctic pack in winter. Nature always finds a way.

In this chapter we learn about the various ways in which ice forms on the ocean surface, grows, deforms and melts again — the set of processes that give rise to the ceaselessly changing landscape of the polar oceans.

2.1. THE STRUCTURE OF THE ICE CRYSTAL

We first have to account for the fact that ice floats on water at all. One of the most extraordinary properties of ordinary ice is that it is one of the few substances where the solid is less dense than its molten form. Geophysically this is of crucial importance

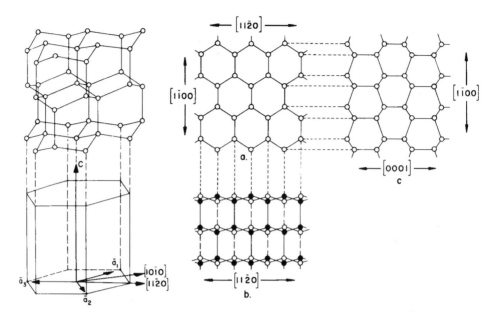

Figure 2.1.   The crystal structure of ice I (after Weeks and Ackley, 1986).

because the fact that ice floats makes it thermodynamically difficult or impossible for a sea to freeze completely. The reason for this behaviour lies in the open structure of ice.

There are at least nine polymorphs (different crystal forms) of solid H_2O under different conditions of temperature and pressure; they are discussed fully, as is the crystallography of ice, by Hobbs (1974) and Petrenko and Whitworth (1999). Under normal conditions found on the surface of the Earth, the only structure encountered is called ice I (fig. 2.1). It is a hexagonal structure. Each oxygen atom is at the centre of a tetrahedron with four other O atoms at the apices, separated by 0.276 nm. The O atoms are concentrated close to a series of parallel planes that are known as the basal planes. The principal axis, or c-axis, of the crystal unit cell lies perpendicular to the basal plane. The whole structure looks much like a beehive, composed of layers of slightly crumpled hexagons.

This structure causes ice to be markedly anisotropic in its physical and mechanical properties. Fracture along the basal plane requires, for a unit cell, the rupture of only two bonds, while fracture along any other plane normal to this requires the breaking of at least four bonds. Thus ice glides and cleaves readily on the basal plane. When an ice crystal grows, it is energetically easier for new atoms to be added to an existing basal plane than to begin a whole new plane. Therefore ice crystals grow more readily along the a- and b-axes (axes in the basal plane) than along the c-axis. These preferred growth directions are the directions of

(a)   the arms of snowflakes growing from vapour;
(b)   the arms of dendritic ice crystals growing on the surface of a newly-freezing sea or lake surface;

(c)   the preferred downward growth direction of sea ice crystals in a sheet;

(d)   internal melt features that form inside ice crystals as a result of absorbed solar radiation.

The net of O atoms is held together by hydrogen bonds. The H atoms lie along these bonds, but in a disordered way. The position of each H atom in the bond is to be closer to one O than to the other. Each O has two H's near it, but there can be only one H along each bond. Subject to these two rules, the H atoms can be arranged in any way. It is the length of the hydrogen bond that creates the open structure of ice; when ice melts, some of the bonds are broken, causing a disordered structure with a higher density. But even in liquid water there is some short-range order, with a few water molecules retaining the crystal-like bonded structure until this is destroyed by thermal motion; this accounts for the curious density behaviour of fresh water, where there is a maximum density at 4°C.

2.2. THE DENSITY AND FREEZING POINT OF SEA WATER

Consider a fresh water body being cooled from above, i.e. a water mass at the end of summer, experiencing subzero air temperatures. As the water cools the density increases (fig. 2.2) so the surface water sinks, to be replaced by warmer water from below, which is in its turn cooled. Thus a pattern of convection sets in such that the whole water body gradually cools. When the temperature reaches 4°C, a fresh water body reaches its maximum density. Further cooling results in the colder water becoming less dense and

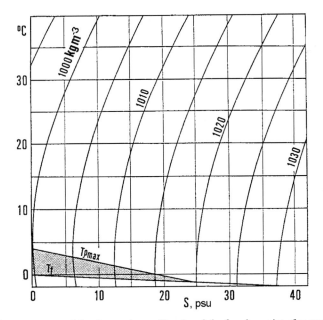

Figure 2.2.   The temperature of density maximum (T_{pmax}) and the freezing point of seawater (T_f), showing how these lines meet at 24.7 psu. The shaded triangle is the range over which cooling at the sea surface can take place without convection. Contours of sea water density are also shown (after Ono, 1995).

staying at the surface. This thin cold layer can then be rapidly cooled down to the freezing point, and ice can form on the surface even though the temperature of the underlying water may still be close to 4°C. Thus a lake or river can experience ice formation early in the autumn or winter, while considerable heat still remains in the deeper parts.

As fig. 2.2 shows, the addition of salt to the water lowers the temperature of maximum density. When the salinity exceeds 24.7 psu, the temperature of maximum density disappears, and so cooling of an ocean by a cold atmosphere will always make the surface water more dense and will continue to cause convection right down to the freezing point - which itself is depressed by the addition of salt to about −1.8°C for typical sea water. It may seem, then, that the whole water column in an ocean has to be cooled to the freezing point before freezing can begin at the surface, but of course there is a density jump at the pycnocline, so convection only involves the surface layer down to that level. Even so, it takes some time to cool a heated summer water mass down to the freezing point, and so new sea ice forms on a sea surface later in the autumn than does lake ice in similar climatic conditions.

The salinity of 24.7 psu is a critical value which is defined as the separation point between brackish water and true sea water. Once again this anomalous behaviour is a product of the way in which molecules in cold liquid water continue to associate in a crystal-like way over short ranges. Most parts of the polar oceans have surface salinities which comfortably exceed 24.7 psu. In the Antarctic they exceed 34 psu in most places, and even in the Arctic Basin, where river discharge is a diluting factor, they are usually higher than 30 psu. There are three important locations, however, where the salinity is less than this critical value (fig. 2.3). They are the Baltic Sea; the Kara Sea off the mouth of the Ob' River; and the Laptev-East Siberian Seas off the Yenisei and other eastern Siberian rivers. Because of its low salinity the Baltic freezes early in winter, and Baltic Sea ice has many of the structural characteristics of lake ice. The low salinity regions of the Russian Arctic shelves also encourage sea ice to form early in the season, and, as we discussed in chapter 1, ice production on these shelves is an important factor in increasing the density of the surface water to the point where it sinks at the shelf edge and ventilates the deep parts of the Arctic Basin.

2.3. THE INITIAL STAGES OF ICE FORMATION

2.3.1. Ice Formation in Calm Water

In quiet conditions the first sea ice to form on the surface is a skim of separate crystals which initially are in the form of tiny discs, floating flat on the surface and of diameter less than 2–3 mm. Each disc has its c-axis vertical, so the disc is growing outwards in a- and b-directions. At a certain point such a disc shape becomes unstable, and the growing isolated crystals take on a hexagonal, stellar form, with long fragile arms stretching out over the surface. These crystals also have their c-axis vertical. The dendritic arms are very fragile, and soon break off, leaving a mixture of discs and arm fragments. With any kind of turbulence in the water, these fragments break up further into random-shaped small crystals which form a suspension of increasing density in the surface water, an ice type called **frazil** or **grease ice.** In quiet conditions the frazil crystals soon freeze together to

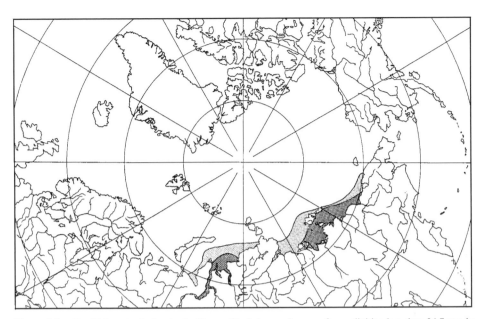

Figure 2.3. Brackish water in the Arctic Ocean. Shaded areas have surface salinities less than 24.7 psu in summer (light shading) and winter (dark shading). Adapted from Ono (1995).

form a continuous thin sheet of young ice; in its early stages, when it is still transparent, it is called **nilas**. When only a few centimetres thick this is fully transparent (dark nilas) but as the ice grows thicker the nilas takes on a grey and finally a white appearance. Once nilas has formed, a quite different growth process occurs, in which water molecules freeze on to the bottom of the existing ice sheet, a process called **congelation growth** which is discussed further in section 2.4. This growth process yields **first-year ice**, which in a single season in the Arctic reaches a thickness of 1.5–2 m and in the Antarctic 0.5–1 m. The details of ice behaviour during these initial stages are discussed by Weeks and Ackley (1986).

We digress at this point to discuss what happens when the initial ice formation occurs in rough water.

2.3.2. Ice Formation in Rough Water: The Frazil-Pancake Cycle

In the Antarctic early studies of the sea ice structure (Gow *et al.*, 1982) showed that most of the thickness of any ice core consists of small randomly oriented crystals characteristic of frazil origin. The mechanism by which the ice is generated was not elucidated until the first expedition was able to work in the pack ice zone during early winter, the time of ice edge advance. This was in 1986, in the Winter Weddell Sea Project, using F.S. "Polarstern". In traversing the ice margin region a careful study of the ice conditions and characteristics was carried out, and the **pancake-frazil cycle** (fig. 2.4) was identified as the source of most of the first-year sea ice seen further inside the pack (Wadhams *et al.*, 1987; Lange *et al.*, 1989).

Ice forming at the extreme ice edge cannot pass through the Arctic sequence into nilas because of the high energy and turbulence in the Southern Ocean wave field, which maintains it as a dense suspension of frazil ice. This suspension undergoes cyclic compression because of the particle orbits in the wave field, and during the compression phase the crystals can freeze together to form small coherent cakes of slush which grow larger by accretion from the frazil ice and more solid through continued freezing between the crystals. This becomes known as **pancake ice** because collisions between the cakes pump frazil ice suspension onto the edges of the cakes, then the water drains away to leave a raised rim of frazil ice which gives each cake the appearance of a pancake. This mechanism was described by Martin and Kauffman (1981). At the ice edge the pancakes are only a few cm in diameter, but they gradually grow in diameter and thickness with increasing distance from the ice edge, until they may reach 3–5 m diameter and 50–70 cm thickness. The surrounding frazil continues to grow and supply material to the growing pancakes, since the water surface is not completely closed off by ice and so a large ocean-atmosphere heat flux is still possible which can dispose of latent heat.

At greater distances inside the ice edge the pancakes begin to freeze together in groups (fig. 2.4b), but in the case of the Antarctic the wave field was found to be strong enough to prevent overall freezing until a penetration of some 270 km is reached. Here the pancakes coalesce to form first large floes (fig. 2.4c) then finally a continuous sheet of first-year ice (fig. 2.4d). At this point, with the open water surface cut off, the growth rate drops to a very low level (estimated at 0.4 cm per day by Wadhams *et al.*, 1987) and the ultimate thickness reached by first-year ice is only a few cm more than the thickness attained at the time of consolidation of the pancakes.

First-year ice formed in this way is known as **consolidated pancake ice** and has a different bottom morphology from Arctic ice. The pancakes at the time of consolidation are jumbled together and rafted over one another, and freeze together in this way with the frazil acting as "glue". The result is a very rough, jagged bottom, with rafted cakes doubling or tripling the normal ice thickness, and with the edges of pancakes protruding upwards to give a surface topography resembling a "stony field" (Wadhams *et al.*, 1987). The contrast between such ice and ice formed in calm conditions in the normal way is shown in fig. 2.5, which depicts profiles generated by drilling holes at 1 m intervals.

The rafted bottom of consolidated pancake ice provides a large surface area per unit area of sea surface, providing an excellent substrate for algal growth and a refuge for krill. The thin ice permits much light to penetrate, and the result is a fertile winter ice ecosystem (see section 8.1).

It is not yet known whether this sequence of ice growth is followed around the entire periphery of the Antarctic, but if so the area occupied by Antarctic pancake ice in early winter could be as great as 6 million km^2, making it an important yet seldom-seen feature of the Earth's surface.

In the Arctic, frazil and pancake ice are found in various parts of the ice marginal zones in winter, but the only area where it forms the dominant ice type over an entire region is the so-called **Odden ice tongue** in the Greenland Sea, discussed in more detail in chapter 6. The Odden (the word is Norwegian for headland) grows eastward from the main East Greenland ice edge in the vicinity of 72–74°N during the winter because of

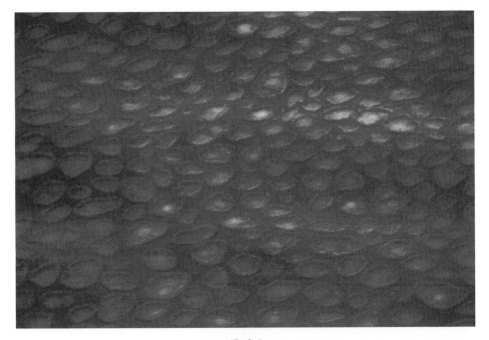

(2.4a)

(2.4b)

Figure 2.4. The frazil-pancake cycle at the Weddell Sea ice edge in winter. (a) Individual small pancakes at the outer ice edge. (b) The pancakes acquire raised rims and freeze together in groups. (c) The pancake groups freeze together into floes, 270 km from edge. (d) The floes have frozen together into consolidated pancake ice, with the pancake rims still visible through a dusting of snow. Photographs by the author.

(← 2.4c)   (2.4d↓)

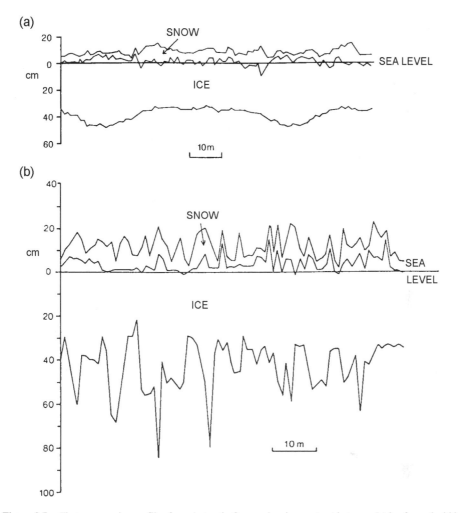

Figure 2.5. First-year sea ice profiles from Antarctic Ocean, showing contrast between (a) ice formed within calm polynyas and (b) consolidated pancake ice ("stony fields").

the presence of very cold polar surface water in the Jan Mayen Current, which diverts some water eastward from the East Greenland Current at that latitude. Most of the old ice continues south, driven by the wind, so a cold open water surface is exposed on which new ice forms as frazil and pancake because of the large amount of wave energy found in the Greenland Sea in winter. Fig. 2.6 shows how this ice tongue appears on passive microwave imagery, while frazil ice at the edge of the tongue can be seen in greater detail in the synthetic aperture radar (SAR) image of fig. 6.10(c) on p.221. The frazil ice, around the edges of the tongue, looks black because the frazil suspension damps down the short ocean waves of a few centimetres wavelength which are effective at scattering back radar radiation to the transmitter. Pancake ice looks brighter on SAR because of strong radar reflections from the raised rims of the cakes.

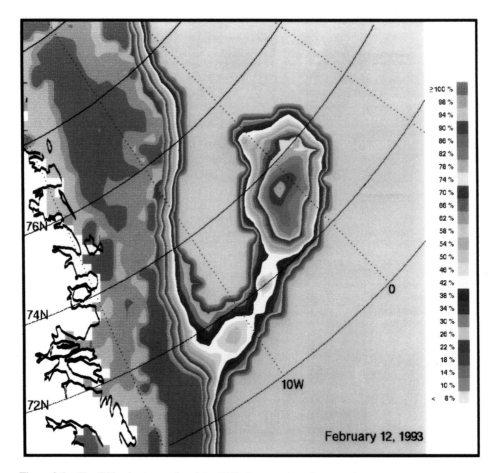

Figure 2.6.   The Odden ice tongue in winter 1993, from passive microwave imagery.

2.4. CRYSTAL GROWTH AND BRINE REJECTION

2.4.1. Crystal Fabric

Once a continuous sheet of nilas has formed, the individual crystals which are in contact
with the ice-water interface grow downwards by freezing of water molecules onto the
crystal face. As we have mentioned, this freezing process is easier for crystals with
horizontal c-axes (such that downward growth can occur along an a- or b-direction) than
for those with c-axes vertical. The crystals with c-axis horizontal grow at the expense
of the others, and as the ice sheet grows thicker crowd them out in a form of crystalline
Darwinism (fig. 2.7). Thus the crystals near the top of a first-year ice sheet are small and
randomly oriented, bespeaking their origin as frazil, or are oriented with c-axis vertical,
which occurs if they grew initially as flat floating discoidal or star-shaped crystals in very
calm water. There then follows a **transition region** which may be 10–20 cm deep, within
which the selection process is taking place and crystals with unfavourable orientations

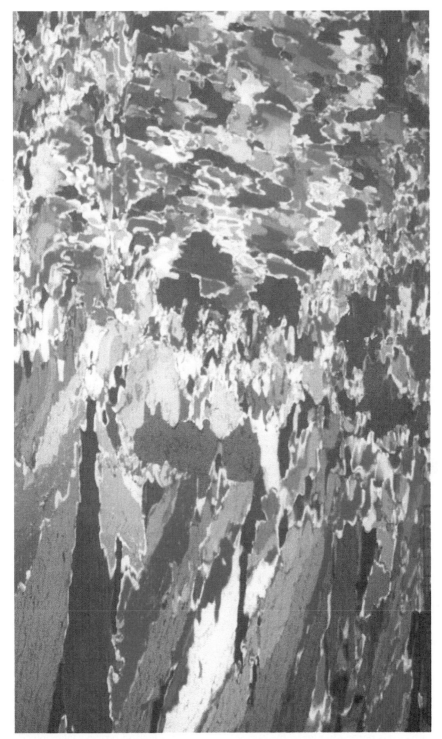

Figure 2.7.   The change of crystal orientation and fabric with depth in a first-year ice sheet. Height of image 30 cm.

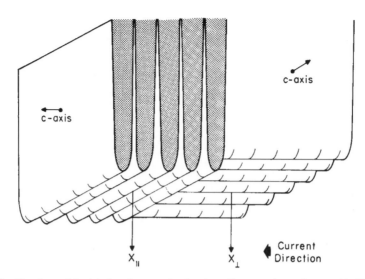

Figure 2.8. The shape of the interface in a growing ice sheet. Two crystals are shown, with differing c-axis orientations relative to the prevailing current (after Weeks and Gow, 1978).

are being eliminated. Below this, the favoured crystals continue to grow downwards, creating a fabric composed of long vertical columnar crystals with horizontal c-axes. This columnar structure is a key identifier of **congelation ice** (i.e. ice which has grown thermodynamically by freezing onto an existing ice bottom), and is a striking feature of first-year ice even when viewed by the naked eye.

It is reasonable to think that the c-axes should be oriented randomly in the horizontal plane, since all the relevant heat exchange processes are vertical. This indeed was found to be the case until some observations were published by Cherepanov (1971) showing large areas of the Kara Sea in which c-axes in the lower part of the first-year ice were highly oriented. Similar effects were then found in the fast ice of the coastal zone of the Beaufort Sea and in other fast ice zones. The observations, and subsequent laboratory experiments (Langhorne, 1983; Langhorne and Robinson, 1986), showed that the alignment occurs when sea ice is growing with a preferred direction of shear current, as in the case of fast ice when the currents (and alignments) are parallel to the shore and in the case of some pack ice zones (e.g. Kara Sea) where although the ice can move, it usually becomes very tightly bound to the surrounding coast and remains stable through most of the winter. The explanation proposed by Weeks and Gow (1978) depends on the fact that the interface between ice and water in a growing ice sheet is dendritic, i.e. corrugated as shown in fig. 2.8. A current perpendicular to the c-axis is parallel to the grooves and can establish a stable boundary layer. A current parallel to the c-axis flows in a turbulent way over the bumpy corrugations that it is encountering, causing mixing of the water nearest the ice, which has a solute-limited boundary layer, i.e. a slight enhancement of its salinity through brine rejection from the ice which slows its rate of growth. The disruption of this brine-rich layer gives such ice a slight growth advantage relative to the ice with c-axis perpendicular to the current, so again a selection process takes place with the parallel c-axis crystals "taking over" from the others. The physics of this mechanism are still not fully determined.

2.4.2. Brine Cells

As fig. 2.1 shows, ice has an open structure and so has a lower density than water. The structure is not so open, however, that other ions can easily find ways to incorporate themselves in "holes" within it. This applies to the ions of the salts in sea water, and so when sea ice grows the salt cannot enter the crystal structure. One might expect it to be rejected, therefore, leading to a sea ice cover composed of pure ice. Such is not the case, however. If you suck on a piece of first-year sea ice it will taste distinctly salty, although multi-year ice may not have any detectable traces of salt. When ice cores are melted, the water from young sea ice may have a salinity of about 10 psu, from first-year ice 4–6 psu and from multi-year ice 1–3 psu. How does this salt get into the ice?

The answer lies in the dendritic structure shown in fig. 2.8. The ice-water interface advances in the form of parallel rows of cellular projections called dendrites. Brine rejected from the growing ice sheet accumulates in the grooves between rows of dendrites. As the dendrites advance, ice bridges develop across the narrow grooves that contain the rejected brine, leaving the brine trapped and isolated. The walls of the "prison" close in through freezing, until the salt is contained in a very small cell of highly concentrated brine, concentrated enough to lower the freezing point to a level where the surrounding walls can close in no further. The cell then remains, a tiny inclusion only about 0.5 mm in diameter. Sheet-like systems of cells develop (fig. 2.9), each vertical sheet corresponding to successive bridgings of the same groove between two growing dendrites. A similar mechanism is responsible for the trapping of **air bubbles** within the growing ice sheet. Air bubbles give the ice a white, matt appearance, and their rate of formation appears to be greatest when the growth rate is high. Air bubble content is responsible for the very variable density of sea ice samples.

2.4.3. Brine Cell Migration and Drainage

The liquid brine retained in the growing ice sheet gives the ice a bulk salinity which is about 10 psu for young ice. However, the brine cells drain out of the ice as the ice sheet grows, by way of a network of **brine drainage channels** which they create, and as the ice sheet ages the brine concentration drops. Fig. 2.10 shows how fast this process is in young sea ice; it slows down as the sheet grows thicker. There are a number of mechanisms by which this drainage can happen.

Firstly, no sooner is the cell encapsulated in the growing ice sheet than it encounters the vertical temperature gradient between the ice-water interface (at −1.8°C) and the upper ice surface, which is at or near the air temperature. This gradient can easily be 30°K m^{-1}, equivalent to a 0.015° temperature difference between the top and bottom of a 0.5 mm diameter cell. The cell as a whole is in thermal equilibrium with the surrounding ice, in that the brine concentration is just enough to prevent further freezing. However, at the top of the cell the ice is colder than the mean, and so freezing occurs. This increases

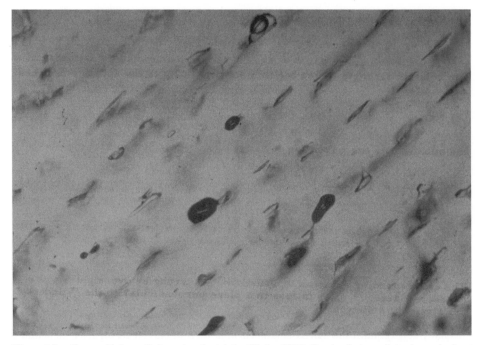

Figure 2.9. Sheets of brine cells in an ice sheet (after Weeks, 1986). Spacing between brine layers is about 0.6 mm.

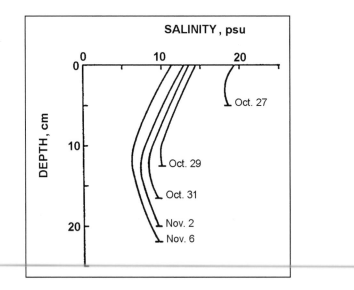

Figure 2.10. The rapid desalination of young ice (after Lewis, 1970).

the brine concentration, allowing melt to occur at the bottom of the cell where the ice is warmer than the mean. The cell as a whole therefore moves downwards; the salt in it is the same, but the water is constantly changing from the melt and freeze of ice along its path. Through this **brine cell migration** down the temperature gradient (in which the brine cell also gets bigger, as it encounters warmer temperatures) the cell eventually drains out of the bottom of the ice sheet. Theoretical explanations of this phenomenon (summarised in Weeks and Ackley, 1986) give migration rates which are very small, only about 1 cm per month, yet it is clear that brine cells migrate much more rapidly than this. One possibility is that internal convection occurs within the brine cell (more efficient with larger cells), which increases the heat transfer rate through the cell and hence the speed of cell migration. Another possibility is that the cell encounters defects in the crystal fabric as it moves, such as screw dislocations, and these provide an easier path for the cell's downward migration. It is a curious aspect of sea ice research that our current knowledge of the large-scale properties and role of sea ice is advancing very rapidly, through remote sensing and climate modelling, while our knowledge of some very basic microphysical properties remains tantalisingly incomplete.

A second mechanism is **brine expulsion**. This occurs when a brine cell experiences cooling of the ice around it, for example as the ice sheet gets thicker so that the cell becomes relatively closer to the upper surface. The thermal contraction of the sea ice around the cell is greater than that of the liquid in the cell, producing a high internal pressure in the cell which may allow the cell wall to rupture. This occurs preferentially along the basal plane (a- and b-axes) driving the brine outwards and downwards. Although micrographs do show cases of thermal cracking around brine cells, this mechanism does not appear to be a dominant one. It was investigated by Untersteiner (1968) and Cox and Weeks (1975).

A third mechanism is **gravity drainage**, and this is believed to be more important than the first two during the winter. As the ice sheet grows, its surface gradually rises higher above sea level to maintain isostatic equilibrium. This means that if a brine cell forms part of an interconnected system rather than being completely encapsulated, it will experience a pressure head which will drive it down and out of the ice (Eide and Martin, 1975). For this mechanism to work we have to postulate an interconnecting system of very fine pores which connects brine cells in such a way that hydrostatic pressure can be transmitted, whilst at other times the pores are too fine to permit easy motion of the brine from cell to cell (otherwise brine drainage would be an instantaneous process). The existence and properties of these pores is a current research problem. Not only does an absolute pressure head exist in the part of the ice sheet above sea level, but also an unstable density gradient exists in the brine cells at all depths, at least during the winter when there is a downward positive temperature gradient. This is because the brine density which allows the brine cell to exist in equilibrium with the surrounding ice is greatest at the lowest temperatures, i.e. near the top of the ice sheet, so if the brine cells are interconnected, the highest densities are at the top and the lowest at the bottom. Brine will therefore flow downwards through the interconnecting pores. This mechanism was investigated by Cox and Weeks (1975) and found to be important if not dominant.

A further loss of brine occurs during the first summer when the top surface of the ice sheet melts and fresh water percolates down through the ice sheet, flushing out much of the remaining brine. This will be considered in section 2.6.

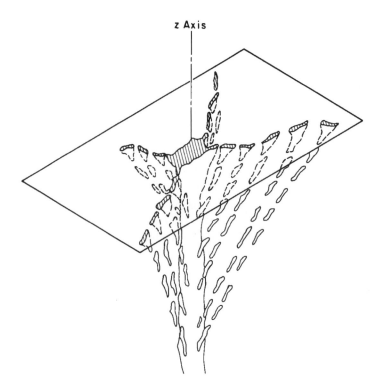

Figure 2.11.   Geometry of a typical brine drainage channel (after Lake and Lewis, 1970).

We might expect that the gravity drainage process would occur in a uniform manner throughout the ice sheet, allowing individual brine cells to drain downwards. However, what is actually observed, and so far has not been explained, is the fact that the brine cells merge to create a system of brine drainage channels, rather like a river with tributaries, through which the brine drainage occurs at a limited number of locations, each with a considerable flow rate. Fig. 2.11 shows the geometry of a typical channel. It is a tube of typical diameter 0.4 cm. The density of tubes is typically one per 180 cm^2 of ice bottom in thick ice (Lake and Lewis, 1970) and one per 33 cm^2 in thinner ice (Saito and Ono, 1980), implying that as the ice grows thicker brine channels merge to create fewer, larger channels. The flow of brine through these channels has been investigated (Eide and Martin, 1975; Niedrauer and Martin, 1979) and found to be oscillatory, with a short period of inflow (8–15 minutes) followed by a longer period of outflow (typically 45 minutes), and a typical drainage rate of 1 litre per minute. Brine drainage channels tend to grow wider with time. This is because as cold brine drains downwards through the tube there is a lateral flow of heat from the surrounding ice into the brine. This warms the brine above its equilibrium temperature and it begins to melt the ice walls around it, which cools the ice and dilutes the brine. Thus there is a very local cooling of the ice around the brine tube, which itself may encourage brine cells to drain into the tube by the brine expulsion mechanism. Brine drainage channels therefore produce a localised variability in the properties of an ice sheet, the implications of which are just beginning to be investigated (Cottier *et al.*, 1998).

Where the brine drainage channel emerges from the ice sheet, a hollow tube of ice can grow out from the ice sheet base, known as an **ice stalactite** (Dayton and Martin, 1971). In stable Antarctic fast ice these have been observed with lengths of 1.5–6 m, but usually they are much shorter and in moving pack ice they are seldom seen at all. They are produced by freezing of sea water around the very cold brine streamer emerging from the bottom of the tube. Once formed, the stalactite grows wider, since the inner walls melt in order to dilute the dense brine, while freezing continues to occur onto the outer walls to allow heat transfer into the cold brine. Under very stable conditions the stalactite will grow both longer and wider throughout the period of brine drainage.

Many other details of brine drainage channels remain to be investigated. For instance, the channels possess a "neck" near the bottom due to freezing of the lower-salinity sea water which rises up the tube to replace the dense brine after each expulsion episode. This helps determine the rate of oscillation of the system. Most important, and least known, is the biological role of brine drainage channels. Phytoplankton have been observed to live on the walls of brine drainage channels, and even larger zooplankton such as amphipods have been observed to crawl up the larger channels. Within a channel there is possibly a higher light level than on the ice bottom, because of the waveguide effect of the channel for light penetrating from above, while the oscillating water flow brings nutrients and oxygen to the resident biological community. In addition the tube provides security from larger browsers. The sea ice ecosystem based on brine drainage channels is especially important in the Antarctic, but is now receiving attention in both hemispheres (Gradinger, 1996; Lizotte and Arrigo, 1998).

2.4.4. Solid Salts in Ice

Brine cells which are isolated in the upper part of a growing ice sheet are subject not only to a temperature gradient, as described above, but also to colder and colder temperatures as the winter progresses and the ice sheet grows thicker (this assumes that the thickening of the ice sheet proceeds more rapidly than downward draining of the brine). Under these circumstances, certain **solid salts** can precipitate out within the brine cell, a function of the water-ice-salt phase diagram. Table 2.1, after Weeks and Ackley (1986), shows the temperatures of initial salt formation. Calcium carbonate precipitates at the high temperature of −2.2°C, but this is only present in very small quantities in sea water. The

Table 2.1. Temperatures of formation of solid salts presumed to exist in sea ice.

Constituent	Temperature of initial formation °C
$CaCO_3.6H_2O$	−2.2
$Na_2SO_4.10H_2O$	−8.2
$MgCl_2.8H_2O$	−18.0
$NaCl.2H_2O$	−22.9
KCl	−36.8
$MgCl_2.12H_2O$	−43.2
$CaCl_2.6H_2O$	< −55.0

first major constituent to precipitate out is sodium sulphate at −8.2°C. Other constitutents precipitate at much lower temperatures. Sea ice that is colder than −8.2°C can therefore be assumed to contain some solid salts within the brine cell system.

2.4.5. Lake and River Ice

We have already shown how lake ice formation occurs more readily than sea ice. Lake ice has no brine inclusions, but does contain air bubbles. Its crystal fabric resembles that of first-year sea ice, with a columnar structure developing if the ice is growing in calm water. River ice often forms in turbulent conditions, and frazil ice is often a dominant component (Prowse and Gridley, 1993). This causes severe problems in blocking inlets to power station cooling systems, and the detection of frazil ice in rivers by remote sensing is an important technical problem. When the ice cover stabilises, the upper layer of the ice sheet comprises randomly oriented crystals of frazil origin, while further growth occurs as columnar crystals. In rivers ice can also grow, or be deposited, on the river bed (especially on vegetation or boulders), when it is called **anchor ice** (Tsang, 1982). This can prevent the exchange of water between the flow in the river itself and the subsurface flow which is often a significant component of the total water flux. In the highest northern latitudes river ice may reach a thickness of 2 m by thermal processes, but the additional deposition of frazil ice by the river flow under a stationary cover can increase the thickness of icing to 5–10 m in the case of Siberian rivers (Shumskiy, 1964). When the ice breaks up in spring, the broken blocks can cause enormous damage to structures, and can also pile up into **ice jams** which can hold up the river flow and cause flooding (Calkins, 1986; Prowse, 1990).

2.5. SNOW LOADING AND METEORIC ICE FORMATION

The Arctic Ocean is a cold desert, and the average snow thickness on sea ice is only a few cm, making a small contribution to the overall thickness or mass. Fig. 2.12 shows the distribution of annual precipitation over the Arctic, showing that it is only near the margins with the open ocean that snowfall really becomes significant. The snow is important in determining the albedo of the surface, and as we shall see in the next chapter, the date at which the snow melts and reveals bare ice is important for determining the thermal balance and the equilibrium ice thickness. However, snow is generally not important as a contributor to the ice structure itself.

   In the Antarctic the situation is quite different. The annual snowfall is greater, and in coastal regions snow is blown onto the sea ice by katabatic winds off the tops of ice shelves. During the July–September 1986 "Polarstern" cruise in the eastern Weddell Sea the mean snow thickness was 14–16 cm on the surface of first-year ice. Since the ice itself is so thin, this was sufficient to bring the ice surface below sea level in 15–20% of cases, leading to the infiltration of sea water into the overlying snow and the formation of either a wet slushy layer on top of the ice or, in the case of freezing, the formation of a "snow ice" layer between the unwetted snow and the original ice upper surface. In

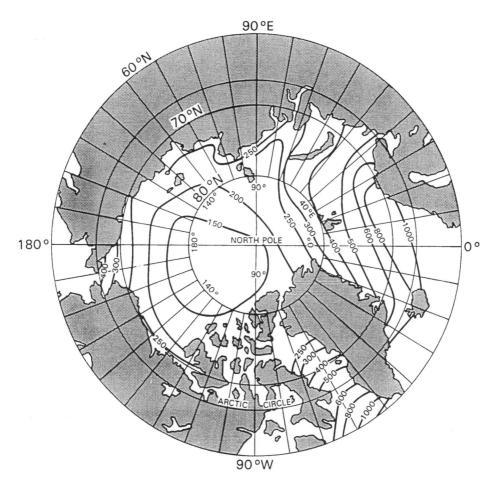

Figure 2.12. Annual snowfall over the Arctic in mm (after Gorshkov, 1983).

September–October 1989 there was also a mean snow thickness of 16 cm over undeformed first-year ice (23 cm over deformed ice, because of piling up of snow against ridges), but in multi-year ice in the western Weddell Sea the snow thickness was much greater. The average was 0.63 m over undeformed ice and 0.7 m over deformed ice. This was sufficient to push the ice surface below sea level in almost every case. Fig. 2.13, from Wadhams and Crane (1991), shows the contrast between the two types of ice cover in this respect. The resulting flooded layer has an effect on passive microwave signatures, making it difficult to unequivocally identify multi-year ice, and also has an impact on the mass and energy balances of sea ice which must be taken into account in modelling efforts. Recent discussions of snow loading and snow-ice formation, from the points of view of field observation and remote sensing analysis respectively, are by Sturm *et al.* (1998) and Markus and Cavalieri (1998).

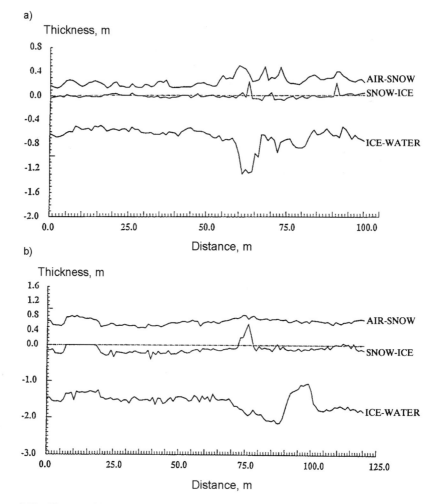

Figure 2.13. First-year (a) and multi-year (b) ice profiles from the Weddell Sea (from Wadhams and Crane, 1991). Note that the snow-ice interface lies below sea level in the multi-year case.

2.6. SUMMER MELT PROCESSES

So far we have considered the development of the fabric of an ice sheet through its first winter. The summer melt period, however, if survived by the ice sheet, is crucial to its further development. In the Arctic, the overlying snow layer typically begins to melt in mid-June and is gone by early July. The meltwater gathers to form a network of **meltwater pools** over the surface. On first year ice, which has a smooth upper surface at the end of winter (except where ridged), the pools are initially very shallow, forming in minor depressions in the ice surface, or simply being retained within surviving snow pack as a layer of slush. As summer proceeds, however, this initial random structure becomes more fixed as the pools melt their way down into the ice through preferential absorption of solar radiation by the water, which has an albedo of 0.15–0.4 compared to 0.4–0.7 for the bare ice (Maykut, 1986). The melt increases the variation in topography of the ice surface, and is one reason why the average summer albedo of the ice is difficult to establish for modelling purposes, since it depends on the fractional area occupied by the melt pools. When pressure ridges exist, they can form a dam for melt pools, and a smooth first year floe surrounded by pressure ridges can appear almost like a uniform shallow lake.

As the melt pools grow deeper and wider they may eventually drain off into the sea, over the side of floes, through existing cracks, or by melting a **thaw hole** right through the ice at its thinnest point or at the melt pool's deepest point. The downrush of water when a thaw hole opens may be quite violent, and on very level ice, such as fast ice, a single thaw hole may drain a large area of ice surface. From the air such thaw holes give the appearance of "giant spiders", with the "body" being the thaw hole and the "legs" channels of melt water draining laterally towards the hole.

The drained melt water forms a temporary low-salinity surface layer a few metres deep which gives the water column in summer a second, shallow pycnocline. This has an effect on sound propagation since it gives a higher-velocity surface sound channel which refracts sound waves downwards.

The underside of the ice cover also responds to the surface melt. Directly underneath melt pools the ice is thinner and is absorbing more incoming radiation. This causes an enhanced rate of bottom melt so that the ice bottom develops a topography of depressions to mirror the melt pool distribution on the top side. In this way an initially smooth first-year ice sheet acquires by the end of summer an undulating topography both on its top and bottom sides (fig. 2.14). Some of the drained melt water may in fact gather in the underside depressions to form **under-ice melt pools**, which refreeze in autumn and partially smooth off the underside, leaving it with bulges but not depressions (Wadhams and Martin, 1990).

A final and most important role of the melt water is that some of it works its way down through the ice fabric through minor pores, veins and channels, and in doing so drives out much of the remaining brine. This process, called **flushing**, is the most efficient and rapid form of brine drainage mechanism, far more rapid than any of the three winter mechanisms that we have described, and it operates to turn the salinity profile of late winter first-year ice into one with much lower salinity, especially in the upper layers (fig. 2.15). The hydrostatic head of the surface meltwater provides the driving force, but once again an interconnecting network of pores is necessary for the flushing process to operate. Given that the strength properties of sea ice depend on the brine volume, this implies that the flushing mechanism creates a surviving ice sheet which during its second winter of existence has much greater strength than in its first winter.

In the Antarctic it is often stated that melt pools do not occur, because the ice of winter continues to move seaward after growth ceases, and hence breaks up and melts completely at the retreating edge. However, recent observations in the NW Weddell Sea by the author (fig. 2.16) have shown extensive areas of melt ponding on thick multi-year floes near the ice edge.

2.7. MULTI-YEAR ICE

In official definitions (WMO, 1970), ice which has survived a single summer season of partial melt is called **second year ice** and ice which is older than this is called **old ice**.

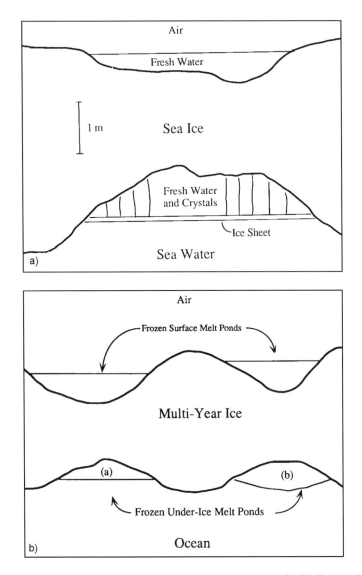

Figure 2.14. Schematic diagrams of top surface and under-ice melt pools (after Wadhams and Martin, 1990). The upper figure shows the situation in summer, with under-ice melt pools forming below surface melt pools, and with freezing going on at the underside fresh water – sea water interface. The lower figure shows the situation after freeze-up, with both sets of melt pools refrozen: the under-ice pools may refreeze as level ice or with a bulge.

In practice, partly because of the difficulty of visually discriminating between second-year and old ice, and partly because all ice older than first-year has certain shared properties, such as greater strength, *any* ice older than first-year is usually simply classed as **multi-year ice.**

In the Arctic, sea ice commonly takes several years to either make a circuit within the closed Beaufort Gyre (7–10 years) or else be transported across the Arctic Basin and

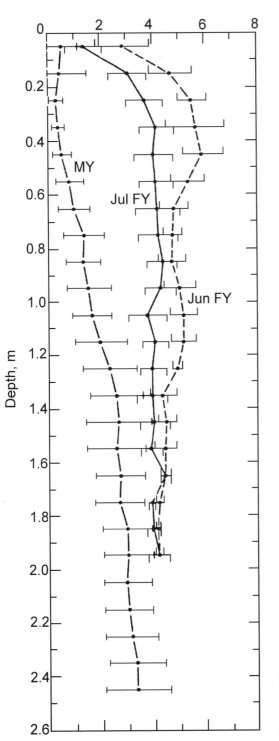

Salinity, psu

Depth, m

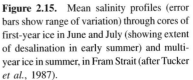

Figure 2.15. Mean salinity profiles (error bars show range of variation) through cores of first-year ice in June and July (showing extent of desalination in early summer) and multi-year ice in summer, in Fram Strait (after Tucker *et al.*, 1987).

Figure 2.16.   Melt pools on multi-year Antarctic ice in NW Weddell Sea during February 1995; (top) from surface, (opposite page) from a vertical aerial photograph. Photographs by author.

expelled in the East Greenland Current (3–4 years). More than half of the ice in the Arctic is therefore multi-year ice. Growth continues from year to year until the ice thickness reaches a maximum of about 3 m (see chapter 3), at which point summer melt matches winter growth and the thickness oscillates through an annual cycle. This old, multi-year ice is much fresher than first-year ice; it has a lower conductivity and a rougher surface. Microwave radiation penetrates further through it than through first-year ice and an incident beam tends to be scattered within the volume of the ice sheet rather than reflected in a specular way from the surface. Its microwave properties are therefore distinct, enabling first- and multi-year ice to be well discriminated by both passive microwave and SAR. The low salinity of multi-year ice makes it much stronger than first-year ice and a formidable barrier to icebreakers.

The melt processes which caused the first undulations to develop in the upper and lower surfaces proceed to enhance this topography in later years. During the autumn following the first summer of melt, any surface melt pools that have not drained refreeze to give thin sheets of fresh ice covering the parts of the ice cover with the lowest elevation. During the second summer these are preferred sites for new melt pools to develop, which melt their way even more deeply into the ice surface. After many years (fig. 2.17) a melt pool may look like a lake with spectacularly mountainous cliff-like sides, and with the unmelted portions standing clear like mountain ridges. Similar continued development of the bottom roughness leads to an underside topography with prominent undulations (most of them positive, i.e. bulges rather than depressions). This is shown in fig. 2.18(b), which is an

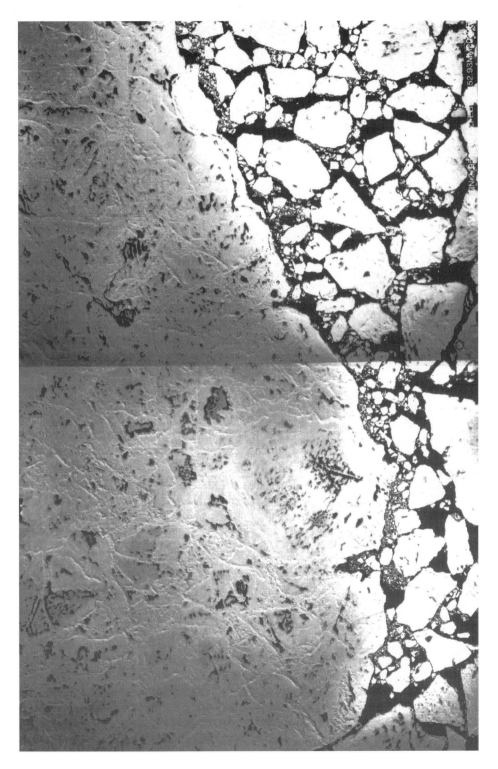

(a)

(b)

Figure 2.17.   Melt pools on (a) first-year (b) multi-year Arctic sea ice. Photographs by author.

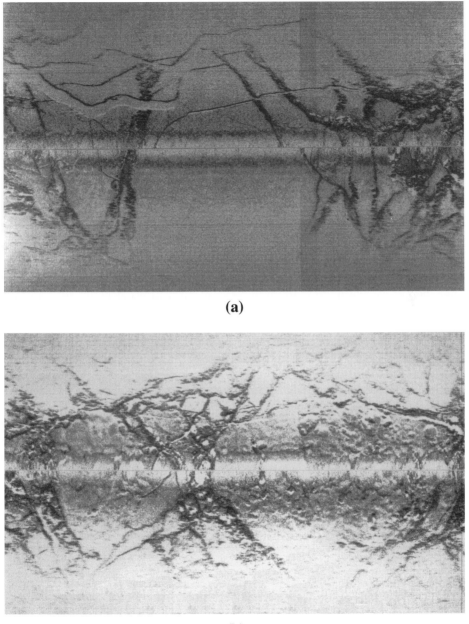

(a)

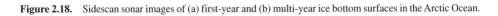

(b)

Figure 2.18. Sidescan sonar images of (a) first-year and (b) multi-year ice bottom surfaces in the Arctic Ocean.

under-ice sidescan sonar image taken by a submarine, showing protuberances as dark and depressions and "shadows" of protruding features as light. The multi-year ice has a bottom landscape of bulges and other roughness elements, while first-year ice (fig. 2.18(a)) shows a smooth bottom except for occasional narrow prominent cracks. Pressure ridges are dark rugged linear features in both cases.

In contrast, Antarctic sea ice normally has a northward component to its motion, so that it is constantly spreading from its source area into lower latitudes, where it eventually melts. Usually this process takes less than a year, so almost all Antarctic ice is first-year ice. Only in the Weddell Sea, the Ross Sea, and to a lesser extent the Bellingshausen Sea, do significant quantities of pack ice survive the summer to become multi-year ice. Usually this occurs because of the existence of a semi-closed gyre. In the Weddell Sea, for instance, ice formed in the eastern part of the Weddell-Enderby Basin is carried west in the southern part of the Weddell Gyre (the Antarctic Coastal Current) into the southern Weddell Sea, and then westward and finally north along the east side of the Antarctic Peninsula (the route followed by the drift of Shackleton's Endurance). This track has been found by the use of drifting ice buoys to take about 18 months. Thus multi-year ice may be found in the western Weddell Sea within the zone of northward motion. In fact it is second-year ice; the only very old multi-year ice to be found is the small amount that has broken out from the fast ice which grows for many years in embayments along the fringes of ice shelves (Wadhams et al., 1987) — this may reach 11 m or more in thickness.

Again, the first opportunity to study such ice during the winter was a comparatively recent cruise, the 1989 Winter Weddell Gyre Study of F.S. "Polarstern", which involved a transect across the Weddell Sea from west (Antarctic Peninsula) to east (Kap Norvegia) (Wadhams and Crane, 1991). Earlier work was done in summer (Ackley, 1979) and early spring (Lange and Eicken, 1991). Multi-year ice was found only west of 40°W, in the region of northward ice drift, and when sampled by drilling had a mean thickness of 1.17 m in regions free of ice deformation. Undeformed first-year ice in the same regions had a mean thickness of only 60 cm. The fact that second-year ice is twice as thick as first-year ice, despite the fact that the growth rate slows radically as ice thickness increases, is evidence of more extreme conditions in the southern Weddell Sea, through which the multi-year ice has passed, than in the lower-latitude circumpolar Antarctic Ocean. These conditions may consist purely of lower mean air temperatures, but it is more likely that lower ocean heat fluxes are also important. Further recent discussions on the structure and thickness respectively of Antarctic multi-year ice are given by Eicken (1998) and Strass and Fahrbach (1998).

2.8. FORMATION OF LEADS AND PRESSURE RIDGES

In this chapter we have been concerned so far with the way in which sea ice forms and changes under thermal processes alone, making no distinction between fast ice and pack ice. Yet we know that pack ice is constantly in motion, driven by the wind, and that this produces many important changes to its appearance and development. Therefore we now give a brief discussion of the two most obvious features of a pack ice zone, leads and pressure ridges. The physical and statistical properties of pressure ridges will be discussed in much greater detail in chapter 5.

The long-term pattern of wind-driven motion in sea ice is the same as the surface current pattern described in chapter 1. However, the wind stress which drives the sea ice through frictional drag is integrated over a large area — it has been estimated that in concentrated pack ice a piece of sea ice responds to wind fields integrated over a distance of 400 km upwind. Therefore a large-scale divergent wind field, created by an appropriate pressure pattern, can also create a divergent stress over a large area of icefield. Since ice has little strength under tension, this divergence can open up cracks which widen to form **leads** (fig. 2.19). In winter leads rapidly refreeze because of the enormous temperature difference between the atmosphere (typically −30°C) and the ocean (−1.8°C). The heat loss from a newly-opened lead can be so violent (more than 1000 W m^{-2}) that the lead steams with **frost smoke** (fig. 2.19(a)) from the evaporation and condensation of the surface water. A young ice cover rapidly forms, within hours, as nilas if the surface is calm, and this cuts out the evaporation. When a subsequent wind stress field becomes convergent, the young ice in the refrozen leads forms the weakest part of the ice cover and is the first part to be crushed, building up heaps of broken ice blocks above and below the water line. Such a linear deformation feature is called a **pressure ridge** (fig. 2.20), the above-water part being the **sail** and the below-water part (more extensive) being called the **keel.** Keels in the Arctic can reach down to 50 m, although most are about 10–25 m deep, with a depth of 30 m seen at least once every 100 km of track (Wadhams, 1978a). The keel is typically about 4 times deeper than the height of the sail, and is also 2–3 times wider, so that apparently undeformed ice near a pressure ridge may have part of the keel underneath it; this occurs because in a compressive deformation it is easier to push ice blocks downwards against buoyancy than upwards against gravity.

Ridged ice in the Arctic makes a major contribution to the overall mass of sea ice; probably about 40% on average and more than 60% in coastal regions. Ridges once formed become **consolidated** through freezing together of the ice blocks, and become permanent features of the winter pack, with strength equal to or greater than the surrounding undeformed ice. However in summer a ridge tends to disintegrate and shed ice blocks to the surrounding ice cover. Mean ice thicknesses in the Arctic range from 2–3 m in the seas north of Russia to 7–8 m north of Greenland and the Canadian Archipelago. The differences are mainly due to differences in the amount of pressure ridging; north of Greenland the wind drives ice towards the coast and creates enormous amounts of new ridging.

In contrast, Antarctic ridging is much less intense. Individual ridges are also much shallower; every ridge sampled during the 1986 and 1989 experiments was less than 6 m in draft. Fig. 2.21 shows a typical ridge cross-section, which resembles that of an Arctic ridge in everything except total depth. Examination of the structure of ridges showed that in most cases the block thickness in the ridge was similar to the thickness of the floes on either side, that is, the ridge has formed by buckling of the floe itself, or by buckling due to the collision between two floes, rather than by the crushing of young ice. This means that the stress due to convergence can be relieved by a large number of individual buckling events rather than by a small number of refrozen lead closures, so that a given ridge need not grow to a great height. There ought to be more frequent ridges than in the Arctic, on this argument, but once again this does not appear to be the case, except possibly in coastal regions. On average, Antarctic sea ice is in a state of divergence as it moves northward into wider spaces of the Southern Ocean, and so it is likely that ridge-

(a)

(b)

Figure 2.19.   (a) Freshly opened leads in winter, with frost smoke. (b) A refrozen lead, with dark nilas covered in salt flowers.

(a)

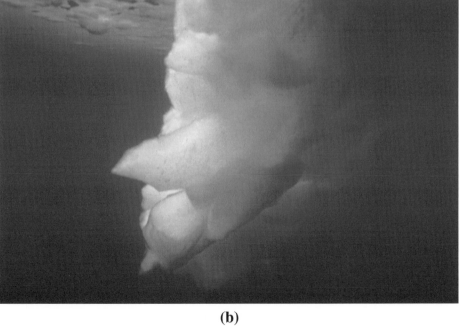

(b)

Figure 2.20. (a) The sail and (b) the keel of an Arctic pressure ridge.

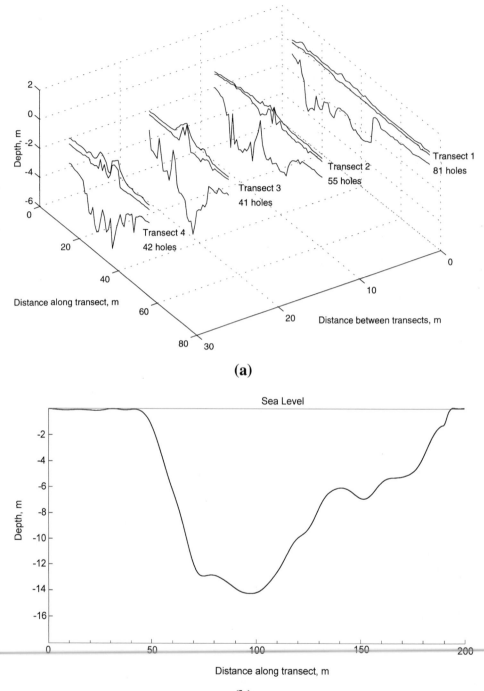

(a)

(b)

Figure 2.21. (a) Set of parallel cross-sections through an Antarctic pressure ridge keel, from holes drilled 1 m apart. (b) Cross-section through an Arctic pressure ridge keel, from sonar profiling. Note similar shape, but difference in vertical scale.

building events occur less frequently than in the Arctic. Furthermore, ridges formed in this way may well disintegrate and revert to brash ice when the pressure is relieved.

It is difficult to assess the overall contribution made by ridging to the mean thickness of Antarctic sea ice. Drilled profiles offer an inadequate data set for such large-scale averages. Nevertheless, the contribution does appear to be important. In 1989, for instance, the mean ice thickness in profiles containing no deformed ice, and some deformed ice, respectively, were 0.60 and 1.03 m (first-year) and 1.17 and 2.51 m (multi-year ice) (see chapter 5). We may expect further insight from the extension of laser profiling surveys over the ice cover (Weeks *et al.*, 1989) and from the installation of moored upward-looking sonar systems recording ice thickness for a year at a time at a given spot (Strass and Fahrbach, 1998). Early results from the Weddell Sea obtained by the latter authors give mean ice drafts of 0.8 m in the central Weddell Gyre, 2.2 m in the eastern inflow and 2.8 m in the western outflow, with ridged ice estimated to contribute 30% to the total ice volume in the interior Weddell Sea and 50% in the boundary regions.

2.9. ICE IN SHALLOW WATER

2.9.1. Grounded Landfast Ice

Ice which is fast both to the shore and to anchoring points on the seabed occurs in winter around the entire rim of the Arctic Ocean as well as in many sub-Arctic areas and around the coasts of Antarctica. We can distinguish this from other types of fast ice, which are afloat but which do not move because they are filling a restricted channel or basin which is too small to allow a wind-driven ice circulation.

The first ice encountered at the shore, which adheres to the beach and which is unaffected by the tide, is called the **ice foot.** It is the first ice to form in autumn, and is separated from the remaining landfast ice by one or more **tidal cracks**. The formation mechanism (Bentham, 1937) involves the freezing of water which is left behind by the ebb tide in pools and puddles on the beach, and a gradual building up of an ice sheet through successive tides. Properties of the ice foot have been reviewed by Feyling-Hanssen (1953) for Greenland and Svalbard, and McCann and Carlisle (1972) for Devon Island. An ice foot occurs even when there is a steep shore or fast currents and large tides which prohibit wider expanses of fast ice from forming, although of course it cannot form on rocky headlands; it is often used in winter for easy sledge travel along a coast. After the ice foot breaks up in spring, remnants often remain buried in beach sediment, so that the ice foot plays a role in shore erosion (Nansen, 1922). Fig. 2.22 shows such a case at Cape Tegethoff, Franz Josef Land, where the beach is being ploughed up and eroded back by the ice foot, which has been pushed up the beach during the break-up season.

The tidal cracks occur because the part of the ice foot which is afloat bends at low tide and fractures. The tidal cracks occur far from shore only if there is little tide or if the shoreline slopes very gently. In this case it is better to talk of **bottomfast ice** rather than an ice foot, to describe ice which is permanently grounded.

Beyond the tidal cracks, floating ice forms along the coast earlier in the winter than further out to sea. This is because the shallow water cools to the freezing point quickly, and often the nearshore water has a lower salinity due to river discharge. A typical

Figure 2.22. Remains of the ice foot on the beach at Cape Tegethoff, Franz Josef Land, in summer (July), showing beach erosion due to ice push.

sequence of events as winter proceeds has been described for the north coast of Alaska by Kovacs and Mellor (1974) and Reimnitz *et al.* (1977), and is illustrated in fig. 2.23. In early winter (2.23a) the fast ice grows seaward beyond the ice foot, only to encounter the offshore polar pack when it is driven landward by the wind. A variety of interactions can take place: a short landward excursion will affect only the outer part of the fast ice, turning it into an area of irregular rafted blocks, while more sustained pressure will lead to the creation of pressure ridges. The deeper ridges created out of deformed fast ice, plus ridges which are already part of the intrusive polar pack, can attain a draft which causes them to ground on the seabed. While the ice is still in motion, the crests of the grounded ridges excavate long narrow troughs in the seabed sediments, a process called **ice scour** (or ice score, or ice ploughing or gouging). Eventually the ridges come to a halt, and provide stable pinning points for the expanding fast ice sheet, possibly assisted in the case of Alaska by grounded fragments of ice islands (see chapter 7) (figs 2.23b,c). When the wind reverses and drives the polar pack out to sea, it leaves a **flaw lead** between the outer edge of the fast ice and the moving polar pack, which in midwinter soon refreezes to create new areas of thin fast ice. This extremely complex set of processes results in a fast-ice region which extends out to slightly further than the water depth limit at which a reasonable density of grounded ridges can occur, in the range of 18–27 m (Stringer, 1974). Within this zone, the portion nearest the coast may be smooth and undeformed, but most of the zone is heavily deformed with many deep ridges, or even fields of continuous ridging, called **rubble fields**, created by prolonged and extensive pressure of the offshore pack on the fast ice.

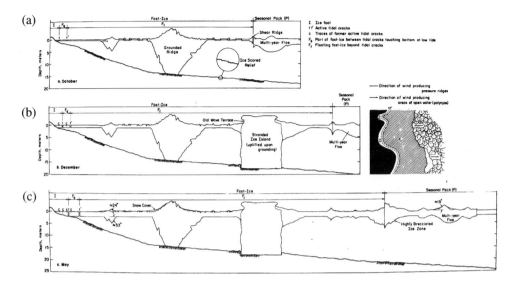

Figure 2.23. Ice-seabed interaction on the Beaufort Sea coast of Alaska during winter (after Kovacs and Mellor, 1974).

In the East Siberian Sea, the waters off Sakhalin, and other shallow seas north of Russia, a characteristic of summer is the existence of isolated grounded ridges called **stamukhi** (Reimnitz *et al.*, 1978; Reimnitz and Kempema, 1984). These are deep pressure ridges or ridge systems which grounded during the winter and became part of the fast ice zone; the remainder of the fast ice melted or broke up around them in spring, leaving the stamukhi to continue to exist as isolated grounded features in the open sea.

Although landfast ice is pinned at its inner edge to the shore, and at its outer edge to grounded pressure ridges and ice islands, it can still make small movements. Cooper (1974) measured displacements of 10 m in Mackenzie Bay, while Tucker *et al.* (1980) used a laser ranging system to measure 10 m displacements in the inner fast ice zone, but up to 160 m near the outer edge. The small displacements are due to thermal expansion of the ice, and the larger to the compressive effect of the offshore pack moving against the fast ice. Such displacements are of engineering significance, for one way of exploring for oil in nearshore Arctic waters has been to place portable drilling rigs on platforms of fast ice (often thickened by pumping water over the surface, sometimes reinforced with wood pulp) during the winter. Typically a drill string can only resist a lateral displacement of 10% of its length, so if the fast ice moves by more than 10% of the water depth there is a danger of breaking the drill string and causing a possible blowout if the rig has reached oil.

The compressive stress of the offshore pack on the fast ice, as well as other forces such as storm surges, can drive the fast ice bodily up the beach, causing the ice foot to plough up the beach sediments as shown in a mild way in fig. 2.22; in more severe cases the sheet can be driven hundreds of metres inland and can overwhelm coastal settlements. This process is called **ice pile-up** or **ride-up**, and was reviewed by Kovacs and Sodhi (1980). In beaches subject to ride-up there is often an escarpment a few feet high (fig.

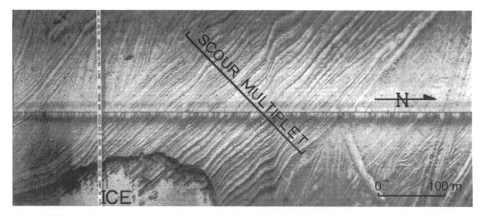

Figure 2.24. A sidescan sonar image of ice scouring in the Beaufort Sea (after Barnes *et al.*, 1984). This shows a multiple scour mark made by a pressure ridge with several deep crests, in a water depth of 25 m.

2.22) marking the limit of erosion by the fast ice during a typical year. An extreme event can eat away and undermine this escarpment, driving it back and causing progressive destruction of the coastline.

2.9.2. Ice Scour

Ice scour was first observed during sidescan sonar surveys of the Canadian Beaufort Sea during the summer of 1970 on the *Hudson-70* expedition (Shearer *et al.*, 1971), which was also the first expedition to circumnavigate the Americas (Wadhams, 1971). It was subsequently observed along the entire Beaufort Sea shelf (e.g. Weeks *et al.*, 1984) as well as in the Russian Arctic and Canadian Arctic Archipelago. A dense network of overlapping plough marks is observed (fig. 2.24) in which recent marks overlie older ones. Repeated sidescan surveys over the same terrain can therefore be used to estimate the rate of scour formation per unit area. Marks can range from single plough marks to multiple comb-like parallel marks, or a comb of marks with a space then another parallel comb. Clearly these correspond to the deepest blocks of an uneven pressure ridge running aground while the more shallow parts of the ridge crest do not touch bottom. Depths of marks typically range up to 2.5 m, with 4.5 m as a maximum value.

One important discovery was that the scour marks do not just correspond with the water depths of fast ice (up to 27 m) but extend out to 40–45 m and even beyond. The deepest observed pressure ridge in the Arctic has a draft of 47 m, while estimates based on extrapolation of ridge draft distributions have suggested that the deepest ridge existing in the Arctic at any given moment is about 55–58 m in draft. The deeper scours can therefore be ascribed to scouring by rare pressure ridges of extreme depth, embedded in the polar pack, and scouring the seabed without bringing the pack as a whole to a halt. The water depths correspond to the **shear zone**, the region where pressure ridging is in fact being created as the polar pack feels the restraining influence of the nearby fast ice. The role of shear zone scouring in reducing ice velocities in this zone has not yet been assessed, but it may be significant enough that it ought to be included in ice dynamics models. Certainly, massive forces are involved, all concentrated at the tip of the gouging ice block; it has been estimated (Palmer *et al.*, 1990) that the ice force required to cut a typical Beaufort Sea gouge is more than 10 MN.

More radically still, some scours have been observed in even deeper water, out to 65 m or more (Barnes *et al.*, 1984). A number of possible explanations exist for this. Firstly, pressure ridges of extreme depth, never observed in the field and created by mechanisms different from those which lead to the conventional negative exponential distribution, may exist in the polar pack. This is unlikely. Secondly, the scours may be due to ice island fragments of greater than expected depth. This is also unlikely. Thirdly, the scours may be "ancient" rather than recent: the seabed in the Beaufort Sea has been sinking over the past few thousand years since the whole Canadian Shield has been tilting through post-glacial rebound. The scours may have been created perhaps 3000 years ago when water depth was shallower, and have not yet filled in with sediment because of the very slow sedimentation rate far from shore. This is more likely, but a conclusive explanation depends on a careful geological examination of individual scours, the sediments contained in them, and the role of bottom currents in transporting sediments into scours.

A similar phenomenon to sea ice scouring is **iceberg scouring,** found on the Grand Banks of Newfoundland, off Baffin Island and in the Antarctic, in fact wherever icebergs move over continental shelves. This phenomenon will be discussed in chapter 7.

Occasionally seabed features are seen in shallow water which are not linear, but more like "potholes". These may correspond to the wallowing and pirouetting motion of grounded icebergs, as discussed in chapter 7, but carried out by an isolated ridge fragment or ice island fragment. In very shallow water a similar type of feature is attributed to **strudel scour** (Reimnitz and Bruder, 1972; Reimnitz *et al.*, 1974). This occurs when the spring run-off from an Arctic river flows out over the fast ice surface and drains away violently through cracks or seal breathing holes; the turbulence from this generates a circular depression in the seabed.

2.10. POLYNYAS AND THEIR ROLE

As the sea ice cover in the Arctic or Antarctic expands in winter, areas of water occur poleward of the ice limit, which do not freeze but which remain open to the atmosphere through all or part of the winter. These regions of long-lived open water are called **polynyas** (WMO, 1970) and are distinguishable by their persistence from the normal short-lived leads discussed in section 2.8. Polynyas in the Arctic have been important to man, historically allowing the Inuit a winter hunting ground (Schledermann, 1980), and have been described as oases within the polar desert allowing biological activity to continue throughout the winter (Lewis, 1990) and serving as habitats for large mammals and birds (Stirling, 1980). Polynyas are of great importance to the heat balance of the Arctic Ocean, roughly one half the total ocean-atmosphere heat exchange for the Arctic Ocean taking place through polynyas and leads (Makshtas, 1991), where the lack of ice insulating the ocean from the atmosphere allows large fluxes of heat and moisture.

A review has categorised two distinct types of polynya (Smith *et al.*, 1990) although most real polynyas have a mixture of both characteristics. A **latent heat polynya** is formed when sea ice is continually removed from the region in which it forms by winds or ocean currents, the heat needed to balance the loss to the atmosphere being provided by the latent heat of fusion of ice which continually forms. A **sensible heat polynya** is formed when a continued source of heat from the ocean prevents ice formation.

In the deep basins of the Arctic and Antarctic the wind-driven ice dynamics usually do not permit a persistent polynya to occur. When it does so it is usually a sensible heat polynya. The most famous Antarctic case was the Weddell polynya (Gordon, 1978; Carsey, 1980; Martinson *et al.*, 1981) which occurred over Maud Rise in the 1970s but which has not developed since. **Coastal polynyas,** however, are a common feature of continental shelves in both polar regions. They are believed to be primarily latent heat polynyas; that is, heat loss from the ocean surface is balanced by the latent heat of new ice formation and the polynya is maintained by wind or current removal of the new ice. The coastal polynyas around Antarctica can be further subdivided into two categories: those off the front of the major bay ice shelves (the Ross, Filchner-Ronne, Amery etc.) and those in the lee of ice tongues or other coastal projections, such as the Terra Nova Bay polynya in the Ross Sea (Bromwich and Kurtz, 1984) or along the Wilkes Land coast (Cavalieri and Martin, 1985). Those off the front of ice shelves involve processes of water circulation under the shelves and katabatic winds whilst inhibition of ice drift into the polynya area is important for those in the lee of ice tongues (e.g. Smith *et al.*, 1990).

Processes inside latent heat polynyas are significant for the sea ice zone as a whole. Being regions of intense heat loss from the ocean to the atmosphere, these polynyas can behave as "ice factories" contributing a significant fraction of total annual sea ice production; it has been estimated, for instance, that the small polynya in Terra Nova Bay contributes 10% of the ice production in the Ross Sea (Kurtz and Bromwich, 1985). Brine rejected during ice growth is concentrated in the polynya areas and can cause intense localised water mass modification as well as significantly increasing the salinity of Antarctic shelf water. The ice-free polynyas also play an important role for Antarctic marine ecosystems and in the control of biogeochemical fluxes. Around the fringes of the Arctic Ocean, coastal polynyas can make a significant contribution to the Arctic halocline, creating 0.7–1.2 Sv of dense water (Cavalieri and Martin, 1994).

Antarctic coastal polynyas which have been studied by remote sensing alone include the coastal polynya in the Cosmonaut Sea (Comiso and Gordon, 1987, 1996) and in Terra Nova Bay in the Ross Sea (Kurtz and Bromwich, 1983; Bromwich and Kurtz, 1984; Bromwich *et al.*, 1990). A summary of recurrent polynyas that have been detected and studied using passive microwave data has been given by Zwally *et al.* (1985). Fig. 2.25 shows the locations of prominent coastal polynyas around the coasts of Antarctica, picked out by passive microwave imagery.

In the Greenland Sea an equivalent type of coastal polynya occurs along the east coast of Greenland. Their role in the overall heat and salt budgets of the Greenland Sea has not yet been addressed, except for the case of one particularly prominent polynya, the **Northeast Water** (NEW). This occurs in winter and spring from just south of Nordostrundingen (81°30′N) extending southwards to about 80°30′N. It has been extensively studied as part of the International Arctic Polynya Programme (Gudmandsen *et al.*, 1995; Lewis *et al.*, 1996; Minnett, 1995; Schneider and Budeus, 1994, 1995). Another large persistent polynyas is the **North Water** in Smith Sound (northern Baffin Bay) between Greenland and Ellesmere Island. Here the southward drift of heavy pack ice

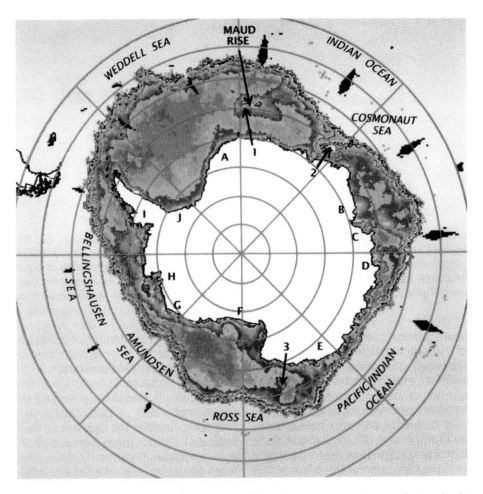

Figure 2.25. Locations of prominent polynyas around Antarctica, from passive microwave imagery in winter 1980 (after Gordon and Comiso, 1988). Coastal polynyas are lettered A–J and polynyas within the pack are numbered 1–3. Polynya 1 is the Weddell Polynya over Maud Rise (at this time it did not contain open water, but a reduced ice concentration), while polynya 2 is the Cosmonaut Sea polynya which contained open water.

through Nares Strait is arrested by a mechanical blockage of the Strait, whereas the surface water continues southward, leaving an open region. In the Bering Sea prevailing northerly winds in winter keep open polynyas on the south side of St. Lawrence Island and in Norton Sound (Pease, 1987). **Flaw lead polynyas**, which develop just off the edge of the fast ice under an offshore wind, occur all around the Arctic, especially in spring. The case of the Laptev Sea has been considered by Winsor (1997) and fig. 2.26 shows a typical development of this polynya. A full description of polynyas and their various origins is given by Smith *et al.* (1990), and we describe the dynamics and thermodynamics of a wind-driven polynya in chapter 4.

Figure 2.26. The Laptev Sea flaw lead polynya in spring 1995, from a visible-band DMSP satellite image (after Brigham, 1996).

Table 2.2. Air-sea heat fluxes and ice formation rates in polynyas (from Smith *et al.*, 1990).

Source	Sea-Air Heat Flux (W/m²)	Time Periods	Mean Ice Production (m/day)	Region	Type
Kurtz and Browich, 1985	713	May–Sept.	0.31	Terra Nova Bay	latent
Cavalieri and Martin, 1985	160–450	June–Sept.	0.05–0.13	Wilkes land coast	latent
Schumacher *et al.*, 1983	535	65 hour event	0.17	St. Lawrence I.	latent
Pease, 1987	150–500	Feb, 82, 83, 85		St. Lawrence I.	latent
den Hartog *et al.*, 1983	329	March 1980		Penny Strait	both
Alfultis and Martin, 1987	446	Dec.–March 1978–1982	0.13	Sea of Okhotsk	sensible
Steffen, 1985	180	Nov.–March		North Water	both

Within the Greenland Sea proper, Koch (1945) identified three recurrent coastal polynyas which occur just beyond the edge of the coastal fast ice in winter and spring, persist for weeks or months, and can open and close repeatedly. These are situated:-

(1) off the southern part of Store Koldeway, from 76° to 78°N,
(2) from Bass Rock southward to Jackson Island (73°30′N to 75°N),
(3) in the mouth of Scoresby Sund.

Wadhams (1986), in discussing these observations, showed satellite imagery of many more, but smaller, winter polynyas which occur on the lee side of protruding capes of fast ice around islands or headlands. In addition, Wadhams discussed another phenomenon identified by Koch, that of **land water.** This is the effect whereby a continuous narrow strip of open water is formed against the fast ice edge by offshore winds. It is the Greenland Sea version of a flaw lead polynya or of the long narrow polynyas formed off ice shelf fronts in the Antarctic. Sensible heat brought by wind-induced upwelling may add to the purely mechanical effect of the wind in keeping the landwater channel open. Landwater has proved to be useful in providing an "inner passage" for supply ships and sealing ships to run northwards inside the pack ice edge in winter and spring. Occurrences of landwater were tabulated and illustrated in Wadhams (1981c, 1986); it tends to be most common in spring (April–June).

In Table 2.2 we present a compilation of air-sea heat fluxes and ice formation rates for selected polynyas in both polar regions (Smith *et al.*, 1990). It has been suggested on the basis of laboratory experiments that the rates of ice production may be greater than shown in this table owing to underwater frazil ice production adding to the production at the surface of a polynya (Ushio and Wakatsuchi, 1993).

A frazil ice skim is the normal initial form of ice that grows at the surface of the water column in a polynya, especially in wind-driven Antarctic coastal polynyas when katabatic winds, funnelling down off the Antarctic ice sheet between mountain peaks, are the force keeping the polynya open. If the ambient wind speed is high enough the frazil ice suspension is ordered into bands by Langmuir circulation in the surface layers (Pease,

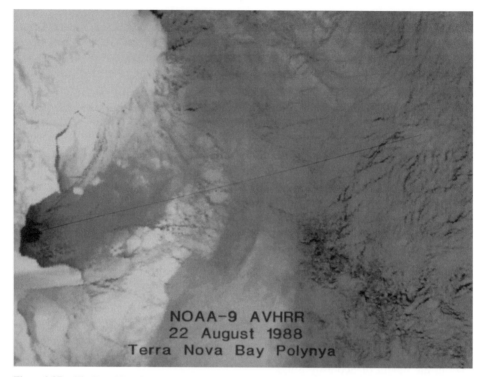

Figure 2.27.   The Terra Nova Bay polynya in the Ross Sea as shown by NOAA AVHRR (visible band) imagery. By courtesy of R. Whritner, Scripps Institution of Oceanography.

1987), leaving open water between the bands. The frazil ice then collects on the downwind edge of the polynya, the depth of the ice layer being a function of both the ambient air temperature and the wind speed (Bauer and Martin, 1983). This layer of frazil ice will eventually damp out the wave energy impinging on the downwind side of the polynya (Martin and Kauffman, 1981; Squire *et al.*, 1995), and the wave radiation pressure acting on this frazil layer turns the frazil into pancakes and ultimately into a first-year ice sheet. This sheet then continues to move downwind and is the product of the polynya "ice factory", a very potent source of ice production for the Antarctic.

A typical case of a small localised polynya occurs in Terra Nova Bay in the western Ross Sea, the place where Captain Scott's Northern Party spent an uncomfortable winter living in a snow bank during 1911–12, and now the site of an Italian scientific base. Fig. 2.27 shows this polynya during winter. It is kept open by katabatic winds blowing down a narrow glacier which drains into the sea as the Hells Gate Ice Shelf. The coastal feature which helps maintain the polynya is the Drygalski Ice Tongue, growing out to sea south of Terra Nova Bay, which shields the bay region from coastal pack ice drifting northwards in the Antarctic Coastal Current. This particular polynya is the one estimated to produce 10% of the ice generated over the entire Ross Sea shelf, so polynyas are clearly important as locations of enhanced ice production. They are also important, however, in their ocean interactions. Such a large amount of ice production is associated with a large amount of brine rejection, and so if the polynya is located in a critical region where mixing is occurring leading to bottom water generation, the additional brine production may have a strong effect enhancing the rate of bottom water production (Zwally *et al.*, 1985), as was discussed in chapter 1. In chapter 1 we also discussed the role of polynyas situated off the edges of ice shelves in producing a circulation under the shelf with associated melting and freezing processes.

3.   THE THERMODYNAMICS OF SEA ICE

You are in a Swedish icebreaker entering Independence Fjord, the great ice-choked fjord of Northeast Greenland that was named by Robert Peary and was the site of the 1906 tragedy in which the explorer Mylius Erichsen perished with all his men. It is a foggy day in summer. The ship comes to a halt in the heavy ice. As the morning mist clears you take a helicopter further up the fjord in search of a type of ice that you have read about but never seen, **sikussak.** *In 1945 Lauge Koch, the revered Danish explorer and surveyor of East Greenland, completed a monumental work on "The East Greenland Ice", a book upon which he had lavished five years of his life whilst interned by the Germans. He had found a few places in the furthest northern reaches of the island where the fast ice remains for decades and becomes immensely thick. The Eskimos, who have a word for every type of ice and snow, call it sikussak,"fjord ice that looks like ocean ice". At last you see it, a wild landscape like the Grand Canyon, formed by the annual cycle of melt and refreezing, with mesas, isolated peaks and spires, and deep gullies filled with running water. A dozen or so melt streams like this radiate out from a thaw hole, like the legs of a spider, but beware of the hole itself — a turbulent mass of fresh water plunges down it into the sea below. If you fell you would never be found. The helicopter drops you and your colleagues and the coring equipment. You begin to drill from a mesa, carefully slicing each core into 10 cm sections which you put into bottles to melt later and measure the salinity. You are 6 m down and nowhere near the bottom. Suddenly the helicopter returns. The ship must leave and sail east to meet the King of Sweden who is flying out by long-range helicopter from Svalbard. You cannot argue with royalty. Reluctantly you pack up and leave. You will never know the thickness reached by sikussak.*

How thick does sea ice grow if left alone for years? Will it grow thicker from year to year or does it reach a limit? If this limit exists, how does it vary in different parts of the world? How is it affected by snow fall and heat flux from the ocean? These are questions which can be addressed by relatively simple models, and which we consider in this chapter.

3.1. THERMOPHYSICAL PROPERTIES OF SEA ICE

Before we start we must examine the basic thermal properties of sea ice, which determine its rates of growth and decay, and especially the phase transitions that occur during melting and freezing. The properties which we need to know are the thermal conductivity, specific heat, latent heat of fusion, and extinction coefficient for radiation. We need to know these properties both for sea ice and for snow, since the snow which covers the sea ice in winter plays an important role in limiting the thickness to which sea ice can grow; its low thermal conductivity makes it act like a thermal blanket which reduces heat loss from the surface of the ice.

Sea ice is a mixture of four components: ice; liquid brine; air bubbles; and solid salts. The conduction of heat through sea ice is influenced both by the porosity (air bubble content) of the ice and by the solid salt content, but the most complex effect occurs with liquid brine. This is because, as we have seen in chapter 2, the brine is contained in tiny cells of concentrated solution; each cell is at its freezing point and is in phase equilibrium with the surrounding ice. When the temperature within the ice rises the ice surrounding the brine cell melts, absorbing latent heat, diluting the brine and raising its freezing point to the new temperature. When the ice temperature falls some of the water in the cell freezes, releasing latent heat and producing a smaller cell containing more concentrated brine with a lower freezing point. The brine cell is thus a **thermal reservoir,** retarding the heating or cooling of the ice. This extra resistance to warming or cooling means that the specific heat of sea ice is a function both of salinity (increasing with increasing salinity) and temperature (increasing with temperature, since the brine volume becomes very large near the melting point).

3.1.1. Thermal Conductivity

The thermal conductivity of sea ice was first investigated by Malmgren (1927) as part of his classic work done during Amundsen's "Maud" expedition. Later work has included theoretical studies by Schwerdtfeger (1963) and Shuleikin (1968) and experimental work by Nazintsev (1964) and Ono (1965), with a review of results by Doronin and Kheisin (1977).

Untersteiner (1961) introduced an approximate formula for thermal conductivity, which was later employed in the model of Maykut and Untersteiner (1971) described in section 3.3. His relationship was

$$k_i = k_o + \text{ß } S_i / T_i \tag{3.1}$$

where k_i is thermal conductivity; k_o is the thermal conductivity of pure polycrystalline ice at temperature T_i (°C); S_i psu is the salinity of the ice; and $ß = 0.13$ W m^{-1}. The thermal conductivity of pure ice was given by Yen (1981) as

$$k_o = 9.828 \exp(-0.0057\ T) \tag{3.2}$$

where T, the ice temperature, is now in °K. k_o has units W m^{-1} °K^{-1}. Yen's formula for pure ice was based on extensive experimentation by many researchers. Equation (3.1), on the other hand, is rather a crude approximation since it has been found that k_i is strongly dependent not only on the salinity of the ice (as expressed in 3.1) but also on its air bubble content. It is a reasonable approximation for use in models which deal only with salinity.

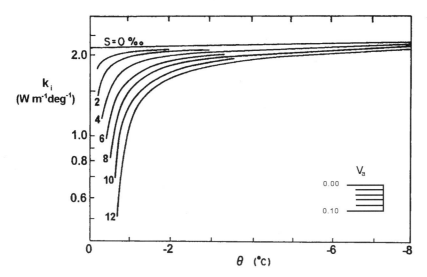

Figure 3.1.   Thermal conductivity of sea ice as a function of salinity and air bubble content (after Ono, 1968).

The question of air bubble content was considered in the studies of both Schwerdtfeger (1963) and Ono (1968). The problem is that the thermal conductivity of seawater brine (the constituent of the brine cells) is about 25% that of pure ice, while the thermal conductivity of air is less than 1% that of ice. The result is that the thermal conductivity of typical sea ice is greatly reduced relative to that of pure ice near the melting point, when the brine volume is high, but as the temperature is lowered reaches an asymptote similar to the thermal conductivity of pure ice *so long as there are no air bubbles.* The air bubble content lowers this asymptote. The first theoretical model, that of Anderson (1958), considered only brine cells, which he allowed to occur either as isolated random spheres or as cylinders or layers oriented as discussed in chapter 2 at right angles to the c-axis. Schwerdtfeger (1963) added randomly distributed spherical air bubbles in the ice to this model. Finally Ono (1968) allowed the air bubbles to be distributed through both the ice and the brine.

We present Ono's (1968) model results (fig. 3.1) as being the most easy to use. The main curves are for bubble-free ice, and it is very clear how salinity depresses the thermal conductivity near the melting point but not at lower temperatures. To account for the effect of air bubbles one takes the bubble-free value and reduces the thermal conductivity by the bubble fraction V_a shown on the inset curve.

Fig. 3.1 shows that very saline, warm ice (e.g. young ice forming in early winter, or melting first-year ice) has a very much lower thermal conductivity than cold or low-salinity ice, by a factor of up to 2 or 3. This is seldom taken into account in discussions of growth or melt rates, but both of these will be significantly retarded in warm salty ice if the model is correct. Of course, none of the models to date take account of possible convection in the liquid or gaseous phases, and we might expect convection in brine drainage channels, for instance, to significantly increase the average thermal conductivity of warm saline ice.

Finally, no measurements have been done on the relationship between k_i and crystal orientation, although the model of Anderson (1958) predicts that the conductivity parallel to the c-axis is much lower than the conductivity perpendicular to the c-axis. If confirmed experimentally, this would provide another reason, in addition to the energetic argument given in chapter 2, why crystals with c-axes horizontal grow faster than their competitors — the latent heat of fusion can be conducted away more easily.

3.1.2. Specific Heat

The specific heat c_i of sea ice was found by Ono (1967) to give a good fit to the following empirical relationship:

$$c_i = c_o + a\, T_i + b\, S_i / T_i^2 \tag{3.3}$$

where $c_o = 2113$ J kg^{-1} °C^{-1} is the specific heat of pure ice, T_i is temperature in °C, S_i psu is ice salinity, $a = 7.53$ J kg^{-1} °C^{-2} and $b = 0.018$ MJ °C kg^{-1}. The third term on the right hand side shows that the effects of temperature and salinity on specific heat are mainly important near the melting point, and are insignificant below −8°C.

A quantity which is easier to measure directly is **thermal diffusivity**, σ_i, defined as

$$\sigma_i = k_i / (\rho_i\, c_i) \tag{3.4}$$

This was calculated by Ono (1968) with results shown in fig. 3.2. It is a directly observable thermal property in that it can be calculated from the rate of change of the temperature profile in an ice sheet. Methods of calculating σ_i are discussed in Ono (1965, 1968) and Yen (1981), and observations made by Ono (1965, 1968), Lewis (1967) and Weller (1968) have agreed quite well with predictions.

3.1.3. Latent Heat of Fusion

The concept of latent heat in the case of sea ice is a complex one, since thermodynamically it is possible for sea ice and brine to coexist at any temperature, and therefore for sea ice to melt at temperatures other than 0°C if it is bathed in a suitably concentrated salt solution, such as occurs at the walls of brine cells when brine cell migration is taking place (section 2.4).

Ono (1968) produced a formula based on thermodynamic considerations and suitable for use at temperatures above −8°C, where complicating factors set in on account of the precipitation of solid sodium sulphate ($Na_2SO_4.10H_2O$) from concentrated brine. His formula is

$$q = 333394 - 2113\, T_i - 114.2\, S_i + 18040\, (S_i/T_i) \tag{3.5}$$

where q is the latent heat of fusion in J kg^{-1}, T_i is temperature in °C, and S_i psu is ice

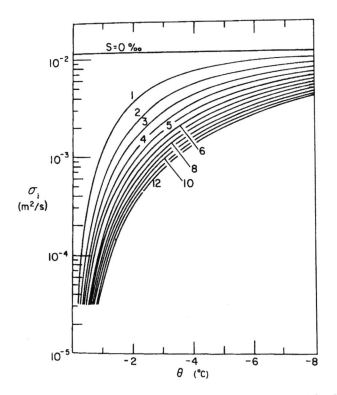

Figure 3.2. Thermal diffusivity of sea ice as a function of salinity and temperature (after Ono, 1968).

salinity. Ono carried out laboratory experiments to test both (3.5) and (3.3), but with results which showed that at times of freeze, melt or rapid temperature change, the brine in sea ice is seldom in equilibrium with the ice itself. These formulae should therefore be applied with caution, and as in so many other aspects of basic sea ice physics, more experimentation needs to be done.

Further discussion of latent heat is given by Doronin and Kheisin (1977), and table 3.1 shows some values of q for temperatures close to 0°C.

Table 3.1. Heat q in kJ required for complete fusion of 1 kg of sea ice.

T_i °C	S_i psu					
	0	1	2	4	6	8
−0.5	335	300	264	194	124	53
−1.0	336	318	301	266	230	195
−2.0	338	329	320	302	284	264
−3.0	340	334	328	316	303	291

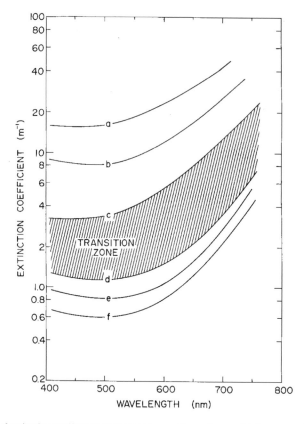

Figure 3.3. Spectral extinction coefficients for various types of ice and snow: (a) dry compact snow, (b) melting snow, (c) the surface layer of melting multi-year ice, (d) the interior of multi-year ice (the area marked "transition zone" representing transitional layers between surface and deep interior), (e) the interior of first-year ice, (f) ice beneath an old melt pond (after Grenfell and Maykut, 1977).

3.1.4. Radiation Extinction Coefficient

When solar radiation is incident on an ice or snow cover, a fraction of it is immediately reflected. This is called the **albedo** (α). The rest penetrates into the interior, where it is subject to absorption and scattering. The rate of absorption depends on the angle of incidence of the radiation and the wavelength as well as the properties of the material. For any wavelength, however, the decay of penetrating energy with distance is found to be exponential (**Beer's Law**), and if we think in terms of penetration through a vertical distance z in the snow or ice cover, can be represented in terms of a **spectral extinction coefficient** $\kappa(z,\lambda)$, i.e.

$$I(z,\lambda) = I_0(\lambda) \exp\left(-\int_0^z \kappa \, dz\right) \tag{3.6}$$

where $I(z,\lambda)$ is the intensity of radiation penetrating to a depth z in the material and $I_0(\lambda)$ is the net radiation penetrating into the surface, given by

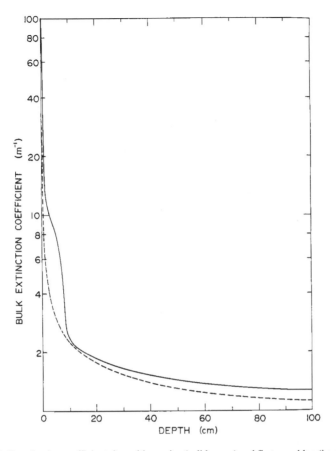

Figure 3.4. Bulk extinction coefficients in multi-year ice (solid curve) and first-year blue (i.e. mainly bubble-free) ice (dashed curve). After Grenfell and Maykut, 1977.

$$I_0(\lambda) = \int_0^\infty (1 - \alpha) \, F_0(0,\lambda) \, d\lambda \qquad\qquad (3.7)$$

Here F_0 is the net total shortwave radiation at the surface.

The spectral extinction coefficient varies enormously between ice types. Fig. 3.3 shows some experimental results. Clearly radiation is extinguished much more rapidly in snow than in ice, and there is a lower extinction rate at the blue end of the spectrum than at the red, hence the tendency for ice to look blue when viewed by transmitted light. If we think in terms of e-folding distances (the distance to reduce the intensity to 1/e, i.e. 37%, of its initial value), we find these varying for typical sea ice from 24 metres at 470 nm (blue) to 8 metres at 600 nm (red) to 2 m at 700 nm (near infra-red) (Perovich, 1998).

For the purposes of thermodynamic modelling, as we shall show in section 3.3, it is an unnecessary complication to have to deal with the variation of extinction coefficients with wavelength, and instead we simply use a **bulk extinction coefficient** κ_z, obtained by weighting the spectral extinction coefficient by the distribution of energy in the shortwave spectrum which reaches penetration z, i.e.

$$\kappa_z = \frac{\int \kappa(z,\lambda)\ I(z,\lambda)\ d\lambda}{\int I(z,\lambda)d\lambda} \tag{3.8}$$

where the integration is carried out over the spectral range which carries all the significant shortwave radiation.

Fig. 3.4 shows bulk extinction coefficients derived in this way for two kinds of ice. Note how the values are highest in the uppermost 20 cm of the ice cover, and then decline to typical values of 1 to 1.5 m^{-1} for interior ice. Much higher values have been found for snow: 4.3 m^{-1} for dense Antarctic snow (Weller and Schwerdtfeger, 1967) up to 40 m^{-1} for freshly fallen snow (Thomas, 1963).

The use of bulk extinction coefficients has been criticised in recent years (e.g. Maykut et al., 1992; Perovich, 1998) because they do not depend entirely on the properties of the ice. They depend on the spectral properties of the albedo, the sky conditions on the day of measurement (sunny days have a relatively greater longwave component in the incident spectrum than cloudy days), and are strongly dependent on conditions in the upper few cm of the ice cover, which therefore take on a critical importance for heat budget estimates.

Albedo, too, should properly be expressed as a **spectral albedo** $\alpha(\lambda)$, and this quantity has been measured for many kinds of ice (Maykut, 1986; Maykut et al., 1992; Perovich, 1998). However, once again for heat budget models it is more convenient to define a bulk albedo in an analogous way to (3.8), i.e.

$$\alpha = \frac{\int \alpha(\lambda)\ I(0,\lambda)\ d\lambda}{\int I(0,\lambda)d\lambda} \tag{3.9}$$

Area-averaged bulk albedo is a highly variable quantity and is particularly difficult to define in summer, when the surface is a mixture of snow-covered ice, bare ice, melting ice and melt pools. At this time of year albedo is most difficult to predict, yet in models it is most important to get the value correct in summer, since incident radiation is greatest then (Curry et al., 1995). Fig. 3.5 shows the wide range of values obtained for different ice types, ranging from 0.87 for new snow down to 0.15 for an old melt pond (and 0.06 for open water, which is also prevalent in summer and which does affect the ice mass budget through lateral melting of floes in water which has undergone surface warming). This variability is one of the greatest problems in ice thermodynamics. The most recent field study to address this problem was SHEBA (Surface Heat Budget of the Arctic), a 1997–8 drift experiment in which a single area of ice was followed through changing seasons (Moritz and Perovich, 1996). It is clear that the additional precision of using spectral albedos in models is not justified while the bulk albedo is so poorly known.

3.2. EARLY MODELS OF ICE GROWTH AND DECAY

Many early attempts were made to develop empirical relationships to predict ice growth from observed air temperatures. The best known were by Barnes (1928), Lebedev (1938) and Zubov (1945). Zubov, in his classic work, *L'dy Arktiki (Arctic Ice)*, found the relationship

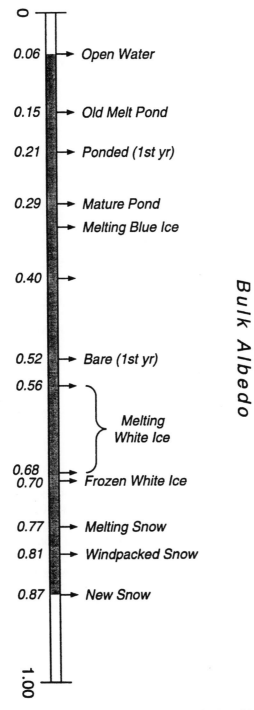

Figure 3.5. Range of observed values for bulk albedo of sea ice (after Perovich, 1998). Data are from Burt (1954), Chernigovskiy (1963), Langleben (1971), Grenfell and Maykut (1977) and Grenfell and Perovich (1984).

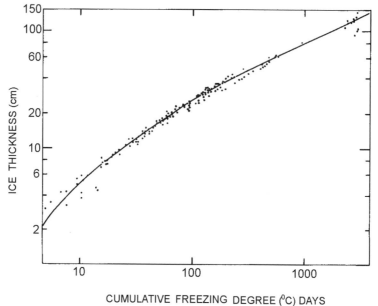

Figure 3.6. Anderson's (1961) relationship for young ice thickness as a function of degree-days of cold.

$$H^2 + 50\,H = 8\,\theta \qquad (3.10)$$

where H is ice thickness in cm, and θ is cumulative freezing degree-days. Lebedev's relationship was

$$H = 1.33\,\theta^{0.58} \qquad (3.11)$$

Both these authors used the concept of cumulative degree-days of cold during the winter. The average daily temperature is subtracted from −1.8°C (the freezing point of sea water) and summed from day to day. The result was found to give a good correlation with the thickness reached by sea ice, with a complicating effect from the snow cover (which inhibits ice growth).

More recent work on such semi-empirical relationships includes Bilello (1961), who developed a statistical method based on observations of fast ice growth, and Anderson (1961), who derived a relationship for young ice in the Arctic with minimum snow cover (fig. 3.6), which fitted the relation

$$H^2 + 5.1\,H = 6.7\,\theta \qquad (3.12)$$

Bilello (1961, 1980) developed a similar technique to predict ice decay in summer for nearshore data, obtaining

$$\Delta H = 0.55\,\theta' \qquad (3.13)$$

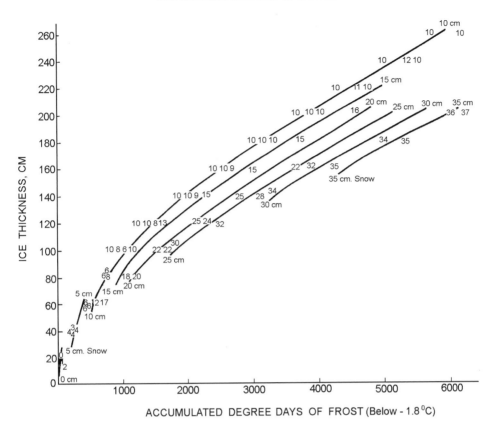

Figure 3.7. Family of curves of ice thickness versus degree-days of cold, for different snow depths (after Bilello, 1961).

where ΔH is the total decrease in ice thickness in cm and θ' is the cumulative degree-days above the freezing point.

These techniques work for fast ice because there is no ocean heat flux involved, since the water is so shallow that it consists of only one layer of polar surface water; they also work well for young ice in refreezing leads because during the early stages of growth the conductive flux in the ice greatly exceeds the oceanic heat flux. They have been used to interpret the best and longest series of fast ice growth data from a whole winter (Nakawo and Sinha, 1981). They are good semi-quantitative indicators of the likely impact of climate change upon the fast ice zone, as we shall discuss in chapter 8. However, they are inadequate for describing thicker ice in the deep ocean, or ice which carries a snow cover. Anderson's (1961) data deviated greatly from his equation at higher thicknesses than about 80 cm because of the variable snow cover, while Bilello (1961) had to develop a family of curves (fig. 3.7) to allow for differences due to snow cover. In addition, we do not expect equation 3.13 for decay rate to be valid anywhere except near shore, where warm winds from land may act, since out in the main polar pack ice zone the summer air temperature always stays very near 0°C. Maykut (1986) showed how relationships of the type (3.10) to (3.12) based on degree-days of cold may be developed on simple theoretical grounds, provided one makes highly simplifying assumptions such as

— a uniform slab of ice with constant thermal conductivity;
— no ocean heat flux;
— crudely parameterised heat exchange at ice surface, proportional to temperature difference between ice surface and air;
— imposed snow layer of constant thickness.

It was clear that a more complete modelling approach based on a full physical theory was required. It is worth mentioning here the very first attempt at such a theory, since it occurred very early, in fact not very long after the radiation law itself was discovered. In 1890 Stefan developed a simple model which showed that the thickness H of young ice should follow the relations

$$H \propto t^{1/2} \tag{3.14}$$

and

$$dH/dt \propto (T_a - T_w) / H \tag{3.15}$$

where t is time and T_a and T_w are the temperatures of the top and bottom of the ice. The growth rate should decrease rapidly as the ice becomes thicker, by about an order of magnitude between ice 10 cm and 100 cm thick in the case of the Arctic Ocean.

This model brings out the main properties of ice thermodynamics: that ice growth slows as the ice gets thicker, so that an icefield in which leads have opened because of divergent wind stress tries to "heal" itself through rapid ice growth in the refreezing leads, and that under thermodynamics alone an icefield composed of ice of varying thicknesses tries to reach a uniform thickness value, with thin ice getting thicker and thick ice getting thinner. However, it still does not take proper account of radiative forcing at the top of the ice, nor of the variability of ice conductivity, nor energy exchanges with the ocean. Clearly a more complete model is necessary.

3.3. THE MAYKUT-UNTERSTEINER MODEL

In 1971 a model was produced for the thermodynamic growth and decay of sea ice in the Arctic Ocean by Maykut and Untersteiner. It is still the basis for understanding the thermodynamics of sea ice, with a simplified version suitable for climate studies being presented by Semtner (1976) and modifications based on more recent data by Maykut (1986). A more recent reassessment and update has been by Ebert and Curry (1993).

3.3.1. Formulation

The model considers the seasonal cycle of short and long-wave radiation fluxes upon a growing or decaying ice sheet, with a seasonally specified snowfall and a given oceanic heat flux. We assume that the ice sheet is an infinite, horizontally homogeneous slab, with

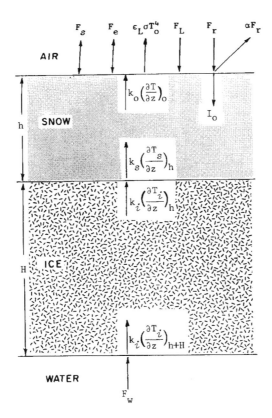

Figure 3.8. Fluxes in a uniform snow-covered ice sheet (after Maykut and Untersteiner, 1971).

no thickness variations due to ice deformation, and that the ocean upon which it floats is at rest so that there is no heat transfer due to friction between the water and the ice. The ice transfers heat between the ocean and the atmosphere. by conduction, but this transfer is affected by brine cells within the ice and by short-wave radiation penetrating the upper surface of the ice during spring, summer and autumn. The snow cover also reduces both the radiative and conductive transfer of heat through the slab and so must be represented as a second layer within the model.

Figure 3.8 shows the geometry and the fluxes involved. A snow layer of thickness h covers a sea ice sheet of thickness H. We assume that mass changes and energy absorption happen only at the snow and ice boundaries. At the top boundary of the snow (or of bare ice in summer), the snow or ice may melt, but mass can be added only through snowfall. At the ice-water interface either ablation (ice melt) or accretion (ice growth) may take place. We consider the heat balance at each boundary in turn, and also the heat transmission in the interior of the snow and the ice. The sign convention is that energy fluxes are considered positive towards a surface and negative away from a surface. The vertical axis z is measured downwards from the snow surface.

Upper surface of snow

We start at the top of the snow layer. Here we have a balance of energy fluxes, where several incoming energy fluxes from the atmosphere are balanced by long-wave radiation from the surface and by heat penetrating into the ice. The incoming radiations are:-

F_r = incoming short-wave radiation from the sun which reaches the surface after penetrating the atmosphere.

F_L = incoming long-wave radiation from the atmosphere and clouds, which themselves have absorbed some incoming solar radiation as it passed through the atmosphere, and have re-emitted it at a lower frequency by virtue of their own absolute temperature.

The outgoing radiations are:-

$\alpha\, F_r$ = the fraction of incoming solar radiation which is immediately reflected by the snow surface. α is the albedo of the surface.

$e_L\, \sigma\, T_o^{4}$ = the outgoing long-wave radiation emitted by the surface. Here T_o is the absolute temperature of the snow surface, σ is the **Stefan-Boltzmann constant** (= 5.671 × 10^8 W m^{-2} K^{-4}) and ε_L is the **long-wave emissivity.** This is simply an expression of the Stefan-Boltzmann law of radiation, which states that the total radiation emitted by a body is proportional to the fourth power of its absolute temperature. σ is the constant of proportionality involved, while the emissivity expresses the way in which the colour or texture of a surface causes it to emit less than the theoretical maximum amount of radiation, called "black body radiation". For snow, the emissivity is usually close to unity, and for sea ice it lies in the range 0.66 to 0.99 (Cavalieri *et al.*, 1981).

There are two additional energy flux terms which can be positive or negative. These are:-

F_s = sensible heat flux to the adjacent air. Sensible heat is the heat actually physically transferred by the snow surface to the overlying air by conduction. If the snow surface is warmer than the air near the surface, the heat transfer is upwards and the sensible heat is negative; if it is colder the heat transfer is downwards and the sensible heat is positive. Sensible heat transfer is a complex process since upward sensible heat involves creating small-scale turbulent convection as the heated parcels of air move upwards and are replaced by colder air parcels moving downwards to interact in their turn with the surface.

F_l = latent heat flux to the adjacent air. This is the energy exchange due to sublimation of snow into water vapour.

If we examine the fluxes downwards from the snow surface into the body of the snow layer, there are two:-

I_o = the flux of radiative energy which penetrates through the snow surface into the body of the snow. This was considered in section 3.1.4.

F_c = the heat conducted downwards into the snow layer (or upwards from the snow layer to the snow surface). This is given by

$$F_c = k_s \, (\partial T / \partial z)_o \tag{3.16}$$

where k_s is the thermal conductivity of the snow and $(\partial T/\partial z)_o$ is the temperature gradient through the snow measured at the snow surface $z = 0$.

If the surface temperature T_o is below the freezing point, these fluxes will all balance, with the balance determining the value of T_o. If, however, the surface of the snow is at the melting point, an imbalance between the incoming and outgoing energy fluxes can be accommodated through the melting of snow, causing a change in h (or of H is the ice surface is bare). Thus there are two possible equations for the energy balance, depending on the surface temperature:-

$$(1 - \alpha) \, F_r - I_o + F_L - \varepsilon_L \, \sigma \, T_o^4 + F_s + F_l + k_s \, (\partial T / \partial z)_o = 0 \quad \text{if} \quad T_o < 273.16$$

or

$$= - \, [q \, d(h + H) / \, dt]_o \quad \text{if} \quad T_o = 273.16 \tag{3.17}$$

where q is the latent heat of fusion of the surface material, whether it be snow or bare ice.

Most of the parameters used in equation (3.17) are external, seasonally varying parameters which have to specified based on our knowledge of environmental conditions in the Arctic or Antarctic Ocean (α, F_r, F_L, F_s and F_l). Other parameters require a knowledge of snow or ice properties (k_s, q, ε_L), while snow deposition, which is a partial determinant of h, must also be specified as an external input to the system.

Interior of snow layer

Inside the snow layer, assuming that there is one (in summer this component of the model is of zero thickness and so is ignored), heat is conducted along the temperature gradient from warm to cold, while at the same time the penetrating solar radiation is gradually absorbed. The absorption of radiation by snow is a complex process dependent on wavelength, angle of incidence, and the physical structure of the snow. It was considered in section 3.4, and is here expressed for simplicity in terms of a single bulk extinction coefficient κ_s, such that the flux I(z) at a depth z is given by

$$I(z) = I_o \, \exp \, (- \, \kappa_s \, z) \tag{3.18}$$

The thermal conductivity of the snow, k_s, is also assumed to be constant.

Consider a small unit cell within the snow, of dimensions δx, δy, δz (figure 3.9). The flux incident on the upper surface of the cell is

$$F_1 = - \, k_s \, (\partial T/\partial z)_z + I_o \, \exp \, [- \kappa_s \, z] \tag{3.19}$$

The flux emerging from the lower surface is

$$F_2 = - \, k_s \, (\partial T/\partial z)_{z+\partial z} + I_o \, \exp \, [- \kappa_s \, (\, z + \partial z)] \tag{3.20}$$

$$F_1 = - k_s (\partial T/\partial z)_z + I_o \exp[-\kappa_s z]$$

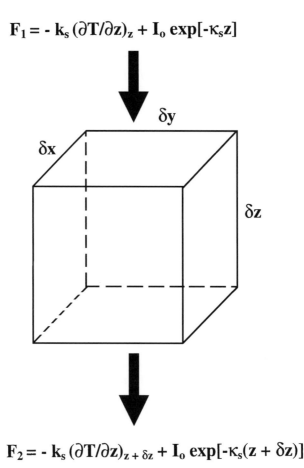

$$F_2 = - k_s (\partial T/\partial z)_{z + \delta z} + I_o \exp[-\kappa_s(z + \delta z)]$$

Figure 3.9.   Schematic of heat conduction through a unit cell of the snow or ice cover.

The net rate of energy gain by the cell is $(F_1 - F_2)\, \delta x\, \delta y$, which can be equated to the rate of temperature rise in the cell by

$$(F_1 - F_2)\, \delta x\, \delta y = \rho_s\, c_s\, \delta x\, \delta y\, \delta z\, (\partial T/\partial t)_z \qquad (3.21)$$

where c_s is the specific heat of the snow and ρ_s is its density. Assuming a slow rate of change of temperature gradient with distance, and neglecting terms above second order, this yields

$$\rho_s\, c_s\, (\partial T/\partial t)_z = \kappa_s\, I_o\, \exp[-\kappa_s\, z] + k_s\, (\partial^2 T/\partial z^2)_z \qquad (3.22)$$

Snow-ice interface

Assuming that a snow layer exists, we can also assume that conduction is continuous

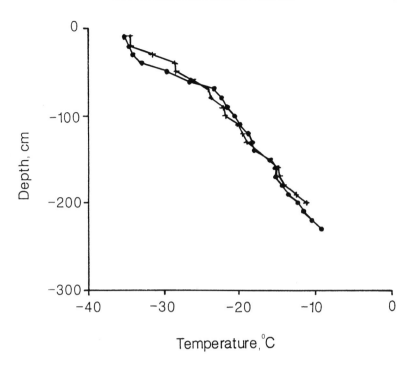

Figure 3.10. Temperature profiles obtained by author using two thermistor chains inserted through the same floe of multi-year sea ice NW of Svalbard, March 19 1993.

through the boundary between the snow and the ice, so that

$$k_s \, (\partial T_s/\partial z)_h = k_i \, (\partial T_i/\partial z)_h \qquad\qquad (3.23)$$

where k_i is the thermal conductivity of the top part of the ice layer and $(\partial T_s/\partial z)_h$ and $(\partial T_i/\partial z)_h$ are the temperature gradients just above and just below the interface respectively. Because snow is a much poorer conductor of heat than ice this means that a temperature profile through snow-covered ice in winter changes its slope at the snow-ice interface, the gradient being much greater above than below the interface. This provides a means to map the thicknesses of snow and ice through a winter automatically. A vertical chain of closely-spaced thermistors is installed through the ice sheet, feeding a data logger. The point at which the temperature profile changes its gradient is the snow-ice interface, while the upper and lower surfaces of the system are the points at which the gradient becomes zero. This does not work in summer when the whole atmosphere-snow-ice-water system is at or near 0°C. Figure 3.10 shows a temperature gradient obtained by the author through multi-year ice in the Arctic Ocean during March; the gradient changes at about 60 cm.

Interior of sea ice layer

Inside the sea ice layer the same conduction and absorption processes occur as inside the snow layer, so the governing equation is the same as (3.22) but substituting ice for snow, i.e.

$$\rho_i \, c_i \, (\partial T/\partial t)_z = \kappa_i \, I_o \, \exp \, [- \, \kappa_i \, z] + k_i \, (\partial^2 T/\partial z^2)_z \qquad (3.24)$$

where κ_i is now the extinction coefficient within the ice.

However, as we have shown in section 3.1, both the specific heat and the thermal conductivity of sea ice are functions of both temperature and salinity. Therefore a single value cannot be used for either $(\rho_i \, c_i)$ or k_i, but both must be expressed in terms of ice salinity $S_i(z)$, itself a function of depth within the ice, and temperature T_i. We have discussed appropriate formulations for these quantities in section 3.1; Maykut and Untersteiner chose to adopt simple approximate formulae due to Untersteiner (1961):

$$k_i = k_o + \beta \, S_i \, / \, T_i \qquad (3.25)$$

and

$$\rho_i \, c_i = (\rho \, c)_p + \gamma \, S_i \, / \, T_i^2 \qquad (3.26)$$

where S_i psu is ice salinity, T_i °C is ice temperature, $\beta = 0.13$ W m^{-1}, k_o is given by (3.2), $(\rho \, c)_p$ is the pure ice value of 1.944 MJ m^{-3} °K^{-1}, and $\gamma = 17.15$ MJ kg^{-1} °K. These relationships should be substituted into eqn. (3.24).

Ice-water interface

At the interface between the ice and the ocean there are only two fluxes, the turbulent heat flux from the ocean into the ice F_w, and the conductive heat flux in the ice close to the boundary. At the interface either freezing or melting may occur, depending on whether the ocean heat flux dominates (melting) or the conductive heat flux dominates (freezing). During winter one expects the conductive heat flux to dominate because the temperature gradient through the ice between the water and the atmosphere is so high, so that ice growth will occur. Nevertheless, if the ice is thick enough the temperature gradient will be reduced to the point where melting may occur instead. Here we see the physical justification for the concept of **equilibrium thickness** of an ice sheet: if some process (such as ridging) generates ice of greater than the equilibrium thickness, the ice will begin to melt at the bottom even in the middle of winter, and even while thinner ice in the vicinity is growing. Recent results show, however, that the Maykut-Untersteiner theory gives too low a melt rate for very thick ice in ridges, which probably ablates through hydrodynamic effects or mechanical erosion.

The equation expressing the heat balance at the ice bottom is:-

$$k_i \, (\partial T_i/\partial z)_{h+H} - F_w = [q \, d(h + H) \, / \, dt]_{h+H} \qquad (3.27)$$

In principle F_w can be expressed in terms of the gradient of water temperature T_w just below the ice-water interface, the density ρ_w and specific heat c_w of the near-surface water, and a so-called **coefficient of eddy diffusivity** K_w in the water column under the ice, a function of its degree of turbulence:-

$$F_w = \rho_w \, c_w \, [K_w \, (\partial T_w \partial z)]_{h+H} \qquad (3.28)$$

However, in practice these quantities are hard to determine while the heat flux itself can be derived from larger scale oceanographic profiling of temperature and salinity, which yields the upward flow of heat from the thermocline into the near-surface layers (e.g. Steele and Boyd, 1998). To a first approximation, this heat flow can be assumed to be lost to the atmosphere (although some heat is stored in the upper layers to give a seasonal temperature variation, most important in the Antarctic) and so can be identified with F_w. It is therefore not necessary to use equation (3.28).

3.3.2. Input Parameters

Maykut and Untersteiner ran their model with energy fluxes for the Arctic Ocean that were a function of time of year but repeated themselves in an annual cycle. The cycle was rerun until a steady annual pattern of temperature and thickness variation was achieved. The fluxes and parameters used were based on data obtained from drifting stations up to that date, and are still regarded as reasonable values for the Arctic, with some exception such as summer albedo, considered later in this chapter. We may summarise them as follows:-

F_r was derived from values proposed by Marshunova (1961) and Fletcher (1965). It is zero from November to February and rises to a maximum of 803 MJ m^{-2} month^{-1} in June.

F_L,   derived from the same sources, is lowest in February and March (431 MJ m^{-2}) and highest in July (799 MJ m^{-2}).

F_s,   derived from Doronin (1963), is from snow surface to atmosphere in winter, with a maximum in January of 49 MJ m^{-2}, and from atmosphere to snow (or bare ice) in spring and summer, with a maximum of -19 MJ m^{-2} in May.

F_l,   also from Doronin, is very slightly negative through the winter, then becomes more strongly negative in summer, reaching -29 MJ m^{-2} in June.

I_o is set at 50 MJ m^{-2} applied evenly through the snow-free period of the year, which means that 17% of net short-wave radiation is assumed to penetrate the ice.

κ_i,   the extinction coefficient, was set at 1.5 m^{-1}, independent of depth (Untersteiner, 1961; Chernigovskii, 1966).

F_w was set at 2 W m^{-2}. This is based on calculations by Crary (1960), Badgley (1961), Panov (1964) and Untersteiner (1964). It has received confirmation (1–3 W m^{-2}), for the Eurasian Basin in the winter months, from Steele and Boyd (1998).

α is set at high values for autumn to spring (0.81–0.85) when fresh snow lies on the ice without melting. The problem occurs in summer, when the albedo is very variable just at a time when the incoming short-wave radiation is highest so that the albedo assumes its greatest importance. The albedo is reduced by the presence of bare ice and melt ponds, but a typical rough Arctic sea ice surface may possess snow patches, bare ice and melt ponds simultaneously in varied proportions. Maykut and Untersteiner used an average value for July of 0.64, but later authors have considered this too high.

Snow deposition, a partial determinant of h, is assumed to comprise a linear accumulation of 30 cm between August 20 and October 30, a linear increase of 5 cm from November 1 to April 30, and an additional 5 cm during May. This was based on

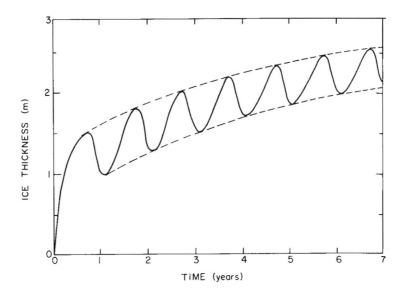

Figure 3.11. Pattern of thermodynamic ice growth in the central Arctic, assuming that an ice sheet begins to grow at the end of summer (time = 0) (after Maykut, 1986).

year-round measurements from US drifting stations (Untersteiner, 1961; Hanson, 1965). Snow density was set to 330 kg m^{-3} during the freezing season, and 450 kg m^{-3} once snow melt had begun.

S_i, ice salinity, affects thermal conductivity, specific and latent heats, and density. Maykut and Untersteiner used a standard profile, a function only of thickness, derived from Schwarzacher (1959).

3.3.3. Results

With heat budget values gathered in this way, the model predicts that a new ice sheet growing from open water at the beginning of winter reaches a thickness of 1.5 m in the first winter, then loses 0.5 m of thickness in summer from bottom and surface melt, with top surface melt making the greater contribution (fig. 3.11). If the ice is left to continue its evolution, subsequent annual growth and decay cycles enable it to approach an **equilibrium thickness** of 2.88 m, an asymptotic value about which the thickness oscillates on a seasonal basis (3.14 m in winter, 2.71 m in summer).

The subsequent annual cycling of the ice once it has reached its equilibrium thickness can in principle go on for ever, with a given parcel of ice gradually moving upwards through the ice sheet from year to year, since there is a net annual growth on the underside to cancel the net annual ablation of the top side. After many years the same ice floe exists but contains none of the ice with which it first started to grow; the skin of ice on the ocean is like human skin in this respect. The rate at which this upward movement happens

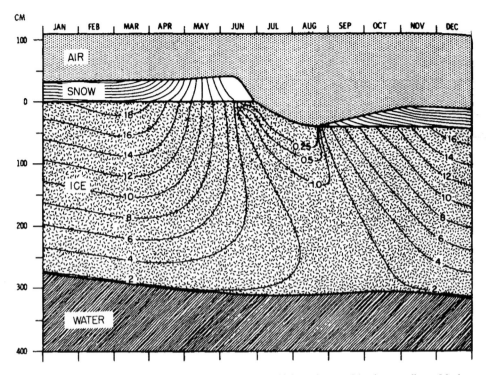

Figure 3.12. Predicted values of equilibrium temperature and thickness in central Arctic, according to Maykut-Untersteiner model. Isotherms in ice are labelled in negative °C; unlabelled isotherms in snow are drawn at 2°C intervals. Melt at upper ice boundary, and melt and growth at ice-water boundary, are shown without accompanying hydrostatic adjustment (after Maykut and Untersteiner, 1971).

is important when we consider how long it takes a sea ice sheet to rid itself of pollutants such as oil introduced at the bottom (see section 8.2.1).

In figure 3.12 we can see that there are some critical dates. Snow melt begins on June 8, and by June 29 all the snow has melted. Ice melt then begins and continues until August 19, by which time 40 cm of ice have melted off the top. This is the ice and the snow which together create the pattern of meltwater pools, and which can percolate down through the ice sheet flushing out remaining brine. New snow begins to accumulate in late August. During summer there is a small amount of bottom melt, but the winter growth of ice at the bottom must exceed this by about 40 cm in order to preserve an annual equilibrium.

This is of course an oversimplified picture. Variations in weather and snowfall from year to year will cause growth rates and equilibrium thicknesses to vary. The patterns for subArctic seas and for the Antarctic are also quite different, although in both cases sea ice is highly unlikely to survive (except as fast ice) for long enough to reach equilibrium thickness. However, the model results do demonstrate some very important sensitivities of sea ice thickness to changes in climatic forcing. In particular, the equilibrium thickness is sensitive to the values used for annual **snowfall** and for **oceanic heat flux**.

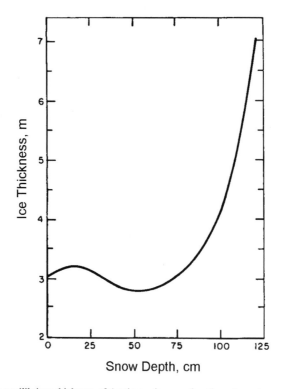

Figure 3.13. Average equilibrium thickness of Arctic sea ice as a function of maximum annual snow depth (after Maykut and Untersteiner, 1971).

Snow depth sensitivity

The sensitivity of equilibrium thickness to maximum annual snow depth is shown in fig. 3.13. The shape of the snowfall curve is due to the competing effects of snow as an insulator - which slows the rate of ice growth in winter - and snow as a covering material which melts off in early summer to reveal bare ice which can then develop a pattern of surface melt pools of low albedo, so enhancing the overall melt rate. The very high ice thicknesses caused by a thick snow cover occur when the snow becomes too thick to melt completely during the brief summer, so that surface melt pools cannot develop. At present the Arctic Ocean sits near the bottom of this curve, with a maximum annual snow depth of 30–50 cm, but if snowfall were multiplied by a factor of about 3 the ice would continue to grow indefinitely, with much of the extra thickness supplied by the snow itself which would grow thicker from year to year.

Ocean heat flux sensitivity

The sensitivity of equilibrium thickness to ocean heat flux is shown in fig. 3.14. The present value of F_w averaged over the Arctic Ocean is about 2 W m^{-2}; this is the value used, for instance, in the Hibler (1979) model of ice dynamics-thermodynamics (see

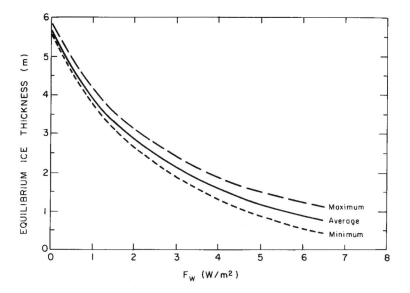

Figure 3.14. Average equilbrium thickness of Arctic sea ice as a function of the average annual oceanic heat flux. The dotted lines show annual maximum and minimum thicknesses (after Maykut, 1986).

chapter 4). Fig. 3.14 shows that if F_w increased to 7 W m^{-2}, with other factors unchanged, the ice would disappear completely. This might happen if there were a big increase in the amount of heat transported into the Arctic by the warm North Atlantic Current, or if the Arctic thermocline were to weaken or disappear because of a reduction in freshwater input from rivers at the surface. Similarly, if there were no oceanic heat flux at all, as can happen in the case of shallow water, the equilibrium thickness would rise to 6 m.

Other sensitivities

It has been suggested that an artificial reduction in albedo, e,g, by sprinkling coal dust on ice, could be used to reduce ice thickness and help clear out ice in summer. The model shows that if the summer albedo were reduced from the 0.64 used in the model to 0.54, the equilibrium thickness would fall to 1 m, and a further 0.1 reduction would cause the ice to melt completely. In practice, however, the artificial darkening of sea ice would probably not have such a drastic effect, as it would merely deepen the melt ponds, with much of the melt water refreezing again in autumn. Any effect big enough to cause basin-wide darkening of the ice, e.g. the fallout of material from an asteroid impact or a nuclear winter, would be so disastrous on a global scale that melting of Arctic sea ice would be one of the more trivial consequences.

A more obvious sensitivity is to air temperature T_a. The Maykut-Untersteiner model, and later heat balance models (Budyko, 1974; Parkinson and Kellogg, 1979), predict that summer air temperatures 4–5°C warmer than at present will result in an ice-free summer Arctic Ocean. Since such temperature rises are indeed predicted for the Arctic within the era of CO_2 doubling, i.e. the next 70 years (Cattle and Crossley, 1995), one might expect

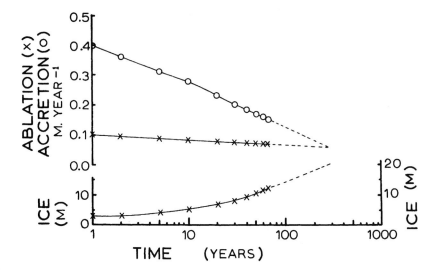

Figure 3.15. Ice thickness (m) and ablation and accretion rates (m yr⁻¹) achieved using Maykut-Untersteiner model with standard Arctic parameters except for an initial thickness of 3 m, an annual snowfall of 1 m, and no ocean heat flux. The extrapolated point at which ablation and accretion rates are equal gives the equilibrium thickness (after Walker and Wadhams, 1979).

that this will remove the ice from the Arctic. However, in reality a much more complex interaction will occur involving evaporation and the stability of the atmosphere over the ice; more complete coupled atmosphere-ocean models, such as that used in the general circulation model described by Cattle and Crossley (1995), predict an ice thickness loss of 1–1.5 m under this degree of warming.

Do special kinds of ice exist?

Given the sensitivity of ice thickness to ocean heat flux, snowfall and other factors, are there special circumstances anywhere in the world which allow ice to grow much thicker than expected? To simulate a coastal Arctic regime with a high local snowfall, Walker and Wadhams (1979) ran the Maykut-Untersteiner model with an oceanic heat flux set to zero and an annual snowfall increased to 1.0 m. The ice thickness (fig. 3.15) reached 12 m in 65 years. Extrapolation of the annual curves of ablation and accretion showed that an equilibrium thickness of 20 m would be eventually achieved after 200–300 years. Can we find ice of this kind?

There are a few isolated observations of thick undeformed floes in the Arctic Ocean. Cherepanov (1964) found that the 80 km² floe on which the Russian drifting station NP-6 had been established was 10–12 m thick, with a crystal structure which was typical of slow congelation growth. A 1 km floe of mean thickness 9.2 m, apparently undeformed, was observed by submarine sonar near the North Pole (Walker and Wadhams, 1979). A 10–12 m thick floe was observed by A R Milne (personal commun.) during an icebreaker voyage west of Prince Patrick Island. However, on the whole, observations of thick floes are scarce in the drifting pack.

A clue to the possible origin of these thick floes came from observations in the Antarctic. During a winter experiment in 1986 aboard FS "Polarstern" a small number of floes of very high freeboard were observed in the pack (Wadhams *et al.*, 1987). Fig. 3.16 shows an example of such a floe. The origin of these floes was found to be bays along the edge of the nearby Fimbul Ice Shelf, from which icebergs had calved. The inlets "healed" themselves by growing multiyear fast ice. Protected from the drifting pack by the shelf edge geometry, the ice could keep growing from year to year with the special circumstances of high snowfall (mainly snow blown onto the ice surface by katabatic winds blowing over the ice shelf) and low oceanic heat flux (an outflow of very cold water from under the ice shelf). In this way the ice could reach 11 m or more, and the occasional small breakouts of ice from these fast ice regions produced the isolated thick floes seen in the drifting pack.

Could this account for very thick floes seen in the Arctic? There is indeed a type of very thick fast ice, first reported by Koch (1945) in north and northeast Greenland and given the Greenlandic Eskimo name **sikussak** ("fjord ice like ocean ice"). Koch reported three areas of sikussak in May 1938 in Peary Land (fig. 3.17), whereas in earlier years it had been more prevalent and had been responsible for holding calving icebergs in place against the edges of glaciers and preventing them from breaking out. In 1980 the author observed ice of sikussak type in the fast ice at the mouth of Danmarks and Independence Fjords in north Greenland (fig. 3.18, Wadhams, 1986). The ice had a highly developed surface drainage system, and a core drilled through 6.1 m of ice failed to reach bottom. A salinity record from the uppermost 4.6 m (fig. 3.19) shows evidence of many years of alternating growth and surface melt. In these fjords similar circumstances prevail as in the Antarctic ice shelf inlets: high snowfall (with high coastal mountains), low ocean heat flux (from the single-layer water structure in the fjord) and intensely cold air temperatures. Sikussak probably exists in other high Arctic coastal locations; for instance there have been observations of fast ice "plugs" of 10 m and 12 m in Nansen Sound and Sverdrup Channel in the Canadian high Arctic (Serson, 1972, 1974). The rare thick floes seen in the pack represent occasional break-outs from these source areas.

An intriguing possibility is that some ice that is conventionally viewed as shelf ice, i.e. ice of terrestrial origin, may really be very old fast sea ice, what may be termed super-sikussak. An Arctic example is the Ward Hunt Ice Shelf on the north of Ellesmere Island. Since the end of the Second World War this ice shelf has been progressively breaking up, giving rise to the famous **ice islands**, several km in diameter and 50 m or more thick, which have drifted in the Arctic Ocean and provided secure bases for research stations. The most famous of all was Fletcher's Ice Island T-3, which left the Beaufort Gyre in 1984 after 27 years and exited through Fram Strait, finally breaking up off SW Greenland. It is also possible that "islands" reported by early explorers north of Ellesmere Island, Peary's "Crocker Land" and Cook's "Bradley Land", were actually ice islands. Ward Hunt Ice Shelf is not actively fed by glaciers, unlike Antarctic ice shelves, and so is either a relic of the last glacial period or else, in whole or part, an accumulation of very thick, very slowly grown sea ice, with fabric properties resembling polycrystalline ice (including zero salinity) because of its very slow growth rate.

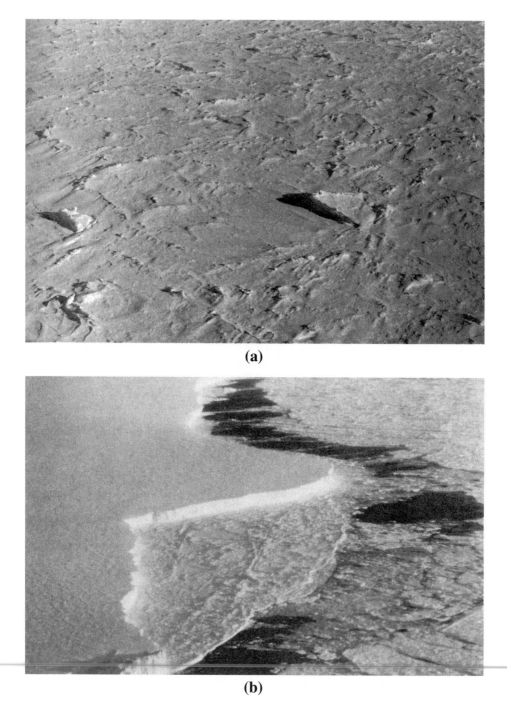

(a)

(b)

Figure 3.16.   (a) Very thick floe observed embedded in first-year pack ice in eastern Weddell Sea during winter of 1986 (after Wadhams *et al.*, 1987). (b) A possible origin of these floes in thick fast ice formed within an embayment of the Fimbul Ice Shelf.

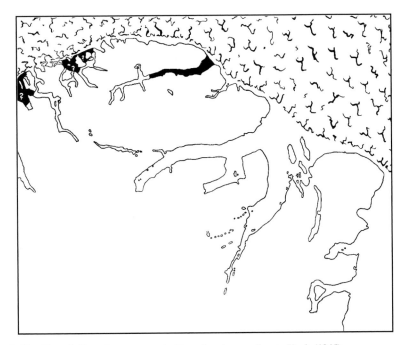

Figure 3.17. Map of sikussak occurrence in Peary Land, according to Koch (1945).

Figure 3.18. Ice of sikussak type observed by author in Independence Fjord, north Greenland (after Wadhams, 1986).

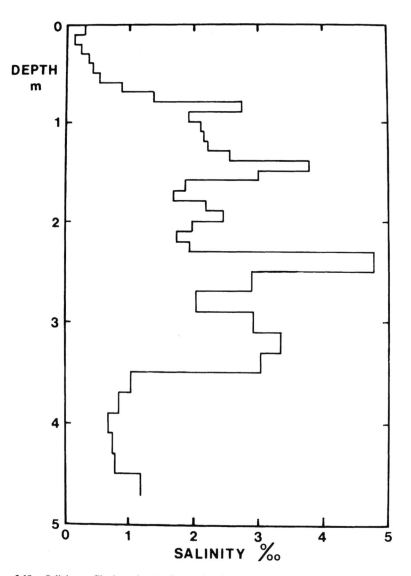

Figure 3.19. Salinity profile through part of core taken in sikussak, north Greenland (after Wadhams, 1986).

4. ICE IN MOTION

You are on the bridge of the German icebreaking research ship "Polarstern" as she ploughs eastward through midwinter pack ice along the Enderby Land coast towards Atka Bay, site of the Georg von Neumayer scientific station. A massive tabular iceberg, several kilometres in width, is dead ahead, drifting westward in the Antarctic Coastal Current close to shore. Should the ship detour around the north of the berg, adding many miles to her journey to relieve the base? Or can she slip through the gap between the iceberg and the coast? The Captain decides to pass between the berg and the land. You are excited. You are on the radar, plotting icebergs and their drift velocities. There are over 100 on the screen, but this one offers the chance of some really accurate readings. But something is wrong. The ship grinds to a halt, and for all her attempts at backing and ramming can make no progress at all. She is trapped and helpless, caught in a terrifying type of deformed ice which has been ground up by shear into a porridge-like consistency of small blocks held together by snow and brash, which clings to the hull and will not allow even this powerful vessel to move. And the iceberg is getting closer. It looms over the ship. You can see every detail of the dark caves in the side. The ship's decks are placed out of bounds; if any part of these ice cliffs collapsed, they would crash onto our deck. The berg is moving west; the pack ice is moving much more slowly and is being sheared and ground up between the berg and the coast. Everyone is scared that "Polarstern" will be driven against the berg and sunk. But your radar plot shows that the berg should just miss the ship. Closer it comes, until it is less than 40 metres away — little more than one ship width. Then, after an agonising interval, it begins to draw past. The pressure is relieved. Suddenly the ship can move again. The heaped-up porridge ice collapses back into a normal first-year ice pack. "Polarstern" sets course for Atka Bay.

At the end of chapter 2 we mentioned the ways in which the morphology of an ice cover is affected by the fact that the ice is not a passive stationary sheet of material covering the sea surface, but is driven by winds and currents. Leads, pressure ridges and coastal polynyas are expressions of the response of the ice to these stresses. We will now go on to consider the mathematical theory required to explain and predict the motion of sea ice. Together with the thermodynamic theory considered in the last chapter, this allows the construction of ice-ocean models which can describe both the velocity and the change of thickness of an ice cover. In their turn, these models can be embedded in General Circulation Models (GCMs) of the atmosphere and ocean which describe the total behaviour of the global atmosphere-ocean system and can predict climate variability and changes in which sea ice plays a role. The formulation of GCMs goes far beyond the scope of this book, but some of their predictions for the future development of sea ice on this planet are discussed in chapter 8.

4.1. THE MOMENTUM BALANCE

To understand sea ice dynamics, we need to deal with four aspects of the air-ice-ocean system:-

(i)   the *momentum balance* — the nature and magnitude of the forces acting on every element of the ice cover;
(ii)   the *ice rheology* — the way in which the ice cover behaves as a material when acted on by different kinds of stress;
(iii) the *ice strength* — what nature and magnitude of stress is needed to make the ice fracture and thus cease to behave as a continuum;
(iv)   the *mass balance* — the dependence of the ice thickness characteristics on growth, decay and ice drift, i.e. how the results of chapter 3 can be applied to ice which is in motion and deforming.

We first derive the momentum balance by considering the forces acting on a unit area of the sea ice cover. Expressed in words, Newton's Third Law of Motion which expresses the force balance for such an element is:-

$$\boxed{\begin{array}{l}\text{Mass} \times \text{Acceleration} \\ \text{of element}\end{array}} = \boxed{\text{Air Stress}} + \boxed{\text{Water Stress}} + \boxed{\text{Coriolis Force}}$$
$$+ \boxed{\text{Internal Ice Stress}} + \boxed{\text{Force due to Sea Surface Tilt}} \quad (4.1)$$

i.e.

$$M\,\mathbf{a} = \tau_a + \tau_w + \tau_c + \tau_i + \tau_t \qquad (4.2)$$

Usually the air stress, water stress and Coriolis force are the dominant forces acting on the ice. Air and water stresses have typical magnitudes of 0.1 N m^{-2}, while a 3-m thick ice floe moving at 0.1 m s^{-1} experiences a Coriolis force of 0.05 N m^{-2}. Internal ice stress is important, however, in close pack ice and in the case of ice in enclosed seas and channels. For instance, the ice in the channels of the Canadian Arctic Archipelago is stationary during winter, despite the fact that it is not grounded on the sea bed, because the internal ice stress transmitted into the ice cover from the coastlines is sufficient to cause the ice to resist the wind stress which is trying to make it move.

To formulate (4.1) and (4.2) in mathematical terms, we look at each force in turn.

4.1.1. Air Stress

It is found from studies of the atmospheric boundary layer that the force exerted by the wind on a surface is proportional to the square of the wind speed relative to the surface. The constant of proportionality linking wind speed to stress exerted on the surface is called the **drag coefficient**, and it is a function of the roughness of the surface. The wind speed, of course, has to be defined as occurring at some arbitrary height above the surface, because the frictional drag of the surface on the wind itself leads to a reduction in wind speed and a change in direction as the surface is approached. The conventional definition is:-

$$\tau_a = \rho_a \, C_a \, |U_a - U_i| \, (U_a - U_i) \qquad (4.3)$$

Here

τ_a = wind stress, i.e. force per unit area exerted by wind on ice;
ρ_a = air density;
C_a = drag coefficient;
U_a = wind velocity, measured conventionally at 10 m above the surface, known as **anemometer height**;
U_i = velocity of the ice surface.

Equation (4.3) takes this more complex vector form because we have to allow for the fact that the ice may be in motion at a substantial fraction of the wind speed, so it is the *relative* velocity of the wind and ice which determines the stress. If $U_i = 0$, (4.3) simply becomes $\tau_a = \rho_a \, C_a \, U_a \, U_a$.

The drag coefficient varies with the roughness of the surface, but for sea ice lies typically in the range 1.4 to 2.1×10^{-3}. When we look at a sea ice surface we see two kinds of roughness, the small scale roughness of the top surface of undeformed ice floes (snow-covered in winter, and a mixture of bare ice and melt water pools in summer), and the larger-scale roughness of the pressure ridges, rafted ice and vertical floe edges which protrude up into the air flow. It is no coincidence that the surface part of a pressure ridge is called the sail, since its vertical surface helps to propel the ice along in the wind. In trying to estimate the total drag due to a composite surface like this, meteorologists often split the drag coefficient into two components, **skin friction drag** due to the carpet of undeformed ice, and **form drag** due to the obstacles to air flow offered by individual structures.

For sea ice, attempts have been made to follow this approach and relate C_a to measurable physical roughness elements of the surface, i.e. to break it down into a skin friction drag coefficient C_{10}, which is more or less constant, and a form drag coefficient C_f which is a function of the density and height distribution of pressure ridges. Thus, for example, Banke *et al.* (1976) proposed the equation

$$C_a = C_{10} + (C_f \, H \, N \, / \, 2) \qquad (4.4)$$

where C_{10} varies over the limited range 0.0013 to 0.0021 for various ice surfaces, C_f lies in the range 0.3 to 0.4 depending on whether the ridges are mostly first-year (higher end of range) or multi-year (lower end of range), H is the mean height of ridges and N is the number of ridges per unit downwind distance.

This was intended to be used in ice models, and is indeed useful in indicating how an icefield is likely to respond much more strongly to the wind after it has become ridged than before. However, today there is less confidence about simple representations such as (4.4), for several reasons. Firstly, even the skin friction drag coefficient can vary greatly; we showed in chapter 2 how consolidated pancake ice, for instance, retains the rough edges of its parent pancakes until the snow cover becomes thick enough to smooth off the surface, and we do know that multi-year ice has an undulating surface which is bound to have a higher skin friction drag coefficient than first-year ice. Secondly, in heavily deformed icefields it is not easy to define individual ridges as in (4.4) since the entire surface may be a rubble field composed of randomly heaped ice blocks. Thirdly, a problem occurs in the marginal ice zone, which has a high drag coefficient due to the raised edges of floes, but where this drag coefficient may vary rapidly as floes get broken up into smaller (and thus rougher, per unit area) cakes by wave action. Finally, even equation (4.3) itself depends on the stability of the near-surface wind flow, i.e. on whether the air temperature is greater or less than the surface temperature of the ice; this affects the way in which the small-scale turbulence, which generates the drag, develops in the flow above the surface.

Because of these difficulties, an empirical approach is normally adopted and for any application C_a is simply measured above the surface of interest using a mast with velocity sensors at different heights to measure the shape of the boundary layer. Some values for different types of ice surface are given in table 4.1. The open water values here of 1.3×10^{-3} at moderate wind speeds and 2.0×10^{-3} at high wind speeds (because of the rougher water surface) are important for comparison purposes. We can see from the table that

- grease ice, nilas and small pancakes actually have a lower drag than open water, because they damp down waves without adding much roughness themselves;
- within each ice type category there is a very large difference between smooth, undeformed ice and rough, ridged ice;
- for any degree of roughness, multiyear ice has a higher drag than first-year ice;
- ice in the Greenland Sea MIZ is rougher than equivalent ice in the Weddell Sea MIZ, because of the higher proportion of thick, multiyear ice floes present.

4.1.2. Water Stress

The force exerted by the relative motion of water and the ice bottom takes exactly the same form as (4.3), i.e.

$$\tau_w = \rho_w \, C_w \, |U_w - U_i| \, (U_w - U_i) \qquad (4.5)$$

where
τ_w = water stress, i.e. force per unit area exerted by water on ice;
ρ_w = water density;
C_w = ice-water drag coefficient;
U_w = water velocity, measured under the ice at the bottom of the logarithmic boundary layer, i.e. about 1–2 m, a depth which must be defined when C_w is derived. The whole oceanic boundary layer is some 30 m thick.
U_i = velocity of the ice.

TABLE 4.1. Air-ice drag coefficients C_a measured for different types of ice surface. Adapted from Guest *et al.* (1994); data from Guest and Davidson (1991), Overland (1985), Smith (1988) and Andreas *et al.* (1993).

Ice type	$C_a \times 10^3$		
	median	minimum	maximum
Grease	0.7	0.6	1.1
Nilas	1.6	1.4	1.9
Pancake			
< 0.75 m diameter	0.9	0.7	1.3
0.75–1.5 m	1.6	1.1	2.2
> 1.5 m	2.4	1.9	2.9
consolidated	1.9	1.5	2.6
Young ice			
smooth	2.3	1.9	2.7
rough	3.1	2.6	3.6
First-year			
very smooth	1.5	1.2	1.9
smooth	2.0	1.6	2.4
rough	3.1	2.2	4.0
very rough	4.2	3.1	5.0
Multiyear			
very smooth	1.5	1.2	1.9
smooth	2.2	1.9	2.5
rough	3.4	2.5	4.1
very rough	4.6	3.6	5.5
extremely rough	8.0	6.7	9.1
Greenland Sea MIZ (U_a < 12 m s^{-1})			
All wind directions	1.8	0.7	3.0
Ice upwind 2–10 km	1.4	1.1	1.8
Weddell Sea winter (Andreas *et al.*, 1993)			
Undeformed ice	n/a	1.1	1.4
Deformed ice	n/a	1.3	1.8
Open ocean steady state (Smith, 1988)			
$U = 10$ m s^{-1}	1.3	n/a	n/a
$U = 25$ m s^{-1}	2.0	n/a	n/a

    The vector form of (4.5) is absolutely vital here, because $\mathbf{U}_w$ and $\mathbf{U}_i$ are usually of similar magnitude. It is just as often the case that surface water is exerting a drag to slow down an ice floe being moved by the wind as it is that a strong current is trying to accelerate a floe.

    A typical value of C_w is about 4×10^{-3}, but again it is a function of the physical roughness of the ice underside, and so varies with ice type (multi-year is rougher than first-year), with ridging density and keel depths, and with the frequency of leads or floe edges which offer vertical obstacles to the flow. Again it can be broken down into skin friction drag and form drag, but normally this is not done. Table 4.2 gives some typical measured values, and shows what a large range of variability occurs. We would expect that an ice cover with a high air-ice drag coefficient will also have a high water-ice drag

TABLE 4.2.   Water-ice drag coefficient C_w measured for different types of ice surface

Reference	Ice type	$C_a \times 10^3$	U_w depth
Johannessen, 1970	10–15 m floes Gulf of St. Lawrence	9–17	2 m
Pease et al., 1983	10–20 m floes Bering Sea MIZ	18–22	1.1 m
McPhee, 1979	Central Arctic pack	20	1 m (adjusted)
Reynolds et al., 1985	Smooth floe Bering Sea	7.8	2 m

coefficient, because surface and bottom roughnesses are so well correlated. We might even hope that, on the basis of equation (4.4) and a knowledge of the ratios of keel depths to sail heights in ridges, we could predict the ratio of C_a to C_w. In fact, this has not yet been found to be possible, but as we shall show later, some interesting characteristics of the ice motion depend on the ratio (C_a/C_w). An important point to note is that the difference between smooth ice and ridged ice should be relatively much greater for water drag than for air drag. This is because the oceanic boundary layer (OBL) under ice is only about 30 m thick while the atmospheric boundary layer (ABL) is some 1000 m thick. Thus a ridge with, say, a sail height of 1 m and hence a keel depth of about 5 m, would protrude through only a negligible fraction of the ABL, having a tiny effect on C_a. Its keel, however, protrudes through a significant fraction of the OBL and has a bigger effect on C_w — it is the equivalent of a hill 150 m high. We expect, therefore, that form drag is a more important component of C_w than of C_a.

In the case of C_w, a complicating factor, discussed by McPhee (1986), is that the whole form of the boundary layer under ice can be greatly affected by melting or freezing. The former increases the density stratification near the surface, hampering vertical momentum transfer and thus reducing drag, producing what McPhee termed "**slippery ice**". The latter decreases stratification and increases drag. C_w values should ideally be measured in stable, neutral conditions where neither melting nor freezing is taking place.

4.1.3. Coriolis Force

The Coriolis force arises because all geophysical measurements of motion on the planet Earth are made relative to the Earth itself as the co-ordinate frame, yet this is an accelerated frame of reference because of the Earth's rotation. The result is that we can apply Newton's laws for the motion of a body (as in eqn. 4.2), but only so long as we introduce an additional force term which takes account of the fact that the body is being accelerated due to the Earth's rotation. This force term is called the **Coriolis force.** It is negligible on the human scale, despite mythical tales of draining bathwater which rotates in a different direction in each hemisphere and of wanderers in deserts who walk in clockwise or anticlockwise circles. However, it is very important on the geophysical scale and plays a vital role in defining the motions of sea ice, icebergs, ocean currents and global winds.

The *magnitude* of the Coriolis force F_c is given by:-

$$F_c = 2 \ m \ \omega \ U_i \ \sin \phi \qquad (4.6)$$

where
m = mass of body (for a unit area of ice cover, m = ρ_i h where ρ_i is ice density and h
    is ice thickness)
ω = angular velocity of Earth in rad s^{-1} = 7.272 × 10^{-5}
U_i = ice velocity
ϕ = latitude

The *direction* of the Coriolis force is 90° to the *right* of U_i in the northern hemisphere
and 90° to the *left* of U_i in the southern hemisphere. The term (2 ω sin ϕ) in (4.6) is known
as the **Coriolis parameter** f_C.

    The Coriolis force is zero at the Equator, and takes its maximum value at the Poles.
This alone makes Coriolis force important for ice motion. At this point we note an
important fact: Coriolis force is proportional to mass. Therefore an iceberg 200 m thick
experiences a Coriolis force per unit waterline area which is 100 times as great as a sea
ice sheet 2 m thick. Thus Coriolis force is a much stronger determinant of iceberg motion
than of sea ice motion, and we shall see in chapter 7 how this leads to different trajectories
for icebergs and sea ice under the same environmental forcing.

4.1.4. Internal Ice Stress

The term τ_i in (4.2) describes the total force which acts on a unit area of ice cover
embedded in a sea ice sheet due to stress transmitted through the ice from the parts of
the ice sheet surrounding the reference area. Such stresses can be due to differential winds
or currents, are very difficult to measure and equally difficult to parameterise in a model.
If sea ice were a rigid material, the internal ice stress on a reference unit area of ice cover
would be a vector sum of the variable environmental stresses acting on all elements of
the ice sheet, and transmitted for great distances through the ice. As it is, ice is not only
not a rigid material, but it also has a thickness interaction in its response — it can be
crushed to form ridges, or open up to form leads. Thus the extent to which a unit area
of ice cover is acted upon by the surrounding ice through internal stress is a function of
the rheology and strength of the ice cover and of the existing thickness distribution, i.e.
factors (ii), (iii) and (iv) mentioned at the beginning of this chapter. Any successful model
of basin-scale ice dynamics must properly account for internal ice stress. The ways in
which it is treated are considered in section 4.3 after we have dealt with the simpler free
drift solution.

4.1.5. Sea Surface Tilt

The final term in (4.2) is τ_t, the stress due to sea surface tilt. This occurs because the
sea surface does not necessarily correspond to the **geoid**, defined as the surface over which

the gravitational potential is constant. If the ocean were uniform and stationary, the geoid is the shape that its surface would take up; it is not quite a smooth surface because varying rock density, due to the distribution of oceanic and continental crust and mid-ocean ridges, causes bumps and hollows. However, the complexity of ocean flow due to factors other than the wind, e.g. the earth's rotation or thermohaline effects (uneven heating, evaporation and precipitation), lead to a piling-up of water in some areas, especially against boundaries, and a depression of the sea surface elsewhere. This tilt of the sea surface relative to the geoid causes a horizontal pressure gradient force which tries to move the surface water. In places where this force is only balanced by Coriolis force, the resulting equilibrium current is called a **geostrophic current.**

Thus sea surface tilt stress can be expressed as

$$\tau_t = - m \ g \ \text{grad} \ H = - \ \rho_i \ h \ g \ \text{grad} \ H \tag{4.7}$$

where H is the elevation of the sea surface with respect to the geoid. It has been found from models and observations of buoy drift (Hibler and Tucker, 1979; Thorndike and Colony, 1982) that sea surface tilt in ice-covered seas is unimportant over a period of a few days, with the resulting steady current accounting for only a few percent of the ice motion, but becomes more important over periods of several months, when wind effects partially cancel out.

4.2. A FREE DRIFT SOLUTION

4.2.1. Inertial Motion

The so-called **free drift solution** to these equations assumes that our unit reference area of ice cover moves as if it were an isolated object on the sea surface (i.e. a single floe), without being affected by collisions with neighbouring floes or by the internal stress transmitted through the ice due to the fact that distant areas of a continuous ice cover are being subjected to different stresses from the test area.

Let us deal with the simplest possible case first. Consider a single ice floe (fig. 4.1a), at rest in the ocean in the Northern Hemisphere and suddenly subjected to a force **F** due to a rising wind. For the moment, let us forget the water drag and assume, improbably, that the ice bottom is perfectly smooth. How will the ice floe move? It will begin by moving off in the direction of **F**, but as soon as it acquires any velocity, the Coriolis force will begin to act and will divert its motion to the right of the wind. So long as the floe retains any component of velocity in the direction of the wind, it will continue to accelerate, but this means that the turning effect of the Coriolis force becomes stronger still and eventually turns the floe so that its velocity now has a component opposed to the direction of **F**. The floe now begins to slow down, and by symmetry it will come to rest at a spot due *east* of the start point, having traced out half a loop. Immediately it will begin the same sequence again, and thus traces out a series of half loops, each of which takes it further to the east of its start point (fig. 4.1a). These loops are called **inertial loops** and are often evident in the trajectories of ice floes and icebergs under varying winds. The shape is that of a cycloid.

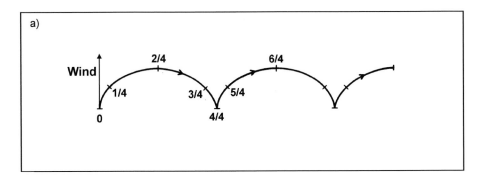

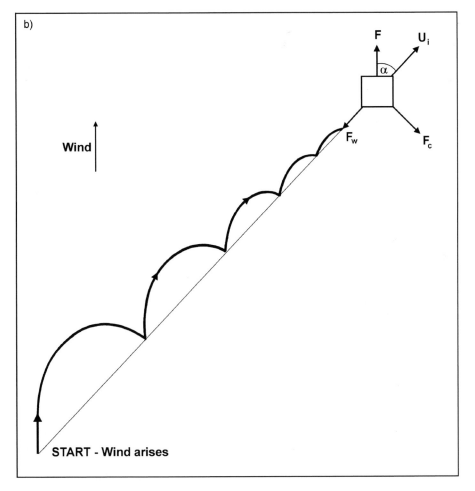

Figure 4.1. (a) Inertial oscillations of an ice floe starting to move under a suddenly-arising wind, in the Northern Hemisphere, without friction. (b) Forces on, and motion of, the ice floe when water drag is included but the surface water is assumed stationary (after Gill, 1984).

If the wind were suddenly removed, then in the absence of water drag the floe would continue to make complete circular loops with a period T given by

$$T = 2\pi / f_C \qquad (4.8)$$

from (4.6), where f_C is the Coriolis parameter. This is called the **inertial period** and is 12 hours at the Pole and infinity at the Equator, where the Coriolis force is zero. In fig. 4.1a the motion of the floe is marked off at intervals of one quarter of an inertial period. The short inertial period at high latitudes has important implications in analysing ice motion in polar seas, since in shallow water there are significant tidal currents (Kowalik and Proshutinsky, 1994), and a 12-hour inertial period is difficult to distinguish from the semidiurnal (M_2 and S_2) tides. Tidal ice motion is discussed further in section 4.5.

Now let us add water drag due to a stationary water surface. The balance of forces on the ice floe is now a triangle as in fig. 4.1b. Once again the floe starts off in the direction of **F** and Coriolis force moves it to the right, but water drag slows its acceleration so that it does not acquire a high velocity as in the first case. Water drag also destroys the symmetry of the motion: as soon as the floe acquires any component of motion opposed to **F** it slows down more quickly, since the water drag is always acting to oppose its motion, and it ends up downwind of, though to the right of, its starting point. This produces a partial inertial loop. When the floe starts on its second loop it already has a component of eastward velocity, and the end result is a set of diminishing loops which eventually lead to a uniform motion governed by the triangle of forces (fig. 4.1b). This motion lies approximately at 45° to the right of the wind, though the exact angle depends on the ratio of air-ice and ice-water drag coefficients, as well as on the thickness of the ice. We recognise in advance that fig. 4.1b is incorrect, since the water drag is shown as due to water which remains at rest, but if we solve the triangle of forces it does give some pointers to the factors involved. The solution for the turning angle α between the surface wind and the ice motion is

$$\tan^2 \alpha = 2 / \{[1 + (2\, \rho_w\, \rho_a\, C_w\, C_a\, U_a / f_C^2\, h^2\, \rho_i^2)^2]^{1/2} - 1\} \qquad (4.9)$$

This demonstrates some interesting results, e.g. that the turning angle increases with f_C (stronger Coriolis force) and ice thickness h, but that it decreases with increasing drag coefficients C_a and C_w and wind speed U_a.

4.2.2. Free Drift in Response to Geostrophic Wind

The above simplified analysis assumes that the surface water remains stationary while the wind pushes the ice, which acquires a final equilibrium velocity by being subjected to the frictional drag of the stationary water. Reality is not so simple, however. In fact the wind drives the ice, and the frictional drag of the ice upon the water itself sets the surface water in motion. This stress is transmitted down through the layers of the water column, diminishing as it goes, so that the near-surface layers are set in motion, each at a different angle to the driving force, because of the Coriolis effect. The full solution to wind-generated motion in the open sea, accounting for these friction-created boundary layers of wind above the ice, and water below it, was derived by Ekman (1905), based

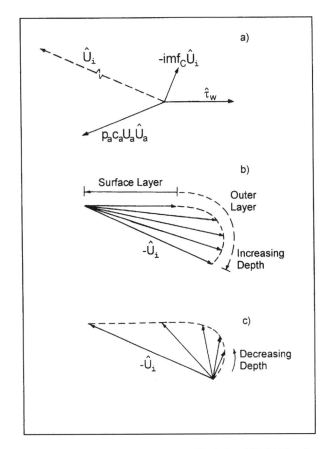

Figure 4.2. (a) Schematic diagram of full force balance on a floe in free drift. (b) Plan view of oceanic boundary layer as seen from the ice. (c) Plan view of velocities in oceanic boundary layer as seen from a frame fixed to Earth. (after McPhee, 1986).

on the data and ideas of Nansen developed during the "Fram" drift of 1893–6. The pattern of changing speeds and directions in the near-surface wind and current is called the **Ekman spiral**.

Let us consider how this analysis applies to sea ice. Consider, as before, a wind acting on the surface of an ice cover in the Northern Hemisphere, with no sea surface tilt or internal ice stress acting, and an ocean beneath the ice cover which is initially at rest (fig. 4.2a). The water drag is no longer directly opposed to the direction of ice drift U_i, because the OBL has itself been set in motion by the traction of the ice on it. Now imagine that we are looking down from the ice at successive levels of the water column, measuring the relative velocity of the water to the ice. Fig. 4.2b shows the result. By analogy with other boundary layers (e.g. flow over the seabed), there is initially a region of shear, with relative velocity varying logarithmically in the same direction as the turbulent stress. This is known as the **logarithmic boundary layer** and is the region marked "surface layer" in fig. 4.2b. A few metres down, the Coriolis force begins to be dominant and the Ekman spiral starts, with a clockwise veering of the mean current relative to the ice with increasing depth. Finally, at a depth of about 30 m we reach the bottom of the boundary layer and the relative velocity takes on its "free-stream" value of $-U_i$. The layer of veering direction in the current is called the **Ekman layer** or (fig. 4.2b) the outer layer. Fig. 4.2c shows how the set of velocities looks when viewed from the bottom of the boundary layer, i.e. at a location fixed relative to the Earth.

The Ekman spiral applies not only to the oceanic boundary layer but also to the atmospheric boundary layer (ABL). The surface wind — or rather, the 10 m wind — used for the definition of wind stress in eqn. (4.3) is actually smaller in magnitude and is turned to the *left* (in the Northern Hemisphere) relative to the wind at the top of the ABL, which is known as the **geostrophic wind**. The **turning angle** θ_a between the geostrophic and the surface wind is found by observation to be about 25° (Brown, 1980; McPhee, 1982). We can thus rewrite (4.3), at the same time simplifying it by ignoring the ice velocity which is a small fraction of the wind velocity, giving

$$\tau_a = \rho_a \, C'_a \, U'_a \, [U'_a \cos \theta_a + k \wedge U'_a \sin \theta_a] \qquad (4.10)$$

where k is a unit vector vertically upwards, U'_a is the geostrophic wind velocity, and C'_a is a new drag coefficient defined with respect to the geostrophic wind rather than the surface wind (i.e. is smaller than C_a). The redefinition of the drag coefficient takes care of the change of magnitude of surface wind with respect to geostrophic wind in the equation.

In a similar way we can rewrite (4.5) as

$$\tau_w = \rho_w \, C'_w \, |U'_w - U_i| \, [(U'_w - U_i) \cos \theta_w + k \wedge (U'_w - U_i) \sin \theta_w] \qquad (4.11)$$

where C'_w is now the water drag coefficient defined with respect to the current U'_w at the bottom of the oceanic boundary layer, with θ_w as the turning angle between the bottom of the boundary layer and the ice underside, again found to be about 25° and positive in the Northern Hemisphere.

We can now derive a better value for the magnitude and direction of the free drift ice velocity. For a steady-state case we insert (4.6), (4.7), (4.10) and (4.11) in (4.2), setting acceleration and internal stress to zero. In the most elementary case we can also set sea surface tilt stress to zero (giving a zero geostrophic current at the bottom of the boundary layer) and even, in the case of very thin ice, Coriolis force to zero (because of h being small). In this simplest of all cases we obtain the result (Leppäranta, 1998):-

$$U_i = \alpha_0 \exp(-i \, \theta_0) \, U'_a \qquad (4.12)$$

where α_0 is the **wind factor**, given by

$$\alpha_0 = [\rho_a \, C'_a / \tau_w \, C'_w]^{1/2} \qquad (4.13)$$

and θ_0 is the turning angle between the ice drift direction and the geostrophic wind direction (positive to the right), given by

$$\theta_0 = \theta_w - \theta_a \qquad (4.14)$$

Since it has been found that θ_w and θ_a are both about 25°, this implies that in the absence of a geostrophic current, the ice drift is in the direction of the geostrophic wind. Since the geostrophic wind blows parallel to the isobars, this result provides an explanation for the empirical **Zubov Law** (1945) of ice drift, which stated that

- Ice drifts parallel to the isobars
- The drift velocity is given by

$$U_i = 13,000 \text{ cosec } \phi \text{ (dp/dx)} \qquad (4.15)$$

where U_i is in km/month, ϕ is latitude and dp/dx is the pressure gradient in millibars per kilometre. Quantitatively, (4.15) is compatible with using reasonable values for (4.13). Zubov's Law, derived empirically from Russian drift station data, was a useful rule of thumb which enabled the general picture of the magnitude and direction of ice drift in polar seas to be derived from perusal of pressure maps — in effect, it was the first ice dynamics model. We have already shown in chapter 1 how the mean pressure fields over the Arctic and Antarctic bear a close relationship to the long-term mean pattern of ice drift. An even earlier rule of thumb, based on surface winds and data from the "Fram", was the **Nansen-Ekman ice drift law** (Nansen, 1902; Ekman, 1902), which stated that ice moves at 2% of the wind speed at 30° to the right of the surface wind in the Northern Hemisphere. Again this is compatible with (4.13).

If we now include Coriolis force and pressure gradient, we obtain a more general solution, which can still be expressed in a form similar to (4.13), i.e.

$$\mathbf{U}_i = \alpha \exp(-i\,\theta)\,\mathbf{U}'_a + \mathbf{U}'_w \qquad (4.16)$$

where α and θ, the more general wind factor and turning angle, are now found to be dependent on a dimensionless quantity

$$R = (\rho_i\, h\, f) / (\rho_a\, C'_a\, U'_a) \qquad (4.17)$$

It is found (Leppäranta, 1998) that the solution is given by

$$\alpha^4 + 2 \sin\theta_w\, R\, \alpha_0^2\, \alpha^3 + R^2\, \alpha_0^4\, \alpha^2 - \tau_0^4 = 0 \qquad (4.18)$$

$$\tan(\theta + \theta_a) = \tan\theta_w + \alpha_0^2\, R / (\alpha \cos\theta_w) \qquad (4.19)$$

We can see from these equations that the wind factor is lower and the turning angle greater than for the thin ice case of (4.13), but that they tend to the same values as $R\rightarrow0$. In the opposite case of very thick ice, as $R\rightarrow\infty$, $\alpha\rightarrow1/R$ and $\theta\rightarrow90°$. Leppäranta (1998) solved these equations numerically to give the results shown in fig. 4.3.

It can be seen from fig. 4.3 that an increase in R, corresponding to an increase in ice thickness or a decrease in wind speed or a decrease in surface drag, causes a decrease in the wind factor and an increase in the turning angle. Hence it is clear that both the Nansen-Ekman and the Zubov Laws are merely rules of thumb applicable to typical Arctic conditions — and even here they work better in the summer, when internal ice stress is

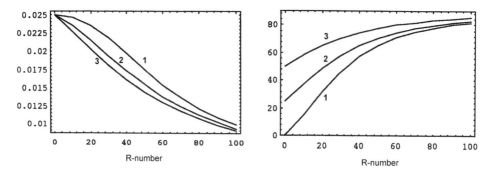

Figure 4.3. The wind drift factor α (left) and the turning angle θ (right) as a function of the parameter R (equation 4.17) for three different turning angles of the oceanic boundary layer: $\theta_w = 0°$(1), $25°$(2) and $50°$(3). (After Leppäranta, 1998).

small, rather than in the winter when it may be dominant (section 4.3). In the Antarctic, conditions can be quite different, because pressure ridge keels tend to be shallow so that bottomside roughness does not scale with topside roughness in the same way as in the Arctic. It has been found for the Weddell Sea, for example (Martinson and Wamser, 1990), that the water drag coefficient was low (1.62×10^{-3}) at a time when the air drag coefficient was as high as in the Arctic, giving a value of 0.8 for C'_a/C'_w and yielding a larger wind factor of 3% for small R as opposed to the 2.5% given in fig. 4.3.

4.3. THE MOTION OF COMPACT ICE

In reality, ice does not move as a whole. Free drift theory works best for isolated pieces of floating ice (e.g. icebergs in the open ocean), for diffuse ice covers such as marginal ice zones, and to some extent for Antarctic sea ice, where the ice is generally divergent as it moves towards an unrestricted ocean edge and where leads are created more readily than ridges. For the central Arctic pack, and for ice in basins and channels restricted by land boundaries, we need to consider how stress is transmitted through the ice and how the ice cover responds to stress when it is not free to move in an unrestricted way.

There is a large body of experimental evidence of the deviation in the motion of compact icefields from free drift. Many measurements of deformation show that, on scales ranging from 10–20 km (Hibler et al., 1974a) to 100–500 km (Thorndike and Colony, 1980), the central Arctic pack experiences strain rates of about 1% per day, uncorrelated with ice velocity. The spatial variability of ice velocity in itself is not linear. When a compact moving icefield has a lateral boundary with fast ice, a discontinuous shear zone develops (fig. 4.4) in which the overall velocity difference is taken up by a number of narrow regions of high shear rate.

4.3.1. Ice Rheology

To develop an ice model which can account for these observed phenomena, it is necessary to derive or propose a **constitutive law**, which relates the strain rate of the ice cover to

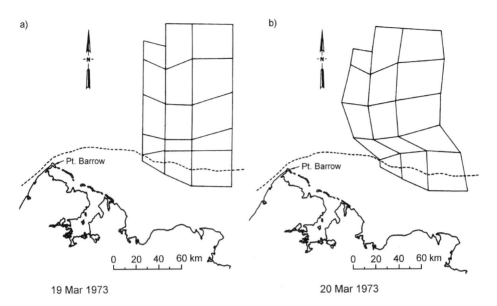

a)

b)

Pt. Barrow

Pt. Barrow

0   20   40   60 km

0   20   40   60 km

19 Mar 1973

20 Mar 1973

Figure 4.4. Observed ice motion in the shear zone off N Alaska (after Hibler *et al.*, 1974b), defined by a grid of identifiable ice features from Landsat images with (b) one day later than (a). The shorefast ice limit is a dashed line.

the applied stress. Some of the characteristics of a sea ice cover which the law should describe adequately are:-

- the ice cover acts as a two-dimensional isotropic continuum (in fact on a small scale it does not, but the law is to be applied to the large scale motion of the entire pack in a model where the grid size is greater than the scale within which the ice acts anisotropically);
- internal stress does not rise far above zero until compactness A exceeds about 0.8;
- small or zero tensile strength, for uniaxial divergence and for two-dimensional dilation — the ice offers little resistance to being pulled apart and dilates easily under a divergent stress;
- high compressive strength — it is difficult to crush an ice cover;
- discontinuous slippage under a shear stress — e.g. in the shear zone near shore, where the velocity differential between the fast ice and the offshore pack is taken up by a series of shear planes, each of which is associated with a system of shear ridges;
- thus, a shear strength which is significant but less than the compressive strength.

The nature of ice pack behaviour suggests that the stresses are largely affected by contact stresses between the floes rather than stresses transmitted by eddy diffusion. Also, early tests showed that linear rheologies could not be made to fit observations of ice dynamics and deformation.

A range of possible rheologies is represented in a simplified way in fig. 4.5, in which stress σ is one-dimensional. The simplest is free drift, where no internal stress is

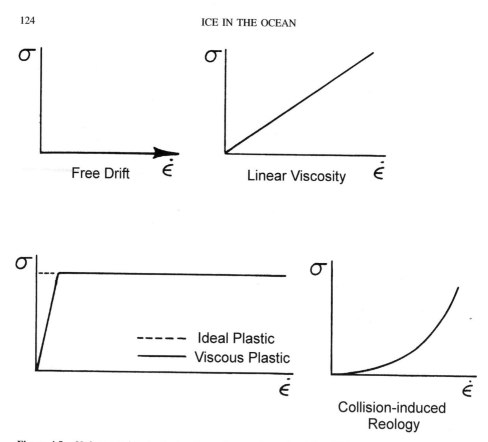

Figure 4.5. Various sea ice rheologies shown in one dimension (after Hibler, 1989). Positive strain rate represents convergence.

experienced and hence the strain rate is zero. In a **linear viscous** rheology strain rate is proportional to stress. An ideal **plastic** rheology was developed by Coon and was the basis for the AIDJEX model (Coon, 1980); when the stress reaches a certain level the material yields. This was based on the perceived similarity of the ice cover to a granular medium, and to the fact that pressure ridge building models (Parmerter and Coon, 1972) showed that, for a given geometry of ice, the forces required to build ridges have a limiting value, which is a property of plasticity. On the other hand Hibler (1979) modified this to a **viscous plastic** rheology, where the stress is proportional to strain rate for small rates, but then reaches a steady level at higher rates. To describe the motion of sea ice in compact MIZ-type icefields composed of discrete floes, various rheologies based on the mechanics of floe collisions have been developed (e.g. Shen et al., 1986). These rheologies typically are equivalent to a stress which rises more rapidly than linearly with the rate of convergence.

In two dimensions the rheology can be represented by a **yield curve** in stress space. Here the stresses lie on the yield curve when flow is occurring, and no stress combination can lie outside the curve. If σ_1 and σ_2 are principal components of the two-dimensional

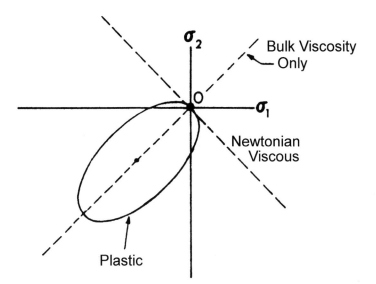

Figure 4.6.   A yield curve showing allowable stress states for viscous and plastic rheologies (after Hiber, 1989).

stress tensor and define the two axes of the yield curve (fig. 4.6), then allowed stress states for a Newtonian viscous rheology (only a shear viscosity allowed) would lie on the line $\sigma_1 = -\sigma_2$, while for a linear viscous rheology with only a bulk viscosity they would lie on the line $\sigma_1 = \sigma_2$. A plastic rheology would have a teardrop-shaped yield curve (shown here as an ellipse), which fulfills the criteria of no tensile strength, moderate shear strength and high compressive strength. Allowed stress states lie inside or on the ellipse.

4.3.2. Sea Ice Models

The first complete sea ice model for the Arctic Ocean was by Campbell (1965) and employed a linear Newtonian viscous rheology, together with a momentum balance and simplified thermodynamics. At that time the only data available to test it were ice station data for dynamics and very sparse submarine data for thickness. Soon afterwards the AIDJEX project (1970–77) rapidly advanced the science of sea ice modelling, since the manned camps produced for the first time an adequate dataset on ice dynamics, deformation and thermodynamics, while an integral part of the project was model development with the aim of ice prediction on time scales of one day and spatial scales of 100 km (Pritchard, 1980). An underpinning advance was the definition by Nye (1973) of the full two-dimensional stress tensor in terms of depth-integrated stresses. The AIDJEX model (Coon, 1980), which was used to interpet AIDJEX data and has since been used for many applications, used a plastic rheology. This required the use of a Lagrangian grid to trace the strain state of a parcel of ice, and this need to retain a memory in the formulation proved numerically complex and computationally expensive.

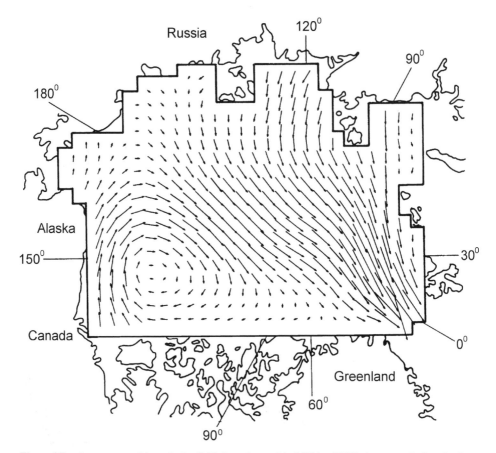

Figure 4.7. Average annual ice velocity field, from the model of Hibler (1979). An arrow the length of one grid cell represents 0.02 ms^{-1}.

An alternative model, which has since become established as the standard approach to sea ice modelling, was by Hibler (1979, 1980), with a more recent update of the numerical scheme by Zhang and Hibler (1997). This employed a viscous plastic rheology, which ovecame the computational problems of the AIDJEX model by enabling an Eulerian grid to be used, and which also proved in many subsequent observational tests to be a very good approximation to actual sea ice pack behaviour. In fact, the most recent set of tests of different model approaches has been that of the SIMIP programme (Sea Ice Model Intercomparison Project) of WCRP, which has concluded, using the same input data to test various models, that a viscous plastic rheology still gives the best fit to real behaviour (Lemke, 1997). The Hibler model was forced with prescribed wind, thermal ocean flux and incoming radiative fluxes. The thermodynamics used was the Semtner (1976) simplification of Maykut-Untersteiner. Fig. 4.7 shows an annual average ice velocity field for the Arctic generated by this model.

The Hibler model was applied initially only to the Arctic, but at the same time a model came out (Parkinson and Washington, 1979) which was applied to both hemispheres. The main problem with this model was that its dynamics were simply free drift, and this showed in the way that it produced fairly realistic results for the Antarctic but erroneous results for the Arctic, where the greatest thickness was given as being in the centre, whereas the Hibler model correctly showed the build-up of thickness around the north Greenland and Canada coastlines due to ridging.

Model development during the 1980s involved improving the parameterisation of coupling between the ice and the ocean and atmosphere, and the use of better forcing data. A first attempt at testing a model against real interannually varying forcing data rather than climatology was by Hibler and Walsh (1982), using pressure and temperature fields from 1973–5 and obtaining good agreement on interannual fluctuations with a poorer fit to results for a single year. Other rheologies were tried out, especially in the light of large new data inputs on marginal ice zone dynamics from the MIZEX project. Rheologies involving interactions between individual floes, giving a yield curve on a statistical basis, were developed by Bratchie (1984) and Shen et al. (1986), while Hopkins et al. (1991) used a similar approach for modelling the ridge-building process by generating and following a rubble build-up. Flato and Hibler (1990) developed a simplified **cavitating fluid** model, where there is no shear strength and ice is in free drift but with the ice velocity corrected for convergence.

The next major development was the production of a full basin-scale **coupled ice-ocean model** by Hibler and Bryan (1987), following a mesoscale model for the MIZ by Røed and O'Brien (1983). Hitherto the sea ice models were fed by prescribed forcing from the ocean, but a fully coupled model enables the joint response of ocean and ice to be reproduced, giving much better predictions of ice edge location and allowing a treament of phenomena such as under-ice convection.

More recent model developments have led in the following directions:

1. Incorporation of ice-ocean models into full General Circulation Models (GCMs), allowing the role of sea ice in climate change to be assessed (see chapter 8);
2. Development of regional sea ice models, for the marginal ice zone, shelf seas or convection regions (e.g. Häkkinen et al, 1992; Backhaus and Kämpf, 1999);
3. Sensitivity studies in which the response of the sea ice to varying key parameters is assessed, e.g. snowfall, air temperature, river runoff, ocean heat flux (e.g. Hibler and Zhang, 1994), or albedo changes (e.g. Ledley and Pfirman, 1997);
4. New formulations for key processes, such as ridging (e.g. Flato and Hibler, 1995);
5. Use of greater computing power to reduce the grid size of a basin-scale model, down to a few km;
6. Development of operational models to provide short-term forecasts, e.g. the UK Meteorological Office's FOAM model (Forecast Ocean-Atmosphere Model) (Alves et al., 1995).

4.4. DYNAMICS AND THERMODYNAMICS OF COASTAL POLYNYAS

We introduced the topic of coastal polynyas in section 2.9. They may play an important role in overall annual ice production in the circumpolar Antarctic, and important regional roles in the thermal balance of Arctic areas such as Smith Sound, NE Greenland and the

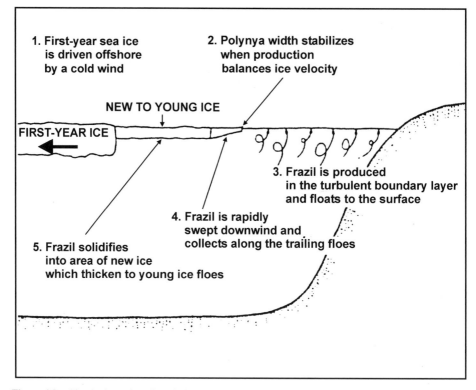

Figure 4.8. Simple dynamics of a wind-driven coastal polynya (after Pease, 1987).

Russian shelves. It is also possible that salt rejection from coastal polynyas is important in bottom water production in the Antarctic, and in the production of dense water on the Russian shelves which runs down the shelf break into deep water and ventilates the Arctic Ocean. It is therefore useful to evolve a simple theory of the dynamics and thermodynamics of a coastal polynya.

A simple theory for a wind-driven coastal polynya was presented by Pease (1987). In this simplest form of a polynya (fig. 4.8) the driving forces are a directly offshore wind and a subzero air temperature. If the coastal water is at the freezing point, frazil ice will be produced in the turbulent boundary layer under the cold offshore wind, and will be rapidly swept downwind. It gathers against the trailing edge of the polynya, where it piles up against the thicker ice which has been consolidating there from earlier ice production and advection. The thicker layer, under pressure from the newly arriving ice, is rapidly consolidating into either pancake ice or a thicker continuous sheet. It is also moving downwind. If no ice production occurred in the polynya, the open water area would steadily get wider, but under a steady wind and a subzero air temperature, the addition of swept-up frazil to the polynya edge builds the edge upwind to offset the bodily motion of the ice downwind. Eventually an equilibrium width of the polynya will be reached, which the theory sets out to estimate.

Let X_t be the width of the polynya at time t, and F_i be the rate of frazil ice production within the polynya, expressed as a thickness of ice added per unit time. This ice is then swept downwind, and is assumed to pile up into a uniform slick of thickness H_i (called the **collection thickness**) against the thicker ice which marks the edge of the polynya. This thicker ice is itself moving downwind under wind stress with a velocity V_i. The rate of widening of the polynya is then given by

$$dX / dt = V_i - F_i X / H_i \qquad (4.20)$$

Under constant wind, and assuming no change in H_i, this is a simple differential equation. If we start with a closed polynya, i.e. $X_0 = 0$, the solution is

$$X_t = V_i H_i [1 - exp (-t F_i / H_i)] / F_i \qquad (4.21)$$

Clearly this implies a maximum width of

$$X_{max} = V_i H_i / F_i \qquad (4.22)$$

which, however, is only reached after infinite time. To be realistic, a polynya can be said to be fully developed when it has reached, say, 95% of its maximum width. The time taken to achieve this width, from (4.21) and (4.22), is given by

$$t_{95} = (H_i / F_i) \ln 20 = 3.0 H_i / F_i \qquad (4.23)$$

Interestingly, this shows that the time required for the polynya to become fully developed depends only on the freezing rate, scaled by the collection thickness H_i. This does not mean that the polynya growth rate is independent of wind speed, since freezing rate and collection thickness are themselves functions of wind speed. It does imply, however, that air temperature is a more critical determinant for the rate of polynya growth than wind.

Pease went on to estimate the freezing rate from a bulk parameterisation of the vertical heat flux at the air-sea interface, using the equation

$$- \rho_i L F_i = (1 - \alpha) Q_r + Q_{ld} - Q_{lu} + Q_s + Q_e \qquad (4.24)$$

where L is the latent heat of freezing for salt water, α is the albedo of the sea surface, Q_r is the short wave radiation flux, Q_{ld} and Q_{lu} are the downward and upward long wave radiation fluxes, Q_s is the sensible heat flux and Q_e is the latent heat of evaporation. Using appropriate values for Bering Sea winter conditions, Pease was able to find relationships between variables as shown in fig. 4.9. She assumed that V_i is 3% of the surface wind speed.

From these diagrams we can draw the following conclusions about polynya behaviour:

1. At very low temperatures polynyas reach maximum size within a reasonable length of time (up to 30 hrs), the persistence of a typical weather system, but at higher air temperatures the opening time increases so most polynyas do not have time to reach equilibrium before atmospheric conditions change.

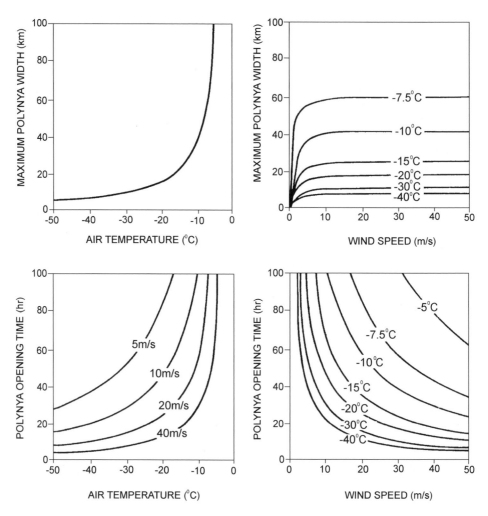

Figure 4.9. Relationships between maximum polynya width, polynya opening time to 95% of maximum width, air temperature and wind speed (after Pease, 1987).

2. Similarly, polynya opening time decreases as wind speed increases.
3. Maximum polynya width is a strong function of air temperature; as the air warms up the polynya width becomes much greater.
4. For a given air temperature, maximum polynya width increases with wind speed, but only up to 5 m s^{-1}, beyond which it stays fairly constant. This is one of the most interesting and surprising results of the analysis.

One empirical parameter in this theory, which must be determined from observation, is the collection thickness H$_i$, which is probably itself a function of wind speed and air temperature. The suggestion of Winsor (1997) is that H$_i$ should be considered a linear function of surface wind speed U_a (m s^{-1}), given in metres by

$$H_i = (1 + 0.1 \ U_a) / 15 \qquad (4.25)$$

The Pease theory has been refined by Ou (1988), who included a finite drift rate for the frazil ice (Pease assumed that the collection process was effectively instantaneous), and showed that rapidly fluctuating winds did not influence the polynya if the period of the fluctuation was shorter than the time needed for a parcel of water or frazil ice to cross the polynya.

Given that the polynya is producing ice to offset its growth, the next problem is to estimate the production rate of ice, and the corresponding production rate of salt in the water column of the polynya. In this way we can estimate the efficiency of the polynya as an "ice factory" and a "salt factory". In the Arctic, one question which can be resolved by such estimates is the role of flaw lead polynyas in generating salt to maintain the so-called **cold halocline** in the Arctic Ocean (Aagaard et al., 1981; Steele and Boyd, 1998). This layer, as mentioned in chapter 1, is the lower part of the polar surface water layer and has a salinity increasing with depth while the temperature remains at or near the freezing point.Where does the salt come from? One possibility is that it comes from the lateral advection of cold, salty water formed on the Arctic shelves through brine rejection from ice formation, a process which is especially effective in polynyas. This is the shelf-slope convection mechanism described in chapter 1. Aagaard et al. (1981) estimated that a flux of 1–2 Sv of shelf water would be needed to maintain the cold halocline, while Björk (1989) estimated 1–1.5 Sv and Cavalieri and Martin (1994) estimated that Alaskan, Siberian and Canadian coastal flaw polynyas could actually supply 0.7–1.2 Sv.

Theoretical models of salt and ice processes in polynyas include that of Winsor (1997). This is basically an extension of the Pease model. The freezing rate is estimated from (4.24), with the polynya width being initialised at zero at t = 0. When the wind speed drops below 5 m s^{-1} it is assumed that the turbulence level is too low to produce frazil ice, and instead the polynya freezes over (with the growth rate being estimated from normal Maykut-Untersteiner thermodynamics as in chapter 3). The polynya width is thus reset to zero and reopening does not occur until the wind speed again exceeds 5 m s^{-1}. The rate of salt production per unit length of the polynya is given by

$$S = F_i \ X \ \rho_i \ (s_w - s_i) \qquad (4.26)$$

where the salinity s_i of the newly forming frazil ice can be expressed empirically in terms of the salinity s_w of the surface water by (Martin and Kauffman, 1981)

$$s_i = 0.31 \ s_w \qquad (4.27)$$

The next question is what this salt does to the water structure in the polynya. Winsor considered the case of the Laptev Sea, or of flaw polynyas on Russian shelves in general, where the water is shallow (about 50 m) and ice production is continuous in the polynya. In this situation he considered that two extreme models could describe the effects of the salt production. In the simplest, non-dynamical, model, the added salt is simply mixed equally over the area of the polynya. In a more complex model, fully developed Ekman

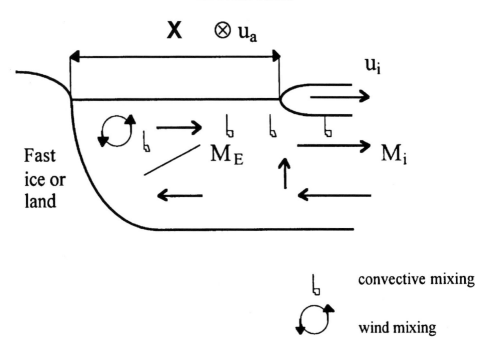

Figure 4.10. Schematic diagram of circulation set up by Ekman transport under the ice edge of a coastal polynya (after Winsor, 1997). The wind U_a is into the plane of the diagram. The mass transport M_i is larger than the Ekman transport M_E within the polynya, resulting in ice edge upwelling.

dynamics occur. In a simplified version of this model we consider a wind u_a directed parallel to the length of the polynya (fig. 4.10). The ice at the polynya edge is advected with a velocity u_i and a direction given by the free drift model, i.e. about 3% of the surface wind speed and has a seaward component. The wind acting on the open water surface in the polynya exerts a stress τ_a with a corresponding Ekman mass transport M_E given by

$$\tau_a = \rho_a\, C_a\, u_a^2 \;; \quad M_E = \tau_a\, /\, \rho_w\, f_C \tag{4.28}$$

where ρ_a is air density; C_a is air-water drag coefficient (taken to be 1.3×10^{-3}); ρ_w is reference water density; and f_C is the Coriolis parameter. The Ekman mass transport is directly seaward for a Northern Hemisphere geometry (fig. 4.10). Under the ice, beyond the edge of the polynya, the direct wind stress on the water is switched off. Instead, wind stress acts on the ice to produce the free drift, then there is an ice-water drag τ_i acting on the underlying water, giving an under-ice mass transport M_i:

$$\tau_i = \rho_w\, C_w\, u_i^2 \;; \quad M_i = \tau_i\, /\, \rho_w\, f_C \tag{4.29}$$

where C_w is the ice-water drag coefficient, taken to be 5.5×10^{-3}. M_i is not directly offshore, because of the additional air-ice turning angle; in addition, as we have already shown in the full theory of wind-driven motion, the first part of (4.29) is a simplification because it does not allow for the water velocity. Nevertheless, the main point of the analysis is that at a sufficiently high wind speed, M_i/M_E becomes greater than unity; Winsor estimated the ratio to be 2.9 for a 10 m s^{-1} wind.

This, surprisingly, implies that the offshore water transport is greater under the ice edge than under the polynya; to maintain continuity there has to be an upwelling at the ice edge as shown in fig. 4.10, associated with a shoreward flow of bottom water. Given that convective mixing is occurring continuously due to the injected salt, the implication is that the salt is being mixed with a greater volume of water than simply the volume initially under the polynya. A more complete analysis suggests that the salt over a season is mixed with about four times the mean volume of the polynya. Thus the outflow of salt-enriched water is greater than predicted by the non-dynamical theory but the salinity enrichment is correspondingly less.

4.5. TIDAL ICE MOTION

It has been known from the times of the earliest explorers that in the shallow seas of the Arctic the ice cover shows a strong tidal response. A ship trapped in concentrated pack ice would often find that after a wait of 6 or 12 hours the pressure would ease and a network of leads would open up. Similarly, in partially enclosed gulfs (e.g. White Sea), channels or inlets there is a tidally periodic motion of the ice cover which can yield a build-up of pressure ridging against opposite coasts; this phenomenon was described for the Russian Arctic by Zubov (1945). Tidal flow on shelves produces a mixing and stirring under the ice and near the bottom which enhances the shelf processes described in chapter 1. Tides also have thermodynamic effects, in that the systems of periodic leads produced result in an enhanced heat exchange between ice and ocean and a greater net rate of ice production in winter. It is therefore important to assess the magnitude of the role of tides in Arctic ice dynamics. In the Antarctic, where the shelves are narrow and fall away quickly to the deep ocean, tidal effects are probably much less important but have been little studied.

The classic work on Arctic tides was by Defant (1924), who established that the semidiurnal tides in the Arctic Ocean are caused by the Atlantic tides, while the diurnal tides are generated internally by astronomical forces. Sverdrup (1926) extended this work for the Siberian shelves. Modern tidal analysis involves modelling both the tidal response of the ocean, in the form of amplitudes and currents due to the different tidal components, and the resulting effect on the motion of the sea ice itself. An example of such a modelling effort is Kowalik and Proshutinsky (1994), the results of which we discuss below.

4.5.1. Tidal Amplitudes and Currents

The first result from models, which matches observation, is that while the tidal *amplitude* due to semidiurnal tides (M_2 and S_2, the principal lunar and solar components) is usually greater than the amplitude of the diurnal tides (K_1 and O_1 solar–lunar and lunar respectively), the *currents* due to these components are often greater for the diurnal tides. In principle, this is clearly because the reversal of the tide is twice as frequent and seabed friction produces a phase lag which does not allow such a large current to grow for a given amplitude. Observations of such effects in the Arctic come from Hunkins (1986), who found a topographic amplification of the diurnal tidal velocities over the Yermak Plateau, and Aagaard *et al.* (1990), who found for the Beaufort Sea shelf that the flow is dominated by the diurnal tide while the amplitude is dominated by the semidiurnal tide. Further analyses of the Yermak Plateau tidal amplification were carried out by Muench *et al.* (1992) and Padman *et al.* (1992), while the effect of these diurnal tides on ice drift was noted by Gascard (personal commun.) in a circular motion induced in drifting ice buoys as they passed over the northern part of the Plateau. Such ice motion would inevitably produce ridging and lead formation. A high-resolution model of the diurnal tides alone (Kowalik and Proshutinsky, 1993) shows that the current enhancement in these specific areas is related to the generation of **shelf waves** by the tide.

The overall tidal model (Kowalik and Proshutinsky, 1994) showed that computed *amplitudes* of the M_2 constituent ranged from 1–10 cm over the deep Arctic Basin to 20–100 cm on the Siberian shelves (greatest near the coast) and in Hudson Bay, with another especially large range in Baffin Bay (up to 100 cm). The S_2 component had lower amplitudes of 0–5 cm in the basin, 5–30 cm on shelves and 10–30 cm in Baffin Bay. The diurnal constituents K_1 and O_1 each had amplitudes of only 1–5 cm in the basin and 2–10 cm on Siberian shelves, but with again a very high amplification to 30–50 cm in Baffin Bay. When these are converted into *currents*, however, and added with appropriate phases to yield a maximum tidal current, the results showed that over the deep Arctic Basin the maximum tidal currents are in the range 5–10 cm s^{-1}. Over the Siberian shelves, Canadian Arctic Archipelago, Hudson and Baffin bays they rise to 10–30 cm s^{-1}, while tidal "hot spots" with velocities exceeding 50 cm s^{-1} are found in only a few specific coastal locations, the largest being the northern end of Hudson Bay, Hudson Strait, the area round Bear Island and the entrance to the White Sea. Considerations of how such currents cause leads to open and close enabled Kowalik and Proshutinsky to estimate the **rate of tidal ice production** per year, which is closely related to the distribution of maximum tidal currents (fig. 4.11). They found that the rate in the central basin was less than 2 cm a^{-1}; on shelves was also typically 2–5 cm a^{-1}; but in the tidal hot spots rose to local values of up to 4 m a^{-1} in northern Hudson Bay, Hudson Strait and off West Greenland, and up to 2 m a^{-1} off Bear Island, off the White Sea entrance and at a few small coastal locations off Siberia, Svalbard and Franz Josef Land. Fig. 4.11 shows the somewhat larger, but still limited, zones where annual ice production exceeds 50 cm, which may be taken as a significant contribution to overall annual ice production. The total additional ice production in the Arctic due to tides was estimated by Kowalik and Proshutinsky at 8×10^{11} m^3 a^{-1}.

4.5.2. Effect of Sea Ice on Tides

The modelling showed that sea ice can alter the amplitude and phase of the tide, but that this effect is small in deep water, amounting to a decrease in amplitude of some 3% and

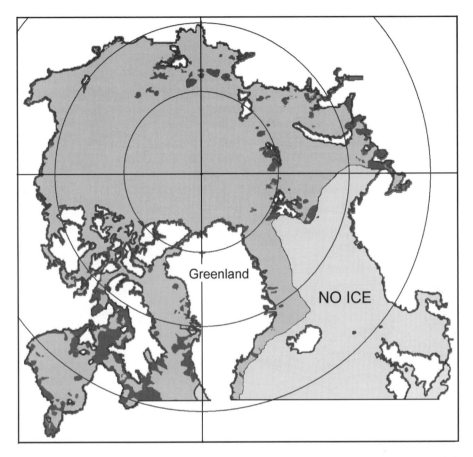

Figure 4.11. Rate of ice production in the Artic due to M_2, S_2, K_1 and O_1 tidal waves. Shaded area is less than 2 cm a^{-1}. Dark areas are greater tha 50 cm a^{-1} (adapted from Kowalik and Proshutinsky, 1994).

a phase lag of 4.5°–5°. The ice acts as a flexible membrane which weakly damps the vertical mode of motion but resists horizontal motion. Only in very shallow water, especially under fast ice, does the ice significantly affect tidal propagation (Murty, 1985). This is because in such conditions the fast ice significantly reduces the water depth, while the tidal range exerts vertical stresses on the fast ice itself, often producing one or more tidal cracks.

4.5.3. Effect of Tides on Sea Ice

The first effect of tides on sea ice, already mentioned and shown in fig. 4.11, is the increase in annual ice production due to tidally-induced divergence and convergence of the ice cover. The tidally-induced currents exert a stress on the underside of the ice cover which

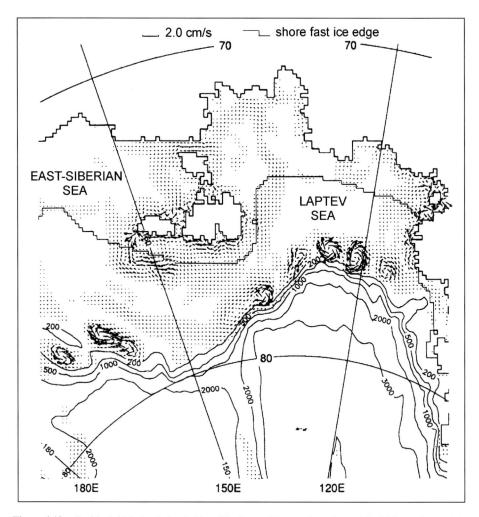

Figure 4.12. Residual tidal circulation in East Siberian and Laptev Seas due to M_2 tidal constituent (after Kowalik and Proshutinsky, 1994). Shorefast ice limit is marked.

has been detected in field measurements (e.g. Tucker and Perovich, 1992) and which causes ice motion. Because the tidal periods are so long, the ice has time to respond to the driving stress, so on the whole the ice velocity is similar in magnitude and sign to the tidally-induced water current, an approximation that was first made by Zubov (1945).

A consequence of the non-linear dynamics of tides is that there is a weak **residual tidal motion** over and above the periodic components. Figure 4.12 shows the residual tidal circulation over part of the Siberian shelf seas due to the M_2 tidal constituent. The currents are quite small except for a number of circulating current systems near the edge of the shelf. These are anticyclonic eddies forming over elevations of the sea floor, and cyclonic eddies forming over depressions, which thus transfer vorticity to the mean field of motion. When translated by the model of Kowalik and Proshutinsky into residual ice motion, the pattern is the same as in fig. 4.12 but the velocity is about half that of the sea current. Thus the pattern of ice motion from fig. 4.12 will generate circulating patterns of motion in the ice as it drifts across the shelf break. In the vicinity of the coastline, the residual tidal ice motion will always be strongly offshore, since during half of the tidal cycle the ice motion is suppressed by the solid boundary. This is one of two ways in which tidal motion influences the role of Siberian shelf polynyas, introduced in chapter 1: the residual tidal motion enhances the offshore drift of ice in the polynya. The other influence is mixing on the shallow shelf caused by the strong tidal currents, which transfers heat to the surface and thus enhances the shelf-slope convection process.

5. PRESSURE RIDGES AND THE ICE THICKNESS DISTRIBUTION

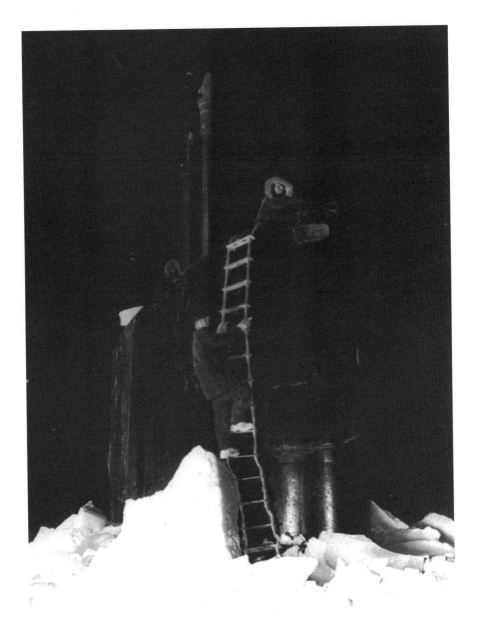

You are in a submarine under the Arctic ice. For days you have been recording the thickness of the ice canopy overhead on upward looking sonar, watching the chart recorder trace out the rough under-ice landscape of steep pressure ridges, hummock fields and stretches of undeformed ice of different thicknesses and ages. It is cosy, crowded and claustrophobic on board, like living in an underground train at rush hour. You long to breathe fresh Arctic air. Now you have reached the North Pole. Here is a polynya overhead, big enough to surface through. But is the ice thin enough? The captain orders a stop trim, then allows the boat to rise slowly until her fin is touching the smooth underside of the refrozen lead. The upward TV camera shows no light penetrating from above; the North Pole has entered its six months of night. The captain orders all main ballast to be blown. A terrible rending, creaking, groaning and cracking noise comes from above, then from all around. Then silence. The ice was too thick. Slowly the boat tilts. Pressed against the ice by her buoyancy, her bow rises to touch the ice, leaving the deck angled steeply upwards. "Pipe all hands to dinner" orders the captain. Why not eat supper under the North Pole? A few hours later you dive, find another polynya, and surface safely. You emerge from the hatch to find darkness all around but stars of dazzling brightness in the sky, with Polaris directly overhead. The floodlights are switched on. A jumble of broken ice blocks lies on top of the casing, with one block perched on top of the fin itself. Out to the edge of the floodlit area stretches the smooth snow-covered ice of the refrozen polynya. You climb down the rope ladder, pick your way over the blocks, slide down the casing and step out onto the young ice. You are taking a walk at the North Pole.

In this chapter we explore the statistical properties of the sea ice thickness distribution and other statistical measures of sea ice roughness in the Arctic and Antarctic. Many of the data from which these properties are derived come from submarines.

5.1. STATISTICAL PROPERTIES OF A SEA ICE COVER

5.1.1. Importance of the ice thickness distribution

In Chapter 3 we found that if we know the history of the air temperature, the snow fall and the ocean heat flux in a region, we can calculate the annual cycle of ice growth and decay. If all the ice in the Arctic and Antarctic were stationary, we could easily calculate how thick it is at any time of the year, and therefore what its role is in heat exchange between the ocean and the atmosphere. The study of sea ice would be very easy. Unfortunately, as Chapter 4 has shown, the sea ice is in motion everywhere except in shallow water and in narrow channels. Because the driving force for this motion, the wind, exerts a variable drag over the ice surface, it induces stresses which tear the ice apart and then crush the weak areas of newly-created young ice into ridges. The result is the highly variable material that we actually see in the polar oceans. Are these processes simply chaotic or can we find ways to describe the distributions of ice thickness and pressure ridge depths that are statistically valid, that tell us something about the processes taking place, and that obey some regular laws? It turns out that we can.

The thickness distribution of sea ice is actually a very fundamental attribute which defines the character of an ice regime. The thickness distribution determines ocean-atmosphere heat exchange, since heat flux is so much greater through thin ice than thick. The shape of the distribution is a measure of the degree of deformation of an icefield. Together with multi-year fraction, the thickness distribution defines the strength and other mechanical properties of the ice cover that are important for ice-structure and ice-vessel interaction. The area-averaged mean thickness h_m, when combined with the ice velocity, defines the mass flux, that is, the rate of transport across a unit width at right angles to the ice drift direction. In Fram Strait, for instance, the ice mass flux is a major component of the overall energy and fresh water exchange between the Arctic Ocean and the Greenland Sea. Finally, and possibly of greatest importance, long-term trends in ice thickness can be interpreted as a response of sea ice to climatic change.

The statistics of pressure ridges, too, are of basic importance to understanding ice processes. Ridges are transported with the surrounding undeformed ice in which they are embedded, and partake of the later melt processes by which the ice gradually loses thickness in sub-polar seas such as the Greenland Sea. The fresh water flux into the Greenland Sea, for instance, is only partially due to thermodynamically grown ice, the rest being ice which has been produced by mechanical means. Pressure ridging is of importance in other contexts: it increases the hydrodynamic and aerodynamic drag co-efficients of the ice surface; deep ridges can generate internal waves in the pycnocline; in shallow water deeper ridges scour the seabed and are responsible for a coastal "stamukhi zone" of highly deformed fast ice; ridges are the determining factor in strength calculations for offshore platforms designed for Arctic use (Sanderson, 1988) and in icebreaker hull design; and ridges cause scattering of underwater sound such that long-range propagation can only be achieved out at very low frequencies. We shall consider all these effects later in this chapter.

5.1.2. Ice Thickness Distribution — Definitions

Let us begin by formally defining the ice thickness distribution, following the terminology of Thorndike *et al.* (1992). Let R define a finite area within the ice cover, centred on a point **x**. Let dA(h, h + dh) represent the area within R covered by ice of thickness between h and (h + dh). Then the probability density function of ice thickness g(h) is given by

$$g(h) \ dh = dA(h, \ h + dh) \ / \ R \qquad (5.1)$$

and has dimensions L^{-1}. Fig. 5.1 (from Wadhams, 1981) shows some typical distributions. g(h) can have a delta-function at h = 0 if open water is present.

When we consider how thickness changes with time, the governing equation is

$$\partial g/\partial t = \text{div}(\mathbf{v}g) - \partial(fg)/\partial h + \mathbf{\Omega} \qquad (5.2)$$

where $f(h,\mathbf{x},t) = dh/dt$ is the thermodynamic growth rate of ice of thickness h at time t

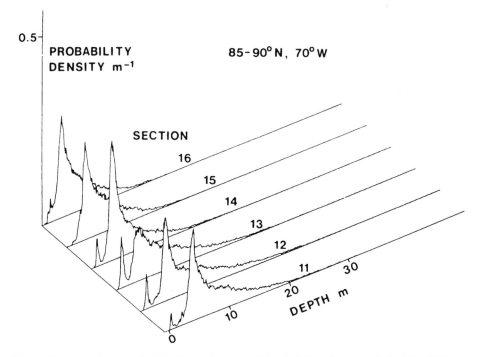

Figure 5.1. Some typical probability density functions of ice draft from the Arctic Basin (after Wadhams, 1981). Each function is derived from 100 km of submarine sonar profile.

and at point x; v is the horizontal velocity vector and Ω is a function that mechanically redistributes ice from one thickness to another, describing the formation of leads and pressure ridges.

The first term in (5.2), divergence within ice, expresses the way in which when the ice cover opens out under the influence of a divergent wind stress, the creation of open water by the formation of leads adds probability at $h = 0$ and subtracts an equal fraction of existing probability from all other thickness categories: it is a source of open water and a sink of ice-covered area. The second, thermodynamic, term causes thin ice to grow thicker and thick ice to grow thinner when averaged over an annual cycle — it represents the results that we derived in chapter 3. The third, and least understood term, describes the ridging process, which simultaneously produces open leads and squeezes thin ice together to form pressure ridges. Thus it is at the same time a source of open water, a sink of thin ice, and a source of thick ice. The relative effects of the three processes on g(h) are shown schematically in fig. 5.2 (after Thorndike *et al*, 1992). It is likely that the g(h) resulting from these processes never reaches an equilibrium state; if you measure g(h) at any one time it will always be in the process of change under the impact of these three processes.

The term Ω is the most difficult to define. In current ice models which take account of ice thickness variations, Ω is treated in a mainly empirical way (e.g. Walsh and Chapman, 1991), and it seems probable that if the physics and mechanics of the ridge-

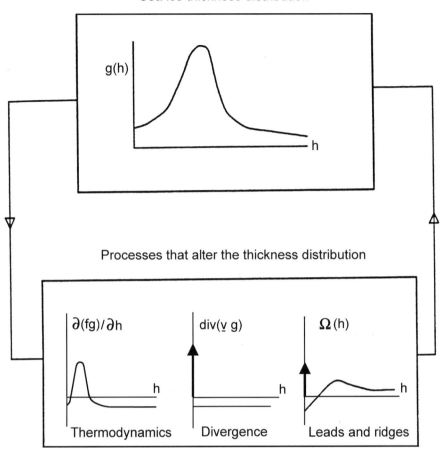

Figure 5.2. The evolution of the ice thickness distribution due to three types of physical process (after Thorndike *et al.*, 1992). For explanation see text.

building process can be more adequately described, it will be possible to improve such models to provide a better description of air-ice-ocean interaction and its role in global climate. We also might expect better predictions of the effect of global warming upon Arctic sea ice: it is not obvious, for instance, that warming will cause a decrease in the mean ice thickness, since thermodynamic thinning may be offset by a greater ease of pressure ridge production.

5.1.3. Ice Thickness Distribution — Typical Shapes

Figure 5.1 demonstrates the typical features of an Arctic Basin ice thickness distribution — actually, in this case, an ice draft distribution since it was measured from a submarine.

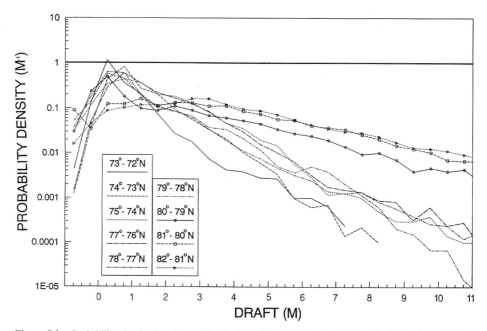

Figure 5.3. Probability density functions of ice draft in 1° latitude bins in the Greenland Sea, from 82°N to 72°N, drawn on a semilogarithmic scale (after Wadhams, 1992).

Firstly, there are one or more small peaks in the range 0–1 m. These correspond to refrozen leads at different stages of development, and young ice which has grown in polynyas during periods of ice divergence. There may also be a peak at zero thickness due to open water. There is usually a relative lack of ice in the 1–1.5 m thickness range, followed by one or more main peaks. The widths and relative heights of these peaks provide important information about the undeformed ice types (first-year and multi-year of different ages) that are present in the ice regime and their relative abundances. Typically the first-year peak occurs in the vicinity of 2 m and the multi-year peak or peaks in the range 3–5 m. Finally there is a tail which contains ice which is thicker than can be attained by thermodynamic growth (> 5 m), and thus represents the flanks and crests of pressure ridges and hummocks. Deformed ice is, of course, found at lower thicknesses as well, but at thicknesses beyond about 5 m it comprises the whole of g(h). To obtain a distribution of this quality requires a submarine to profile over a distance of 100 km, so that in this case g(h) is defined over a length scale rather than an area scale R as in equation (5.1).

The shape of g(h) as shown in fig. 5.1 has the interesting statistical property that the tail of the distribution gives a good fit to a negative exponential distribution. Figure 5.3 demonstrates this for the case of a set of ice thickness distributions obtained at different latitudes in the Greenland Sea, plotted on a semilogarithmic scale. Although the slope of the tail increases as the latitude decreases (evidence that thick ice is melting preferentially as the ice stream moves southward), each distribution is a good approximation

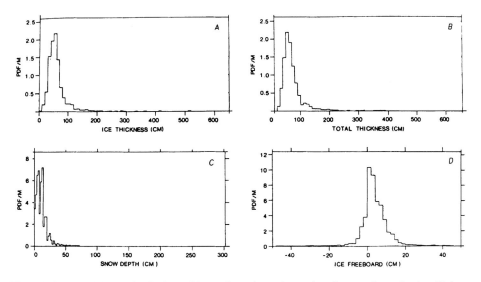

Figure 5.4. The thickness distributions of Antarctic sea ice and snow in a first-year ice region in midwinter (after Wadhams *et al.*, 1987).

to a negative exponential. Since the tail of g(h) represents the ice contained in pressure ridges, we conclude that the probability density function of the thickness of deformed ice is a negative exponential.

In the Antarctic our knowledge of g(h) is much less extensive. Systematic data have been obtained mainly by repetitive drilling, which can never achieve the data density required for a distribution of the quality of figure 5.1. Figure 5.4 shows the ice and snow thickness distributions obtained from drilling 4400 holes in a region of first-year ice (the eastern Weddell-Enderby Basin) in mid-winter of 1986 (Wadhams *et al.*, 1987), showing that the modal ice thickness is only 50–60 cm, that maximum observed keel drafts are only about 6 m, and that the snow cover (measured independently) is sufficient in many cases to push the ice surface below water level and cause flooding and the formation of a new ice type, snow-ice (Lange *et al.*, 1989). The data set is too crude for us to tell whether the thick end of the distribution fits a negative exponential. More recently data which have been retrieved from moored upward sonar in the Weddell Sea (Strass and Fahrbach, 1998) show tails which appear to be close to negative exponentials, although not explicitly analysed for this property.

5.1.4. The Distribution of Morphological Features

In considering how the character of an icefield is revealed by the thickness characteristics of the ice, we have to deal not only with g(h) but also with the distribution of thicknesses and spacings of pressure ridges and refrozen leads, i.e. we must consider the distribution of morphological *features* in the ice as well as the thickness of the ice as a whole.

There have been a number of attempts to model ridge building forces in order to predict sail and keel dimensions of pressure ridges under different conditions, but none of the models yet proposed adequately describes the complete process. The earlier models described limits to size and rates of development by equating kinetic energy loss from the colliding ice floes with the potential energy gained by the ice debris to calculate the ridge-building force (Zubov, 1945; Parmerter and Coon, 1972). Later models have attempted to equate the limits of ridge development with the weight, density and plasticity of the ridge building material and the maximum possible ridge slope (Kovacs and Sodhi, 1980; Sayed and Frederking, 1986). A more recent model (Hopkins, 1994) considers the development of an array of broken ice blocks in a statistical way, whereby the ice in a refrozen lead buckles and fractures, then the ice blocks pile up randomly, governed by gravity and frictional force; realistic ridge shapes can be achieved.

Pressure ridges have been observed to adopt a wide variety of shapes and sizes (Wadhams, 1978a; Weeks *et al.*, 1971; Rigby and Hanson, 1976). It is difficult to define pressure ridges in terms of dimensions such as ridge width or ridge slope (angle of repose) since multiyear keels can be eroded to semi-elliptical or even semi-circular profiles by seasonal melt and ocean currents, and sails are sculpted by the wind or obscured by snow fall. The simplest representation for a ridge cross section is an isosceles triangle with rounded crest, and many ridges fall into this category. In addition, many authors have assumed that pressure ridge cross-sections are symmetrical, but an examination of published ridge profiles appears to refute this assumption (Kovacs, 1971; Dickins and Wetzel, 1981).

Pressure ridges have keel draft to sail height ratios of the order 3–4:1, with larger ratios for first-year ice (Tucker, 1989). The deepest ridge keel recorded had a draft of 47 m (Lyon, 1961) and the highest recorded free-floating sail was 13 m high (Kovacs *et al.*, 1973). The draft:height ratio is lower than the volume ratio because keels are usually wider than sails. Studies of the composition of the blocks in first-year ice (e.g. Tucker, 1989) indicate a relationship between the thickness of the parent ice blocks and ridge height, implying that there is a limiting height to which ice of a certain thickness will build, given sufficient force. In first year ice, keel slopes have been measured in the range 20°–55°, averaging at 33° with sail slopes ranging from 15°–30° and averaging 24° (Kovacs, 1971). For multiyear ice, keel slopes of almost zero to 51° have been recorded (Wadhams, 1978a) with a concentration around 16°–28° averaging at 23.9°, whereas sail slopes are of the order of 14°–19° (Weeks *et al.*, 1971). The Parmerter-Coon model (1972) predicts typical ridge slopes of 25° and 35° for sails and keels respectively; these are in reasonable agreement with the findings of Kovacs (1971) for first year ice, but not so well matched by the multiyear ridges reported by Wadhams (1978a).

5.1.5. Pressure Ridge Depths

Our knowledge of ridge distributions is empirical and derived mainly from observations of ice profiles. When we come to analyse a submarine ice profile we frequently find a rugged surface in which successive ridges appear to overlap one another, so that we need a systematic criterion for defining an "independent" ridge. The definition of individually identifiable ridges in one-dimensional sonar or laser profiles is commonly based on the

Rayleigh criterion for separating spectral lines in optics, stating that an independent ridge must have a crest elevation or draft relative to the local undeformed ice which is more than double that of the troughs which bracket it. Using this method for the identification of ridges in the analysis of laser profiles (Leppäranta, 1981; Tucker *et al.*, 1979; Wadhams, 1976, 1981; Wadhams and Lowry, 1977), the distribution of sail heights has been found to fit a negative exponential with a high degree of accuracy. Sonar profiles of the ice underside obtained using a narrow beam instrument (McLaren *et al.*, 1984; Wadhams and Horne, 1980; Wadhams *et al.*, 1985; Wadhams, 1992), show a similar distribution.

We can therefore define the distribution of keel drafts (or sail freeboards) thus:-

$$n(h) \ dh = B \ \exp \ (-bh) \ dh \ |_{h > h_0} \tag{5.3}$$

where $n(h)$ is the number of keels per km of track per m of draft increment, and B, b are derived in terms of the experimentally observed mean keel draft (h_m), the mean number of keels per unit distance (n_k) and a low level cutoff draft (h_0):-

$$b = (h_m - h_0)^{-1} \tag{5.4}$$

$$B = n_k b \ \exp \ (bh_0) \tag{5.5}$$

Wadhams and Davy (1986) confirmed the validity of the negative exponential relationship by a careful examination of keel depths involving the use of order statistics to test for a negative exponential.

The mean slope angle of pressure ridges can be inferred from $g(h)$ if it is assumed that pressure ridges of all depths are, on average, geometrically similar and can be approximated in shape by an isosceles triangle. Each ridge then contributes equal quantities of ice to $g(h)$ down to its maximum depth. It is easy to show that the relative slopes of log $n(h)$ and of the tail of log $g(h)$ are related to the mean along-track slope angle ∂ of pressure ridges by the relationship

$$g(h) \ dh_{[h > 5]} = [2 \ B \ \cot \partial \ \exp \ (-b \ h) \ / \ b] \ dh \tag{5.6}$$

The best values for ∂ calculated from these relationships appear to lie in the range 11–14°. Thus the problems of why the pressure ridge draft distribution and the tail of the ice draft distribution are both negative exponentials can be reduced to a single problem in ice mechanics if it is assumed that ridges are geometrically congruent. The reason for quoting a 5 m minimum value for h in (5.6) is that the relationship is valid only when the ice draft exceeds any value that can be reached thermodynamically.

5.1.6. Pressure Ridge Spacings

The spacing of pressure ridges was also initially reported as fitting a negative exponential distribution (Mock *et al.*, 1972), such that

$$P_r(x) \ dx = n_k \ \exp \ (-n_k x) \ dx \tag{5.7}$$

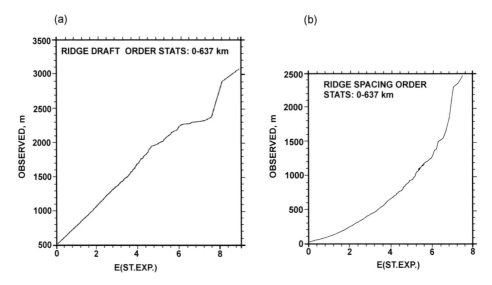

Figure 5.5. Order statistics for ridge drafts (a) and spacings (b) from an ice profile in the southern Beaufort Sea (after Wadhams and Davy, 1986).

This relationship is expected if the positioning of ridges is assumed to occur randomly over an initially undeformed ice cover. However, such appears not to be the case, since the work of Wadhams and Davy (1986) showed that a lognormal distribution offers a more accurate description of the spacing distribution. Later observations by Key and McLaren (1989) confirmed these results.

Wadhams and Davy reanalysed all available datasets on ridge depths and spacings. A specially sensitive test is so-called "order statistics". We take the observed drafts (or spacings) $S_1 \ldots S_n$ in order of size, starting with the smallest. In fig. 5.5 we have plotted the draft against the expected value of the corresponding order statistic from a negative exponential distribution with mean unity and the same sample size. If $S_1 \ldots S_n$ is a random sample from a standard negative exponential distribution and $S_{(1)} \ldots S_{(n)}$ are the corresponding order statistics, then

$$ES_{(i)} = [1/n + \ldots + 1/(n + 1 - i)] \tag{5.8}$$

where ES is the expectation of S (Cox and Snell, 1981). If the observed data fit a negative exponential, the resulting graph should be a straight line, with some expected random variability at the top end because of the small sample size for the greatest drafts. As fig. 5.5 shows, pressure ridge drafts in the sample considered (637 km of narrow-beam sonar data from the southern Beaufort Sea) do indeed fit a negative exponential very well. However, pressure ridge spacings do not, and the curves for spacings have a smooth convexity which strongly suggests that an alternative distribution would provide a better fit to the data. It was found that an excellent fit to the spacing distribution for a large number of datasets (narrow and wide beam) is provided by the lognormal distribution.

The lognormal distribution (Aitchison and Brown, 1957) is defined as follows. Let X be a random variable. If there is a number θ such that the random variable $Z = \ln (X - \theta)$ is normally distributed, then X is said to have a lognormal distribution. If the mean and standard deviation of Z are μ and σ, then the probability density function of X is

$$f(x) = [(x - \theta) \sigma (2\pi)^{1/2}]^{-1} \exp [-\{\ln (x - \theta) - \mu\}^2 / 2 \sigma^2] \qquad (5.9)$$

The mean of X, equivalent to the reciprocal of the number of keels per unit distance, is

$$1/n_k = \theta + \exp (\mu + \sigma^2/2) \qquad (5.10)$$

and the variance of X is

$$\sigma_X^2 = \exp (2\mu + \sigma^2) [\exp (\sigma^2) - 1] \qquad (5.11)$$

The mode of X is

$$m_x = \theta + \exp (\mu - \sigma^2/2) \qquad (5.12)$$

which expresses the most likely spacing of ridges and which is therefore useful as a physically meaningful "characteristic length" for the deformed ice field.

Since only logarithms of positive numbers are defined, x must be greater than θ, which is called the **threshold**. The lognormal approaches a normal when $\sigma^2 << \mu$, which is the case for variables such as the heights of adult males. When $\sigma^2 = \mu$, there is a strong positive skewness; such distributions fit variables as diverse as incomes, age at first marriage, particle sizes in a soil, and the number of words in sentences written by George Bernard Shaw (Williams, 1940).

The recommended approach to estimating the parameters of a lognormal (e.g. Pollard, 1977) is to guess a suitable value for θ (e.g., slightly less than the lowest observed x_i) and to replace x_i, the sample keel spacings, by $z_i = \ln (x_i - \theta)$. The estimators of μ and σ^2 are then the sample mean Z and the sample variance σ_z^2 of the z_i values.

The best graphical way to test a fit to a lognormal is to use logarithmic probability paper, in which the cumulative probability

$$L(x) = \int_0^x f(x) \, dx \qquad (5.13)$$

is plotted as its equivalent normal deviate against ln x. A two-parameter lognormal gives a straight line. Figure 5.6a shows the results for keel spacings in a Beaufort Sea data set obtained by the US submarine *Gurnard*. It can be seen that the raw data give a good fit to a lognormal, except for a deficit of spacings at 100–600 m and an excess of 10 m. By replacing x with (x − 3), i.e., putting $\theta = 3$ m, we obtain a fit which is now excellent throughout the spacing range up to 400 m (the 94th percentile). The physical reason for trying $\theta = 3$ is that the profile was digitised with a horizontal spacing of 1.5 m between data points; it is therefore impossible for two independent ridges to be closer than 3 m apart.

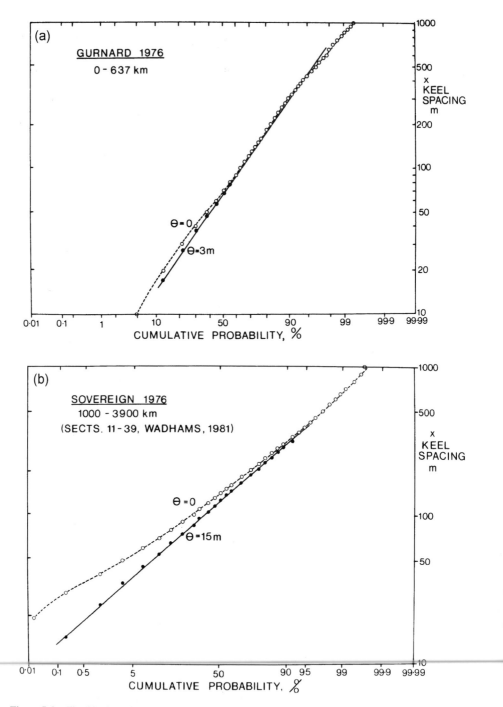

Figure 5.6. Fit of keel spacing data to a lognormal. (a) Narrow-beam sonar in the Beaufort Sea: open circles have a threshold of zero, closed circles of 3 m; both sets converge for spacings beyond 100 m. (b) Wide-beam sonar in the Eurasian Basin: closed circles have threshold of 15 m (after Wadhams and Davy, 1986).

When wide beam datasets were examined (e.g. fig. 5.6b, from a British submarine) it was found that they still gave excellent fits to a lognormal so long as the threshold parameter was increased to a figure, typically 15 m, which approximated to the surface beam diameter of the sonar. A simulation exercise with a set of artificially generated ridges with negative exponential drafts and two-parameter lognormal spacings showed that it is indeed the "smearing" effect of the wide sonar beam which causes a larger threshold to be needed, since closely spaced ridge crests can no longer be distinguished.

Thus we conclude that the best fit to keel spacings is a lognormal which would be a simple two-parameter lognormal for a hypothetical profile obtained with an infinitely narrow beamwidth of sonar and with data points infinitely close together. For real under-ice profiles the appropriate distribution is a three-parameter lognormal with a threshold set either as twice the spacing of successive data points, or as the surface beam diameter, whichever is the greater.

5.1.7. Pressure Ridge Slopes and Widths

Most under-ice profiles from which statistics can be derived have been obtained from upward-looking sonar, which yields accurate data on ridge depths and along-track spacings, but tells us nothing about the true slopes of individual ridges, because of the unknown angles at which the ridge axes cut the submarine track. To derive ridge slope distributions it has hitherto been necessary to assume that ridge orientations are random (Wadhams, 1978a).

The statistical treatment developed by Wadhams took account of

(a) the beamwidth of the sonar, which gives ridges an apparently shallower slope angle and an apparently greater width than would be seen with narrow-beam sonar;
(b) the random angle at which the submarine track crosses the ridge axis.

Figure 5.7 shows how, by the use of this method, an observed distribution of ridge slopes was converted into a "true" slope distribution based on the assumption of random orientation. To show how difficult it is to actually define a ridge slope, figure 5.8 shows some observed profiles of deep keels (all deeper than 30 m) drawn without vertical exaggeration. Clearly a systematic technique is needed to define what we mean by "slope angle" of a keel.

Davis and Wadhams (1995) studied ridge keel morphology, using upward looking (UL) and sidescan (SS) sonar records obtained simultaneously during a 1987 cruise, focusing on pressure ridges from six geographically distinct Arctic regions. They examined 729 independent ridges which could be identified on the UL and also on SS, and thus for the first time it was possible to convert the observed keel slopes (from UL data) into real keel slopes, using ridge orientation information provided by the SS. Thus we have the first very large dataset on real underwater ridge shapes, enabling us to draw statistically valid conclusions about the slopes and widths of ridges. A wide range of keel widths (defined as the distance between the troughs which bracket the keel crest) was seen, clustered around 50–150 m. The widths fitted a log normal distribution at all latitudes with correlation coefficients better than 0.95. Keel slopes were in good agreement with

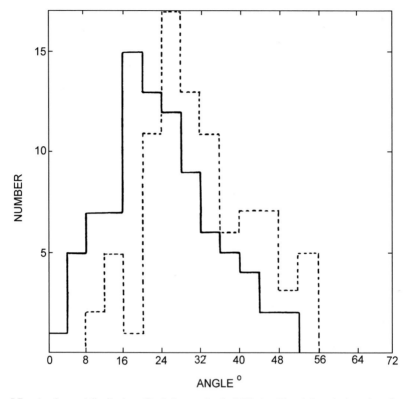

Figure 5.7. An observed distribution of keel slope angles (solid line), with an inferred orientation of true slopes (dotted lines) based on an assumption of random orientation (after Wadhams, 1978).

the earlier study of Wadhams (1978a), and there was a similar good fit to a log normal distribution (correlation coefficient 0.991) with no significant difference between leading and trailing slopes. The ratios of half width to draft were similarly distributed (correlation coefficient 0.994) and variations of slope with latitude appeared to be slight. A tentative relationship between slope and half width/draft ratio was found, of the form

$$\text{Slope} = a + b.c^{\text{ratio}} \quad (R = 0.93) \tag{5.14}$$

where a,b,c are parameters which are best fitted by a = 8.71, b = 76.3, c = 0.585 when the keel slope is expressed in degrees. This purely empirical relationship requires further investigation.

5.1.8. Lead Widths and Spacings

Finally, as another piece of the puzzle of ice deformation mechanisms we must mention leads, which when refrozen are the raw material out of which ridges are created. Leads

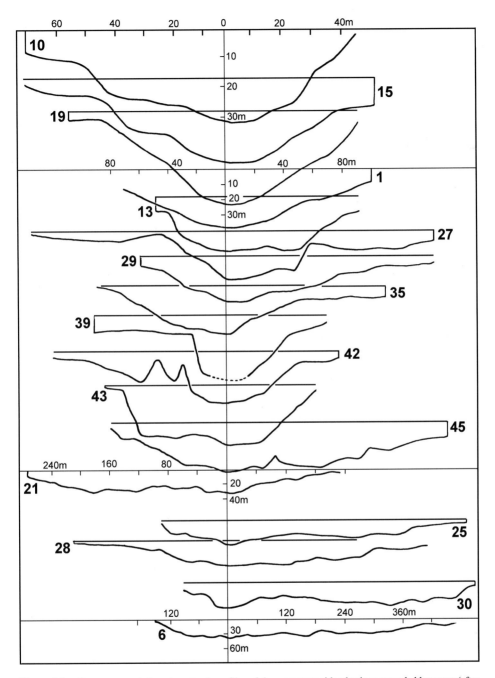

Figure 5.8. Some representative along-track profiles of deep pressure ridge keels as recorded by sonar (after Wadhams, 1978a).

and polynyas are transient features of the ice cover, reflecting its present state of strain, as opposed to ridges which represent the history of stress experienced by the ice.

An early study of lead widths (Wadhams, 1981) found a tentative power law relationship between lead frequency and width. This was confirmed in a later study (Wadhams, 1992) which showed that the distribution of lead widths fits a power law of form

$$P(w) = K \, w^{-n} \qquad\qquad (5.15)$$

where P(w) is the probability of width w per unit increment. The best fit to an exponent (Wadhams, 1992) was 1.45 for leads less than 100 m wide and 2.50 for wider leads, using the criterion that ice in a refrozen lead cannot exceed 1 m in draft (1.67 and 2.76 for a 0.5 m criterion) and the lead must be at least 5 m wide. Lead spacings fitted a negative exponential at moderate spacings (400–1500 m), but with an excess of lead pairs at small and very large spacings.

Other studies have not shown such a simple relationship, and in these cases a "trafficability diagram" (fig. 5.9) is often used to represent lead concentrations. This gives a mean distance between encounters with leads of more than a critical width, and is thus useful for such practical problems as assessing the distance between landing sites for aircraft or surfacing sites for submarines. Fig. 5.9 shows the great contrast between lead frequencies in summer and winter in the same region, with many more leads in summer.

Typically leads (defined by the 1 m criterion) occupy 1–5% of the ice cover in the central Arctic Ocean in winter, with the percentage rising to 10–20 for the marginal ice zone and for divergent drift currents such as the East Greenland Current. The lead fraction is a vital parameter of the ice cover, since most of the ocean-atmosphere heat flow occurs through ice less than 1 m thick (Maykut, 1986). Leads and ridges tend to occur together; it has been found (Wadhams, 1981, 1990a) that lead frequency is high in the heavily ridged zone north of Greenland, where massive convergence and ridge-building are taking place.

5.1.9. Fractal Properties of Ice Surfaces

When the first laser and sonar profiles of sea ice surfaces were obtained, it was natural to try spectral analysis of these surfaces to test whether significant wavelengths of roughness were present in the ice and to examine the spectral shape of the surface. Examples of such work were studies by Hibler (1972) and Hibler and LeSchack (1972), who examined significant spectral peaks from profiles taken in different directions in order to test for anisotropy in the deformed ice cover. No clear conclusions emerged from these studies.

As the theory of fractals (Mandelbrot, 1977, 1982) was developed as an alternative way of analysing random surfaces, it was applied to sea ice by Rothrock and Thorndike (1980), who studied the fractal properties of the ice underside in relation to spectral characteristics. The two spectral properties considered were the autocorrelation function $R_{yy}(\tau)$, defined as

$$R_{yy}(\tau) = E[y(x) \, y(x + \tau)] \qquad\qquad (5.16)$$

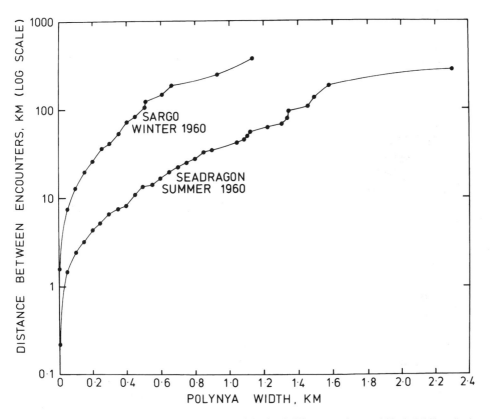

Figure 5.9.   Trafficability diagram for encounters with leads of different maximum widths in M'Clure Strait in winter and summer (after McLaren *et al.*, 1984).

where y(x) is the ice draft and τ is a lag; and the power spectral density function S(k), defined as the Fourier transform of $R_{yy}(\tau)$, which expresses the variance present in the under-ice profile at a wave number k, per unit increment in k.

In most cases, at small lags, Rothrock and Thorndike found that the autocorrelation function for ice profiles varied as a power of the lag, i.e.

$$R_{yy}(t) = R_{yy}(0) - c \; \tau^{2\alpha} \tag{5.17}$$

Under these circumstances, the surface is said to obey the Lipschitz condition, with α as the Lipschitz exponent. It can be shown that a set of points in a plane can be assigned a dimension, the so-called Hausdorff or fractal dimension D, which depends on the extent to which the set of points resembles an area (dimension 2) or a smooth line (dimension 1). The fractal dimension is related to α by

$$D = 2 - \alpha \tag{5.18}$$

It can also be shown that, if a function obeys the Lipschitz condition, then in most cases the power spectral density varies as k^p, where

$$p = -2\alpha - 1 \qquad\qquad (5.19)$$

If $p < -3$ the profile is smooth, i.e. differentiable, while if $-1 < p < -3$ the profile is rough. When $p = -1$ the profile can be described as "fully rough".

Thus the shapes of the autocorrelation function at small lags and the power spectrum across the whole wavelength range offer two ways of testing whether sea ice obeys the Lipschitz condition and, if so, of deriving the Lipschitz exponent and the fractal dimension of the surface. It has been found (Wadhams, 1990b; Key and McLaren, 1991; Wadhams and Davis, 1994) that sea ice surfaces do obey the Lipschitz condition reasonably well, and that D varies significantly over the Arctic, but not in the same way as physical roughness. Some studies of individual ice keels and short sections of profile of particular ice types have also been carried out (Bishop and Chellis, 1990; Connors et al., 1990).

Figure 5.10 shows the autocorrelation function and power spectrum of a typical section of under-ice profile from the Eurasian Basin, obtained by the author in May 1987. This demonstrates that the ice surface has a fractal dimension which is intermediate between that of a "smooth" and a "fully rough" surface. In a study of the whole of this profile, encompassing both the Arctic Ocean and the Greenland Sea (Wadhams and Davis, 1994) it was found that the fractal dimension remained very consistent in the Arctic Basin, in the range 1.40 to 1.48, but that it increased to 1.61–1.77 in Fram Strait and the Greenland Sea, despite reduced mean ice drafts and less ridging. The variation of D is therefore counter to the variation of "roughness" as we understand it physically. The two techniques for calculating D were found to give very similar results, with 8–20 m being used as the small lag autocorrelation range (avoiding lags less than 8 m because of the smoothing effect of beamwidth) and 10–200 or 300 m being the typical range of wavelengths over which a single power law fitted the power spectrum.

One problem with the simple unifractal approach to characterising an ice surface is that it rests on the implicit assumption that the ice draft or elevation is a single random process in which pressure ridges, for instance, arise as statistically random maxima in the same way as mountain peaks in the fractal landscapes of Mandelbrot (1977). In fact, we know that the ice underside is actually an alternation between two types of physical process, i.e. congelation growth (producing refrozen leads and undeformed first- and multi-year ice) and mechanical deformation (producing pressure ridges). This suggests that further progress in the fractal analysis of sea ice requires a multifractal approach, i.e. an algorithm is used to separate deformed from undeformed ice and the two types are treated as separate fractal processes.

5.1.10. Summary

Summarising, we can say that what is known about the statistics of Arctic sea ice morphology at present is:

- The thick end of the draft pdf fits a negative exponential;

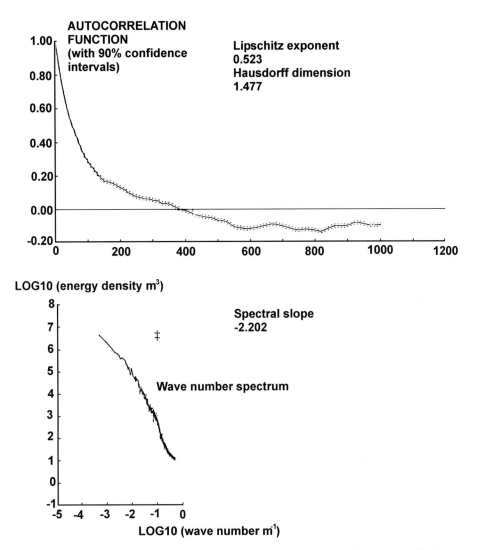

Figure 5.10. Autocorrelation function and wave number power spectrum for a 50 km sonar profile from the Eurasian Basin (after Wadhams, 1990b).

- Keel drafts and sail heights fit a negative exponential;
- Ridge spacings fit a two-parameter lognormal, unless they are obtained from a wide-beam sonar, in which case a threshold parameter must be used which is approximately the same as the surface beam diameter;
- Ridge slopes are variable; keels are wider than sails;
- Keel widths fit a lognormal;
- Keel slopes fit a lognormal;
- There is a relationship between slope and (width/draft) ratio.

There is still insufficient evidence to know whether these relationships hold for the Antarctic.

5.2. THE MEASUREMENT OF ICE THICKNESS

5.2.1. Current Techniques

We can learn a great deal about sea ice from satellite and aircraft surveys — its extent, its type, its surface features. But its thickness is hard to measure by remote sensing, because the brine cells in the ice give it a high electrical conductivity such that electromagnetic waves do not easily penetrate. The radio echo sounding methods which have been used to measure the thickness of terrestrial ice sheets and glaciers cannot therefore be used for sea ice.

So far, five techniques have been commonly employed for measuring ice thickness distribution. In decreasing order of total data quantity, they are:

1. Submarine sonar profiling
2. Moored upward sonars
3. Airborne laser profilometry
4. Airborne electromagnetic techniques
5. Drilling.

Submarine sonar profiling

Most synoptic data to be published so far have been obtained by upward sonar profiling from submarines. Beginning with the 1958 voyage of "Nautilus" (Lyon, 1961; McLaren, 1988), which was the first submarine to the North Pole, many tens of thousands of km of profile have been obtained in the Arctic by US and British submarines, and our present knowledge of Arctic ice thickness distributions derives largely from the analysis and publication of data from these cruises. Problems include the necessity of removing the effect of beamwidth where a wide-beam sonar has been employed (Wadhams, 1981), and the fact that the data are obtained during military operations, which necessitates restrictions on the publication of exact track lines. For the same reason the dataset is not systematic in time or space. Figure 5.11 shows a typical under-ice profile obtained by a British submarine.

Moored upward sonars

The second technique is the use of upward sonar mounted on moorings, so as to obtain a time series of g(h) at a fixed location. Experiments have involved bottom-mounted systems in shallow water in the Beaufort Sea (Hudson, 1990; Pilkington and Wright, 1991; Melling and Riedel, 1995) and Chukchi Sea (Moritz, 1991), and systems in deeper water in Fram Strait and the southern Greenland Sea (Vinje, 1989). Work of this kind has been taking place since 1991 using lines of sonars which span the East Greenland Current in Fram Strait, at 75°N and in Denmark Strait. In conjunction with the use of AVHRR or ERS-1 SAR imagery to yield ice velocity vectors, this technique permits the time

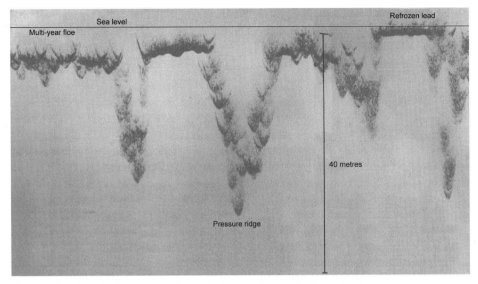

Figure 5.11. An under-ice sonar profile obtained by a submarine in the Arctic Ocean.

dependence of ice mass flux to be measured, and hence the fresh water input to the Greenland Sea at different latitudes (Vinje *et al.*, 1998; Kwok and Rothrock, 1999). The observations involve mounting the sonars at 50 m depth (together with a current meter) in water up to 2500 m deep on a taut wire mooring. Similar work has been done in the Weddell Sea, Antarctica, using a greater mooring depth and an armoured cable to protect against iceberg damage (Strass and Fahrbach, 1998).

The chief advantage of submarine sonar surveys as a way of generating ice thickness distributions is that upward sonar (mobile or moored) is still the only direct and accurate means of measuring the draft of sea ice, from which the thickness distribution can be inferred with very little error, while submarine-mounted (as opposed to moored) sonar allows basin-scale surveys to be carried out on a single cruise, giving the geographical variation in ice thickness characteristics. Submarine-mounted sonar also permits the shape of the ice bottom to be determined more accurately than moored sonar, including pressure ridges and the roughness of undeformed multi-year ice, allowing spectral and fractal studies to be undertaken and an understanding of the mechanics of the ridge-building process to be gained. By the use of additional sensors and concurrent airborne studies, a submarine can be used as a powerful vehicle for validating remote sensing techniques, including laser, passive and active microwave. The chief drawbacks of submarines are that, unlike moored sonar, a submarine cannot generate a systematic time series of ice thickness at a point in space. Submarines also cannot carry out surveys safely in very shallow water, so that many interesting aspects of ice deformation near shore cannot be studied; for instance, the dataset obtained by USS *Gurnard* north of Alaska in 1976 and analysed by Wadhams and Horne (1980) began at the 100 m isobath and so did not cover the whole of the Alaskan shear zone. Finally, it is unlikely that a military submarine would be available in the Antarctic, both because of remoteness and because the Antarctic Treaty requires that military vessels used in the Antarctic must be available for international inspection. The Antarctic is, however, a very suitable region for the use of sonar on an autonomous underwater vehicle (AUV).

Airborne laser profilometry

Laser profiling of sea ice in the Arctic Ocean has been carried out extensively during three decades (e.g. Ketchum, 1971; Weeks *et al.*, 1971; Wadhams, 1976; Tucker *et al.*, 1979; Krabill *et al.*, 1990), while limited studies have also been carried out in the Antarctic (Weeks *et al,* 1989). The aim has been to delineate the frequency and height distributions of pressure ridge sails and the spatial distribution of surface roughness. On only two occasions has it been posssible to match a laser profile against a coincident profile of ice draft over substantial lengths of joint track. The first was a joint aircraft-submarine experiment (Wadhams and Lowry, 1977; Lowry and Wadhams, 1979; Wadhams, 1980, 1981) while the second was a similar experiment in May 1987 which involved a NASA P-3A aircraft equipped with an Airborne Oceanographic Lidar (AOL) and a British submarine equipped with narrow-beam upward-looking sonar (Wadhams, 1990a; Comiso *et al.*, 1991; Wadhams *et al.*, 1991). Only the more recent experiment permitted a direct comparison of the pdf's of draft and freeboard to be made, since the AOL has a superior capability over earlier lasers in the removal of the sea level datum from the record. Comiso *et al.* (1991) found from this experiment that over a 60 km sample of track the overall pdf's of ice freeboard and draft could be brought to a close match across the entire range of data by a simple co-ordinate transform of the AOL data based on the ratio of mean densities of ice and water. Specifically, they showed that if R is the ratio of mean draft to mean freeboard, then matching of the freeboard pdf with the draft pdf is achieved by expanding the elevation scale of the freeboard pdf by a factor of R, and diminishing the magnitude of the pdf per m by the same factor. This is equivalent to saying that if a fraction F(h) of the ice cover has an elevation in the range h to (h + dh), then the same fraction F(h) will have a draft in the range R h to R(h + dh). R is related to mean material density (ice plus snow) ρ_m and near-surface water density ρ_w by

$$R = \rho_m / (\rho_w - \rho_m) \qquad (5.20)$$

The success of this correlation prompted an analysis of the entire 300 km of coincident track (Wadhams *et al.*, 1992) divided into six 50 km sections, all from north of Greenland, within the zone 80.5–85°N, 2–35°W. The results of the analysis were as follows:

(i)   Despite variations in mean draft from 3.6 to 6 m, the six values of R all lay within a narrow range, of mean 8.04 ± 0.19. This corresponds to a mean material density of 910.7 ± 2.3 kg m^{-3}.

(ii) When each section was subjected to a co-ordinate transform based on its own value of R, the pdf's matched the sonar pdf's extremely well when plotted on a semilogarithmic scale (fig. 5.12).

(iii) When plotted on a linear scale, the agreement was less good, in that mid-range depth probabilities are enhanced by the transformation, while very thin and very thick ice probabilities are reduced (fig. 5.13). This is comprehensible on the basis of

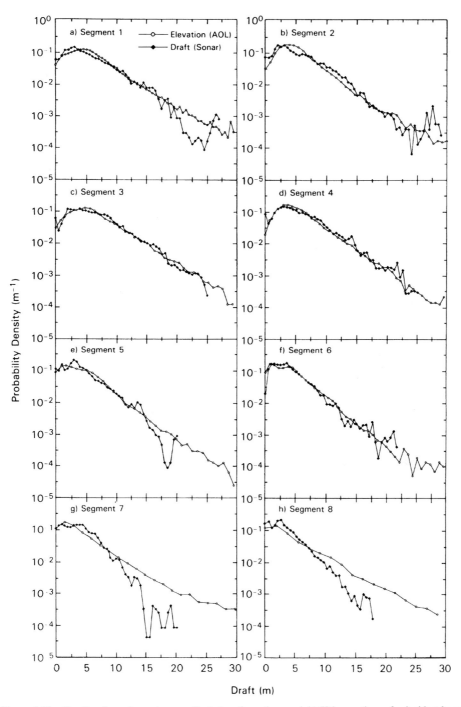

Figure 5.12. Results of carrying out a co-ordinate transformation on eight 50 km sections of coincident laser and sonar track, using the mean draft-elevation ratio. Segments 7 and 8, with a poor fit, did not comprise coincident data (after Wadhams *et al.*, 1992).

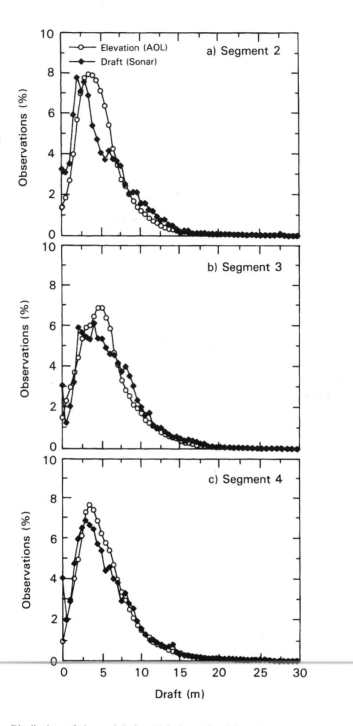

Figure 5.13. Distributions of observed drafts and drafts predicted from laser elevations using a mean draft-elevation ratio of 7.89 (after Wadhams *et al.*, 1992).

considering a uniform snow cover, which would give a low value of R for thin ice and a high value for thick ice. The use of a single average value for the transformation causes thin ice to be moved into thicker categories, and thick ice into thinner, thus making the converted distribution more narrow and peaked than the real ice draft distribution. In principle, a freeboard-dependent R could be used for the transform, but this would involve a sacrifice of simplicity.

The question arises of how R might vary with time of year and location. It is difficult to develop even a simple model, since snow thickness on Arctic and Antarctic sea ice is poorly known, and there are few systematic measurements of ice density, especially of any fundamental difference between the densities of first- and multi-year ice. Near-surface water density must be known to high accuracy, since R depends on the difference between water and ice density, and we know that it diminishes during summer because of dilution by meltwater. One set of results, applicable to the Arctic Basin, is shown in figure 5.14, based on the seasonal snow depth assumptions that were used in the Maykut-Untersteiner model (section 3.3). We see that there is a large seasonal variation in R, mainly due to snow load. R is at its highest value on August 20, at the start of the snow season, with bare ice. R falls rapidly during the autumn snow falls of September and October, then diminishes only very slowly between November and the end of April, when little snow falls. A further onset of spring snow brings R to its lowest level at the beginning of June. As soon as surface snow melt begins R rises rapidly until by the end of June it has risen again almost to its August value. The final slow drift is due to surface water dilution. In fact there will be one or more higher peaks for R during the summer period, as meltwater pools form on the ice surface (increasing R), then drain (decreasing R), but no useful data exist on mean meltwater pool coverage, mean depth, or draining dates. In any case the results show that snow load is a critical parameter, and that the best time to carry out surveys is when dR/dt is at its lowest value, i.e. between November and the end of April.

We can conclude that it is indeed worthwhile to consider the use of an airborne laser as a way of surveying both the mean ice thickness and the thickness distribution over the Arctic. The model suggests that a survey during early spring would be most useful, with annual repetition in order to examine interannual fluctuations and trends (possibly climate-induced) in the mean ice thickness or the form of the distribution. However, better data are required on the seasonal and spatial variability of snow load, ice density and surface water density before full confidence can be attached to the results. Even in the absence of improved background data, the present results suggest that in regions with mean ice thickness in the range 4–6 m, the ratio R will lie in the vicinity of 8.0 in spring, and can be estimated to an accuracy of $\pm$ 2.4% in 300 km of track. This yields an accuracy of about $\pm$ 12 cm in mean thickness over 300 km of track (30 cm in 50 km), neglecting other sources of error. This is an acceptable accuracy for detecting and mapping variability. In the Antarctic much more validation would be needed before a laser could be employed in this way, since snow has a much stronger influence on mean density than in the Arctic, and there is a major difference between snow load on first-year and multi-year ice (since the snow cover does not all melt during summer). However, if such validation were done, this technique could be a highly effective mapping tool for the vast expanse of Antarctic sea ice cover in winter. Most valuable of all for synoptic purposes would be a laser mounted in a satellite. At the time of writing such a system is being designed by NASA.

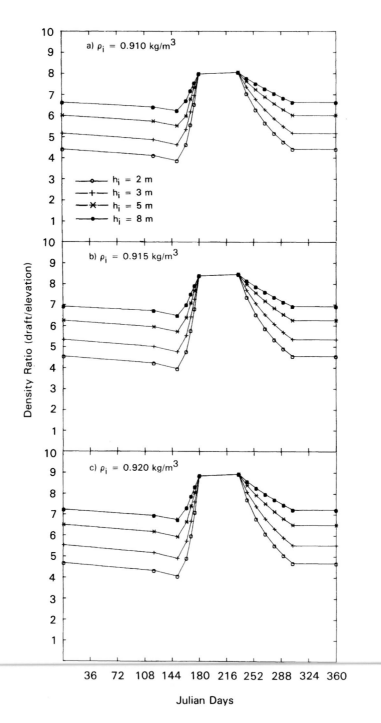

Figure 5.14. Results of a model of the variation of draft / freeboard ratio R with season (after Wadhams *et al.*, 1992).

Airborne electromagnetic techniques

The first electromagnetic technique to be applied to sea ice was impulse radar (Kovacs and Morey, 1986), in which a nanosecond pulse (centre frequency about 100 MHz) is applied to the ice via a paraboloidal antenna on a low- and slow-flying helicopter. The technique was found to have many limitations besides the slowness of data gathering. Superimposition of surface and bottom echoes means that it does not function well over ice less than 1 m thick, while absorption and scattering by brine cells means that better results are obtained over multi-year than first-year ice, with in any case a fading of the return signal at depths beyond 10–12 m. Thus the full profile of deeper pressure ridges is not obtained. Such instruments can therefore be regarded as of restricted application to local surveys in ice regimes of a favourable kind.

A more recent development is the use of electromagnetic induction. The technique was devised by Aerodat Ltd. of Toronto, and has subsequently been developed by CRREL (Kovacs and Holladay, 1989, 1990) and by Canpolar, Toronto (Holladay et al., 1990). The technique involves towing a "bird" behind a helicopter flying at normal speeds. The bird contains a coil which emits an electromagnetic field in the frequency range 900 Hz – 33kHz, inducing eddy currents in the water under the ice, which in turn generate secondary EM fields. The secondary fields are detected by a receiver in the bird; their strength depends on the depth of the ice-water interface below the bird. The bird also carries a laser profilometer to measure the depth of the ice-air interface below the bird, and the difference gives the absolute thickness of the ice. The method appears promising but requires further validation. Figure 5.15 illustrates the principle and shows some results obtained in the Labrador Sea during the LIMEX experiment (Holladay et al., 1990). A similar system has been fitted into a Twin Otter aircraft of the Finnish Geological Survey and used in the Baltic (Multala et al., 1995).

Drilling

Drilling is the traditional method of measuring ice thickness, and most polar scientists have improved their muscles by extensive application to this art. In fact the first systematic measurements of ice thickness in the Arctic were made by Nansen (1897), who drilled through undeformed ice during the drift of "Fram". Many methods have been used: manual drilling is the most painful, since in very thick ice a large number of extensions are needed to the drill, and if the bit gets stuck there is little that the driller can do except spend many hours chipping downwards with a chisel to free it. The use of a gasoline-powered head is an improvement, while the most rapid technique is the hot water drill, where water is heated in a boiler and pumped through a hose and out as a jet through a heavy bronze probe, which thus melts its way quickly down through the ice. A tape with a self-opening set of scissors at the bottom is sent down the hole and the draft, ice thickness and snow thickness read off as quickly as possible before the hole refreezes.

Drilling as a technique was considered statistically by Rothrock (1986), who estimated that 62 independent random holes would give a mean thickness with a standard deviation of 30 cm, while 560 holes would give a 10 cm error. Eicken and Lange (1989) followed this approach and obtained a reasonable approximation to an Arctic ice thickness distribution. In general, however, we must conclude that the technique is not bad as a means of estimating mean ice thickness, but is poor as a way of giving the shape of g(h). It is an essential validation for any other technique.

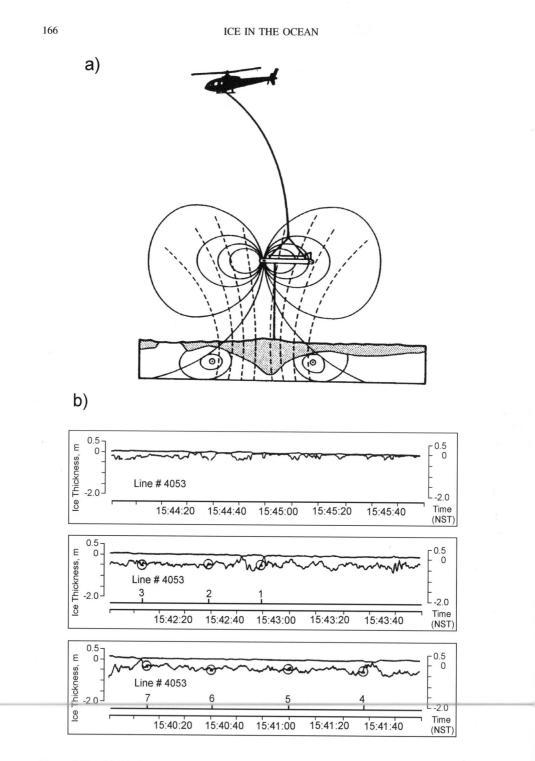

Figure 5.15. (a) Principle of electromagnetic induction sounding of ice. (b) Results from electromagnetic induction sounding over first-year ice in the Labrador Sea (after Holladay *et al.*, 1990). Circles indicate validation of the profiles by drilling.

5.2.2. Possible Future Techniques

There are several new techniques which show some promise for measuring g(h), or parts of it, under certain conditions. We may mention the following:

1. Sonar mounted on AUVs or neutrally buoyant floats
2. Acoustic tomography or thermometry
3. Inference from a combination of microwave sensors.

Sonar on AUVs and floats

The purpose of mounting upward sonar systems on mobile platforms other than military submarines is to obtain systematic datasets along a repeatable grid of survey tracks, enabling interseasonal and interannual comparisons to be carried out for identical geographical locations. Possible platforms with synoptic potential include autonomous underwater vehicles (AUVs), of which several are under development including long-range systems with basinwide capability; long-range civilian manned submersibles, of which the Canadian-French "Saga I" is an example (Grandvaux and Drogou, 1989); and neutrally buoyant floats, on which an upward sonar could be mounted to generate and store a pdf which can then be transmitted acoustically to an Argos readout station on a floe. To date only AUVs have been used successfully, and only for local surveys with control by acoustic beacons or divers.

Acoustic tomography

Acoustic tomography is a technique for monitoring the structure of the ocean within an area of order 10^6 km^2 by measuring acoustic travel times between the elements of a transducer array enclosing that area (Munk and Wunsch, 1979). Guoliang and Wadhams (1989) showed in a theoretical study that the presence of a sea ice cover should measurably decrease travel times by an amount which is dependent on the modal ice thickness. The way that this works is that in polar seas sound is always refracted upwards towards the surface, since sound velocity increases monotonically from the surface downwards. The sound rays are reflected downwards, then arch back upwards and undergo a number of such "bounces" before reaching the receiver. If the sea surface is covered by ice each "bounce" occurs at a depth equal to the ice draft, reducing the travel path for the sound ray. There are complications caused by the phase change suffered by the ray at reflection, but on the whole the travel time should decrease as the ice thickness increases.

Data were obtained during the 1988–9 Scripps-WHOI tomography experiment in the Greenland Sea (Jin et al., 1993), and this technique may have a more general application within the Arctic Ocean itself, although the additional scattering due to under-ice roughness has been found to have an important limiting effect on resolution. A development of tomography is called ATOC (acoustic thermometry of ocean climate), which uses much lower frequencies (e.g. 57 Hz) to achieve greater transmission lengths between a single transmitter and receiver (Johannessen et al., 1993), and this may allow the technique to be used to estimate a modal ice draft for the Arctic Basin. The first experiments on transmitting sound across the Arctic Basin from north of Svalbard to an ice camp in the Beaufort Sea were conducted successfully in April 1994 (Michelevsky et al., 1999).

The use of microwave sensors

Possible ice sounding techniques involving microwave sensors have been reviewed by Wadhams and Comiso (1992). In principle, after sufficient validation has been done, we might hope to find empirical relationships between the distribution of passive microwave brightness temperatures, of SAR backscatter levels, and of aspects of g(h) such as the mean thickness. To date, however, the only quantitative validation of a microwave sensor against ice thickness has been a 1987 joint survey between a sub-marine, an aircraft equipped with the STAR-2 X-band synthetic aperture radar system, and a second aircraft equipped with a laser profilometer and passive microwave radi-ometers. The SAR system operates at HH polarisation and 9.6 GHz, with a swath width of 63 km and resolution of 16.8 m. This provided an opportunity to examine correlations between ice draft as measured by the sonar, and backscatter level along the same track measured by the SAR. The question which can be addressed is to what extent can SAR brightness variability be used to infer the shape of the ice thickness distribution. The results were discussed in Comiso *et al.* (1991) and Wadhams *et al.* (1991). We can summarise them as follows.

Firstly, a qualitative examination of the profile of SAR backscatter along the tracks of the submarine and aircraft showed that there was a clear positive correlation with both the draft and elevation profiles. This is to be expected since pressure ridges in particular give strong returns on account of their geometry (Onstott *et al.,* 1987; Livingstone, 1989; Burns *et al.*, 1987). Next 125 km of matched SAR and laser data were examined. With each set of data averaged over a window length of 1 km the correlation coefficient between elevation and SAR backscatter reached 0.51. Figure 5.16 shows the mean elevation and mean backscatter using this window. Clearly there are regions of both good and bad correlation, the best correlation often appearing to occur in areas of low elevation. This is probably because these correspond to open water, young ice and first-year ice, which on X-band SAR offer a high contrast with multi-year and ridged ice. Evidence indicates that SAR backscatter for ridges is dependent on look-angle and angle of incidence (Leppäranta and Thompson, 1989), while other results (Holt *et al.*, 1990) show that lower-frequency SAR (0.44 GHz) may well be superior in its resolution of ridges to high-frequency SAR such as X-band.

The correlation between SAR backscatter and sonar ice draft was carried out over a more restricted 22 km section of track (Wadhams *et al.*, 1991) where there was excellent matching between tracks. Once again it was found that the correlation coefficient depends on the window length, but in fact the highest correlation was obtained with a lower window length. At an averaging length of 252 m (15 SAR pixels) the correlation between draft and backscatter reached 0.68, which is better than the best correlation with elevation and which implies that 46% of the backscatter variance can be explained by draft differences. Figure 5.17 shows the scatter plot of draft versus SAR backscatter.

It is clear that SAR brightness alone cannot be used to infer the complete shape of the ice pdf. Firstly, less than half of the variance of the SAR brightness can be explained by draft variations, so it cannot be used alone as a predictor. Secondly, this correlation is developed over averaging lengths of 252 m, indicating that to some extent we are relating a mean ice draft to a mean SAR backscatter, rather than obtaining an algorithm which can generate a fine-resolution pdf as in the case of the laser technique.

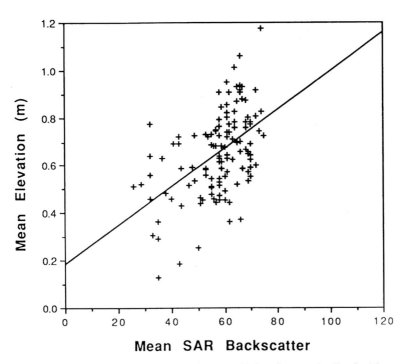

Figure 5.16. Mean elevation versus mean SAR backscatter with best fit regression line for 1-km averaging window.

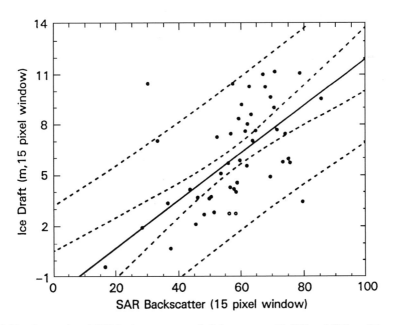

Figure 5.17. Scatter plot of SAR backscatter versus draft from sonar with 95% and 99% confidence limits.

Thirdly, the experiment described above was specific to X-band SAR, whereas the types of satellite SAR which produce routine synoptic data from the polar oceans are of different frequencies (e.g. C-band, 5.3 GHz, on ERS-1). Nevertheless the results show some promise for the use of SAR, in conjunction with other sensors, as an element in an empirical scheme for determining ice thickness using microwave backscatter or emission properties as a proxy. A recent approach towards such a scheme, by Kwok *et al.* (1999), uses SAR to track ice velocity and define ice types, and an ice model to predict growth and melt rates, in this way obtaining an ice thickness distribution and following its evolution.

A final and more direct method of extracting ice thickness from microwave data comes from the radar altimeter. S. Laxon (Wingham, 1999) has suggested that the apparent difference in elevation between altimeter returns coming from consolidated ice and those coming from ice with refrozen leads (which give strong echoes from, essentially, sea level) may be a genuine measure of mean ice freeboard, in which case a procedure such as that used for laser profile analysis could yield mean thickness. Again much validation is needed.

5.3. PRESENT KNOWLEDGE OF ICE THICKNESS DISTRIBUTION

5.3.1. Mean Ice Thickness in the Arctic

Our knowledge of the regional variability of g(h) in the Arctic comes almost entirely from upward sonar profiling by submarines. Therefore our level of knowledge of g(h), and of the mean regional thickness h_m, depends on whether submarines have been able to operate in the area concerned. To date, data have been obtained mainly from British submarines operating in the Greenland Sea and Eurasian Basin, and from US submarines operating in the Canada and Eurasian Basins, with some earlier cruises in the Canadian Arctic. Very large new datasets have been obtained from the US SCICEX civilian submarine programme (Rothrock *et al.*, 1999).

Moving northwards from subpolar regions, the ice in Baffin Bay is largely thin first-year ice with a modal thickness of 0.5–1.5 m (Wadhams *et al.*, 1985). In the southern Greenland Sea, too, the ice, although composed largely of partly melted multi-year ice, also has a modal thickness of about 1 m (Vinje, 1989; Wadhams, 1992), with the decline in mean thickness from Fram Strait giving a measure of the fresh water input to the Greenland Sea at different latitudes. Over the Arctic Basin itself there is a gradation in mean ice thickness from the Soviet Arctic, across the Pole and towards the coasts of north Greenland and the Canadian Arctic Archipelago, where the highest mean thicknesses of some 7–8 m are observed (LeSchack, 1980; Wadhams, 1981, 1992; Bourke and McLaren, 1992). These overall variations are in accord with the predictions of numerical models (Hibler 1979, 1980) which take account of ice dynamics and deformation as well as ice thermodynamics.

The temporal variability of g(h) has been much less extensively measured. On only a few occasions have submarine sonar tracks been obtained over similar regions in different years or different times of year. Results show that in the Eurasian Basin far from land, in the region between the Pole and Svalbard, the mean ice draft is remarkably stable between different seasons and different years (Wadhams, 1989, 1990b; McLaren, 1989), although McLaren *et al.* (1992) found considerable interannual variability at the Pole itself. In the region between the Pole and Greenland, where a build-up of pressure ridging normally occurs due to the motion of the ice cover towards a downstream land boundary, very considerable differences in g(h) and in mean draft (more than 15%) were observed between October 1976 and May 1987 (Wadhams, 1990a). In the Beaufort Sea in summer, also, considerable differences were observed between records taken several years apart (McLaren, 1989). Bourke and Garrett (1987) used all available data to attempt to construct seasonal contour maps of mean ice draft. Figure 5.18 shows Bourke and Garrett's maps for summer and winter, while figure 5.19 shows the variability observed by Wadhams between 1976 and 1987. It should be noted that the data used for the Bourke and Garrett map did not include open water, so these maps are over-estimates of the mean ice draft. The maps were updated by Bourke and McLaren (1992), who also gave contour maps of standard deviation of draft, and mean pressure ridge frequencies and drafts for summer and winter, based on 12 submarine cruises.

In order to assess reliably whether ice thickness changes are occurring in the Arctic it is necessary to obtain area-averaged observations of mean ice thickness over the same region using the same equipment at different seasons or in different years. Ideally the region should be as large as possible, to allow us to assess whether changes are basin-wide or simply regional. Also the measurements should be repeated annually in order to distinguish between a fluctuation and a trend. Because of the unsystematic nature of Arctic submarine deployments this goal has not yet been achieved, but a number of comparisons between pairs of datasets have been carried out.

McLaren (1989) compared data from two US Navy submarine transects of the Arctic Ocean in August 1958 and August 1970, stretching from Bering Strait to the North Pole and down to Fram Strait. He found similar conditions prevailing in each year in the Eurasian Basin and North Pole area, but significantly milder conditions in the Canada Basin in 1970. The difference is possibly due to anomalous cyclonic activity as observed in the region in recent summers (Serreze *et al.*, 1989). Another possibility is that since August is the month of greatest ice retreat in the Beaufort Sea, the difference is simply due to differences between the extent to which the ice retreated in the Chukchi and southern Beaufort Seas during the respective summers. The extent and duration of the open water season in the Beaufort Sea is known to have a high interannual variation, and an unusually open southern Beaufort Sea would lead to more open conditions within the pack itself.

Wadhams (1989) compared mean ice thicknesses for a region of the Eurasian Basin lying north of Fram Strait, from British Navy cruises carried out in October 1976, April–May 1979 and June–July 1985. All three datasets were recorded using similar sonar equipment. It was found that a box extending from 83°30′N to 84°30′N and from 0° to 10°E had an especially high track density from the three cruises (400 km in 1976, 400 km in 1979 and 1800 km in 1985), and this was selected for the comparison. It is a region far from any downstream boundary, and represents typical conditions in the Trans Polar Drift Stream prior to the acceleration and narrowing of the ice stream which occurs as it prepares to enter Fram Strait. The mean thicknesses from the three cruises were remarkably similar: 4.60 m in 1976; 4.75 m in 1979; and 4.85 m in 1985. It should be remembered that these datasets were recorded in different seasons as well as different years.

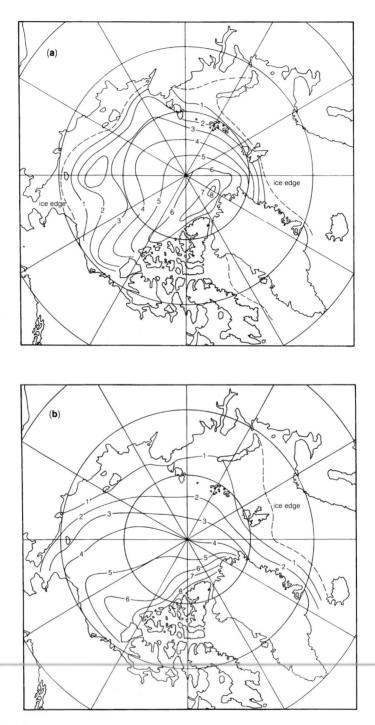

Figure 5.18. Contour maps of estimated mean ice drafts for (a) summer and (b) winter in the Arctic Basin (after Bourke and Garrett, 1987).

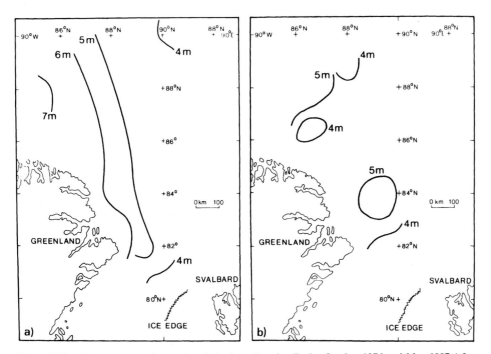

Figure 5.19. Contour maps of mean ice drafts from Eurasian Basin, October 1976 and May 1987 (after Wadhams, 1990a).

Later Wadhams was able to compare data from a triangular region extending from north of Greenland to the North Pole, recorded in October 1976 and May 1987 (Wadhams, 1990a). Mean drafts were computed over 50 km sections, and each value was positioned at the centroid of the section concerned; the results were contoured to give the maps shown in figure 5.19. There was a decrease of 15% in mean draft averaged over the whole area (300,000 km^2), from 5.34 m in 1976 to 4.55 m in 1987. Profiles along individual matching track lines (figure 5.20) show that the decrease was concentrated in the region south of 88°N and between 30° and 50°W. From fig. 5.19 it appears that the build-up of pressure ridging which gave the high mean drafts near the Greenland coast in 1976 was simply absent in 1987, but in fact the situation is not that simple. Table 5.1 shows a comparison between the probability density functions of ice thickness from the pairs of profiles shown in fig. 5.20 (strictly, two 300 km sections from the southern part of the N–S transect and two 200 km sections from 40–50°W in the E–W transect). In 1987 there was more ice present in the form of young ice in refrozen leads (coherent stretches of ice with draft less than 1 m) and as first-year ice (draft less than 2 m). There was less multi-year ice (interpreted as ice 2–5 m thick) and less ridging (ice more than 5 m thick) in 1987 in the E–W transect, although slightly more ridged ice in the N–S transect. The main contribution to the loss of volume appears to be, then, the replacement of multi-year and ridged ice by young and first-year ice.

To determine how this may have occurred the tracks of drifting buoys from the Arctic Ocean Buoy Program (Colony *et al.*, 1991; R.L. Colony, pers. commun.) were examined.

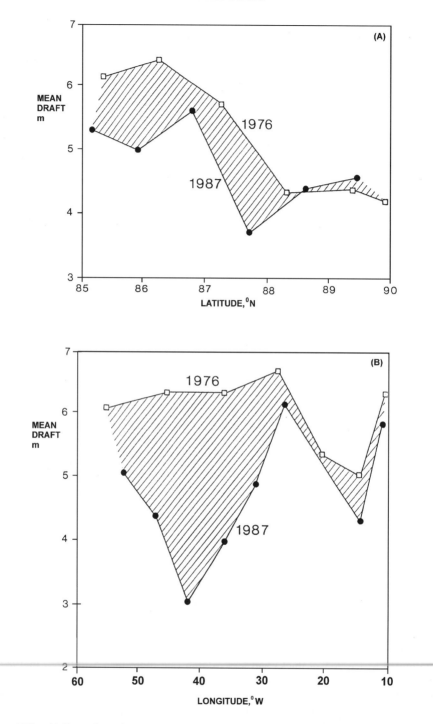

Figure 5.20. (a) Comparison of mean ice drafts from 1976 and 1987 along a N–S transect from North Pole to 85°N. (b) Comparison of mean ice drafts for transect across north of Greenland from 60°W to 10°W.

Table 5.1. Comparisons of ice statistics from 1976 and 1987 datasets

	N–S transect		E–W transect	
	1976	1987	1976	1987
Mean draft (m)	6.09	5.31	6.32	4.07
Ice < 2 m draft (%)	11.6	16.7	7.9	29.9
Ice 2–5 m draft (%)	48.7	38.6	46.5	39.6
Ice > 5 m draft (%)	39.7	44.7	46.0	30.5
Refrozen leads (%)	4.0	7.9	3.7	15.6

Four buoys were in the region during the months prior to the 1987 cruise (fig. 5.21). The three in the westernmost positions, corresponding to a portion of the Beaufort Gyre, remained almost stationary during the period January–May 1987, while buoy 1897, in the Trans Polar Drift Stream, moved towards Fram Strait at an average speed of 2 km per day. The result of this anomalous halting of the motion of part of the Beaufort Gyre would be a divergence within the experimental region, as the Trans Polar Drift Stream ice continued its motion towards the SE. This would lead to the opening up of the pack, creating areas of young and first-year ice. Thus the indications are that it is an ice motion anomaly rather than, or as well as, an ice growth anomaly that is responsible for the observed decrease in mean ice draft.

Further observations are clearly necessary from other years (past, recent and future), and a closer study should be made of pressure fields over the Arctic and the variations that they may be capable of creating in the geographical distribution of mean ice drafts. One of the surprising aspects of the results described above is that the build-up of mean ice draft towards Greenland was once thought to be a very stable aspect of the ice climatology of the Arctic, appearing consistently in the model predictions of Hibler (1980) and in the seasonal climatology of Bourke and Garrett (1987). It appears that the ice cover, like the ocean, possesses a weather as well as a climate, and it is vital that sufficient data be examined to resolve details of this weather such that the underlying trend in basin-wide mean ice thickness can be revealed.

A final regional comparison which the 1987 dataset has made possible occurs in the region immediately north of Fram Strait, between 82°N and 80°N (fig. 5.22). Here it was possible to compare data from the four years 1976, 1979, 1985 and 1987. This is a region which is ice-covered in most years, and where mixing occurs between the various ice streams preparing to enter Fram Strait, notably the streams of old, deformed ice moving S from the North Pole region and SE from the region north of Greenland; and the stream of younger, less heavily deformed ice moving SW from the seas north of Russia. Figure 5.22 shows all available mean drafts from 50 km sections (from Wadhams, 1981, 1983, 1989 and unpublished analyses). There is very good consistency among these four datasets, regardless of year or season of generation; fluctuations appear to be random in character, and where centroids from different experiments lie close to one another, the mean drafts are usually similar. Only the 1976 data points appear somewhat thicker than their neighbours.

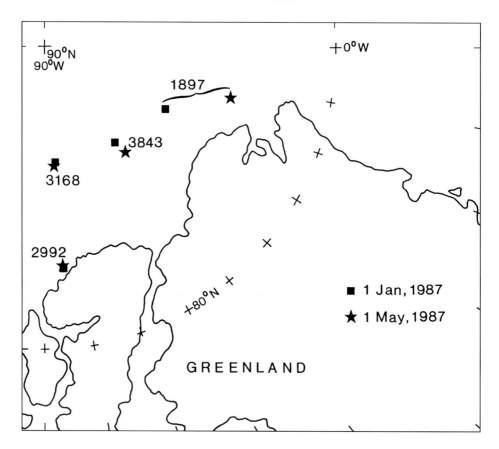

Figure 5.21. Positions of drifting buoys north of Greenland on 1 January and 1 May 1987.

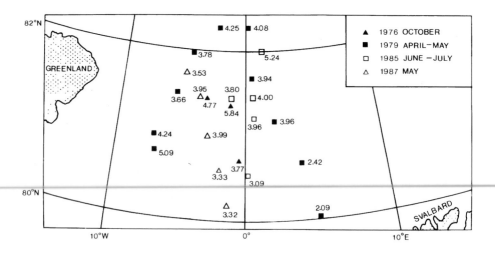

Figure 5.22. Comparisons of mean ice drafts measured in the region north of Fram Strait.

McLaren *et al.* (1992) analysed 50 km and 100 km sections of ice profile centred on the North Pole from 6 cruises from 1977 to 1990. They found that the mean ice draft from 50 km sections in the late 1970s (1977, 1979) was 4.1 m (4.2, 4.0 m respectively), while the mean draft for the late 1980s was 3.45 m (2.8, 4.1, 3.3, 3.6 m for 1986, 1987, 1988, 1990). The difference of 0.65 m is 15%. Using a t-test they showed that this difference of means is significant only at the 20% level, i.e. non-significant although possibly indicative.

Most recently Rothrock *et al.* (1999), by comparing US submarine datasets from 1993–6 with those from 1958–76, found a significant decrease in mean draft for the end of summer, from 3.1 m to 1.8 m, which occurs throughout the Arctic Basin. The latest data from a 1996 UK submarine in the Eurasian Basin (Wadhams and Davis, in preparation) also show a significant decline in mean draft of more than 40% relative to 1976.

From the data comparisons made so far in the Arctic, we can thus draw the following tentative conclusions:

1. Ice reaching Fram Strait via the Trans Polar Drift Stream along routes where it is not heavily influenced by a downstream land boundary shows great consistency in its mean thickness from season to season and from year to year, at latitudes from 84°30′N to 80°N and longitudes in the vicinity of 0°.
2. Ice upstream of the land boundary of Greenland can show great changes in its mean ice draft, notably a significant decline between 1976 and 1987, but it is possible to trace these to anomalies in the balance between pressure ridge formation through convergence (the normal source of a high ice draft in the region) and open water formation through divergence (the anomalous situation). A deeper knowledge of wind field anomalies is needed to understand these changes fully, together with more adequate datasets.
3. Data obtained from the North Pole region in 1977–90 show evidence of a decline in mean ice thickness in the late 1980s relative to the late 1970s.
4. Ice in the Canada Basin in summer also appears to show great variability in mean ice draft, but here there is a free boundary with ice-free marginal seas, permitting relaxation of the ice cover into a less concentrated state under certain wind conditions.
5. Elsewhere in the Arctic Basin there is increasing evidence of a significant decline in the ice thickness in summer, amounting to more than 40%.

5.3.2. Mean Ice Thickness in the Antarctic

In the Antarctic our knowledge of g(h) is much less extensive, since systematic data have been obtained only by repetitive drilling except for the use of moored upward sonar at certain sites in the Weddell Sea. Figure 5.4 shows a typical distribution in a region of first-year ice in midwinter. In the limited regions of the Antarctic where multi-year ice occurs, there is a preferred ice thickness of about 1.4 m (western Weddell Sea, winter) for this ice type, but with no apparent increase in maximum ridge drafts (Wadhams and Crane, 1991; Lange and Eicken, 1991). Given the paucity of data, it is in the Antarctic that the development of new techniques for synoptically monitoring g(h) would yield the greatest impact.

The only spatial profiling data of any kind from the Antarctic are airborne laser profiles of the upper surface (Weeks *et al.*, 1989) and some limited thickness profiling by impulse radar from a helicopter (Wadhams *et al.*, 1987). Laser data are valuable in delineating the occurrence of pressure ridging and in assessing its likely contribution to the overall thickness distribution, but cannot be used reliably to infer ice thickness directly by isostasy, largely because of the effect of the thick and variable snow cover. The usefulness of impulse radar in the Antarctic is limited by the fact that the first-year ice is thin and has a high salinity, and by the unconsolidated nature of ridges, which present misleading water horizons to the electromagnetic radiation; thus, such radar has only been used as an auxiliary to drilling efforts.

The winter pack ice in the Antarctic is of real global importance because of its vast extent and large seasonal cycle, so it is of the greatest importance to know the thickness of the ice which forms and melts annually over such a large area of ocean. The first penetration of the pack during Antarctic spring was the cruise of the "Mikhail Somov" in October–November 1981, which reached 62.5°S in the eastern Weddell Sea, but it was not until 1986 that the first deep penetration into the circumpolar Antarctic pack during early winter, the time of ice edge advance, took place. This was the Winter Weddell Sea Project (WWSP) cruise of FS "Polarstern", during which systematic ice thickness measurements were made throughout the eastern part of the Weddell–Enderby Basin, from the ice edge to the coast, covering Maud Rise and representing a typical cross-section of the first year circumpolar Antarctic pack during the season of advance (Wadhams *et al.*, 1987; Lange *et al.*, 1989). After a spring cruise in 1988 (Lange and Eicken, 1991) a second winter cruise was carried out in 1989: the Winter Weddell Gyre Study (WWGS) of Alfred Wegener Institute involved a crossing of the Weddell Sea from the tip of the Antarctic Peninsula to Kap Norvegia in the east during September–October, and thus allowed the multi-year ice regime of the western Weddell Sea to be studied in midwinter (Wadhams and Crane, 1991). During the later part of this cruise, and in collaborative work carried out by "Akademik Fedorov", ice conditions were studied in the same region of the eastern Weddell Sea as was covered by "Polarstern" in 1986, permitting the first interannual comparison of winter ice thicknesses in the same region of the seasonal pack to be carried out.

Ice thickness measurements in the cruises were made by drilling holes at 1 m intervals along lines of about 100 holes. Typically, during a daily ice station, two such lines would be laid out at right angles to one another, with an attempt at a random choice of location so that ridges were properly represented in the results. Clearly very thin ice and open water could not be sampled, but their contribution to the overall pdf was estimated from the results of aerial photography and video recording. Ice thickness, snow thickness and freeboard (hence ice draft) were measured in each hole. Cores were taken at each site to examine ice composition and character.

The end product of the dominant frazil-pancake cycle process described in section 2.3, which occurs in the outer part of the advancing Antarctic pack, is an ice thickness distribution like that of fig. 5.4 (after Wadhams *et al.*, 1987). Note the peak at the very low value of 50–60 cm, with a peak in snow cover thickness at 14–16 cm. The snow cover was sufficient to push the ice surface below water level in some 17% of holes drilled, and this leads to water infiltration into the snow layer and the possibility of formation of a new type of ice, snow-ice, at the boundary between ice and overlying snow. The mechanism of formation of most of the ice out of frazil and pancake explains structural observations (Gow *et al.*, 1987) which showed most of the thickness of ice cores from the Antarctic to be composed of small randomly oriented frazil-like crystals, rather than the long columnar crystals characteristic of freezing onto the bottom. It is reasonable to suppose that these mechanisms are typical of the entire circumpolar advancing ice edge in winter (neglecting embayments such as the Ross and Weddell Seas). This implies that the first-year ice, which makes up the bulk of the winter pack, has a low mean draft of less than 1 m.

Multi-year ice was virtually absent in the region covered by the 1986 cruise. Only a small number of thick "islands" were seen (up to 11 m thick), which are thought to be very old fast ice broken out from sites along ice shelf fronts. Only in 1989 was it first possible to sample both first- and multi-year ice in an intimate combination, in the western Weddell Sea. The Weddell Gyre carries ice from the eastern Weddell Sea deep into high southern latitudes in the southern Weddell Sea off the Filchner-Ronne Ice Shelf, and then northward up the eastern side of the Peninsula. This journey takes about 18 months, and so permits much of the ice to mature into multi-year (strictly, second-year) ice. Such ice was seen only west of 40°W in the crossing of the Weddell Sea, the zone which experiences the northward drift regime.

Multi-year ice could be identified by its structure in cores, and by the very thick snow cover which it acquires, which is almost always sufficient to depress the ice surface below the waterline. Ice profile lines were divided into four types on the basis of this identification: undeformed first-year ice; first-year ice profiles containing deformed ice; undeformed multi-year ice; and multi-year profiles containing deformed ice. Table 5.2 shows the mean thickness results from these classes. It can be seen that:

1. The mean thickness of undeformed multi-year ice (1.17 m), is about double that of first-year ice (0.60 m), indicating conditions in the southern Weddell Sea which permit much more rapid growth for an existing ice sheet than those experienced in the eastern Weddell Sea in 1986;
2. The mean thickness of first-year ice throughout the Weddell Sea is very similar to the thickness observed in 1986 in the Maud Rise area;
3. The presence of ridging roughly doubles the mean draft of the 100-m floe sections in which it occurs (0.60 to 1.03 m in first-year; 1.17 to 2.51 m in multi-year);
4. Snow is very much deeper on multi-year ice (0.63–0.79 m) than on first-year (0.16–0.23 m).

If we consider only those ice stations in the passage northwards out of the eastern Weddell Sea, i.e. the region covered in the 1986 results, we find mean ice thicknesses as shown in fig. 5.23. The results have been plotted over similar data from 1986. At first sight it appears that the thicknesses tend to be greater in 1989, but a number of factors should be noted. Firstly, the 1986 plot was made to examine the thicknesses reached by the natural growth process; profiles consisting largely of ridging were excluded, and those with some ridging were represented with brackets around them, while in 1989 all profiles from the region are included. Secondly, the 1986 northward decline in mean thickness in the latitude range 62° to 58° represented the approach to a stationary, but compact, ice edge. In 1989 the ice edge lay much further north (the last ice was seen at 53°44′S),

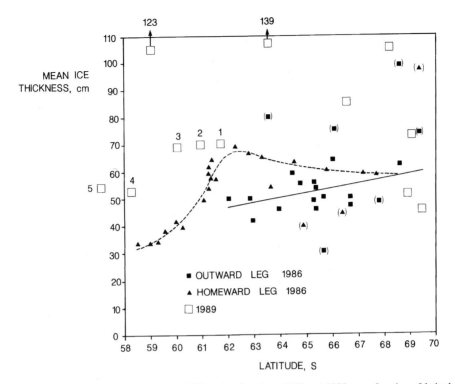

Figure 5.23.   Mean ice thicknesses from drilling sites in winter 1986 and 1989, as a function of latitude.

although its outer regions were diffuse and composed of wide bands, characteristic of off-ice winds. Consolidated pack ice could be studied as far north as 57°, at which point fig. 5.23 shows that there were some signs of melting (stations 4 and 5 relative to stations 1–3). The five ice stations north of 62° in which undeformed ice was profiled can therefore be seen as reproducing the trend of ice thickness against latitude seen in 1986, but with a bodily northward shift of about 3°. We conclude that there was no evidence of a change in the thickness to which first-year ice grows by congelation in the eastern Weddell Sea between the 1986 and 1989 winters. The difference lay in the fact that the ice edge at this longitude lay much further north in 1989.

Drilling from ships has the three advantages that data can be obtained from many locations, that first-year ice and multi-year ice can be clearly discriminated, and that the detailed structure of ridges can be measured. In other respects, however, moored upward-looking echo sounders (ULES) give far more information. Strass and Fahrbach (1998) collected data from six such ULES systems moored across the Weddell Sea (fig. 5.24a) during a two-year period from December 1990 to December 1992. The results for mean drafts (fig. 5.24b) are in good agreement with drilling data for winter, but also reveal the annual cycle ("effective draft" in this figure is true mean draft, i.e. including the open water component). It can be seen that the westernmost ULES, in the Weddell Sea outflow (207), has a cycle ranging from just over 1 m in summer to about 3 m in winter, in good

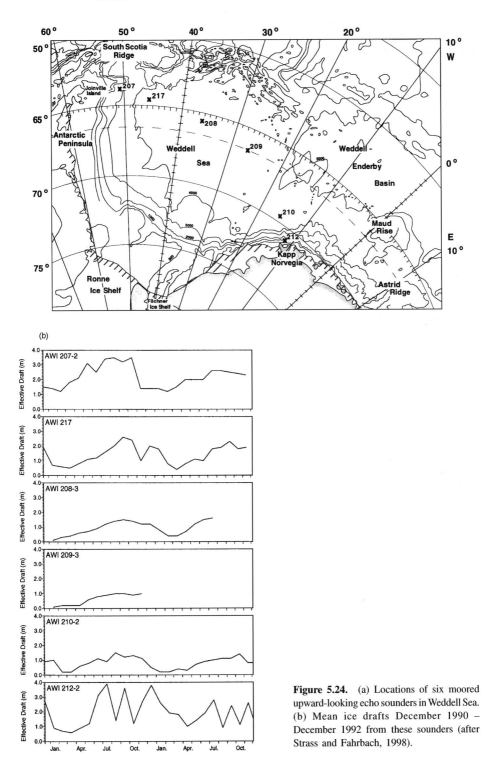

Figure 5.24. (a) Locations of six moored upward-looking echo sounders in Weddell Sea. (b) Mean ice drafts December 1990 – December 1992 from these sounders (after Strass and Fahrbach, 1998).

Table 5.2. Mean ice thicknesses for different categories of ice floe drilled during 1989 Winter Weddell Gyre Study (courtesy of M.A. Lange).

Class	Holes		Snow	Ice	Draft	Total
			Mean thickness			
Undeformed	2034	Mean	0.16	0.60	0.60	0.76
First-year		St. Dev.	0.10	0.21	0.20	0.25
Deformed	2195	Mean	0.23	1.03	1.00	1.26
First-year		St. Dev.	0.17	0.60	0.55	0.64
Undeformed	349	Mean	0.63	1.17	1.25	1.80
Multi-year		St. Dev.	0.18	0.35	0.36	0.45
Deformed	282	Mean	0.79	2.15	2.48	3.30
Multi-year		St. Dev.	0.23	1.08	1.05	1.09
All ice	5339	Mean	0.26	0.97	0.96	1.23
		St. Dev.	0.23	0.73	0.70	0.86

agrement with the "ridged multi-year ice" component of table 5.2 from drilling. The thickness diminishes considerably over the central Weddell Sea (208, 209) but then rises again near the Enderby Land coast (212) to a very variable mean value. This last ULES, very close to the coast, is in a shear zone where much ridging can occur as well as deformation around grounded icebergs (see introduction to Chapter 4).

5.3.3. The Differences Between Arctic and Antarctic Sea Ice

We can summarise the major differences between the morphology of Arctic and Antarctic sea ice as follows:

1. Antarctic sea ice of a given age is much thinner on average than Arctic sea ice. Undeformed first-year sea ice is about 0.6 m thick and second-year ice is 1.2 m thick, whereas in the Arctic first-year ice reaches 1.6–2 m and multi-year ice 3 m or more.
2. The overlying snow cover, however, can be thicker. On first-year ice it averages 16-23 cm and on second-year in the NW Weddell Sea 63–79 cm. The reasons for the much greater thickness in the second year are that the snow does not necessarily melt during the first summer, while during its second year it enters the inner part of the Weddell Sea where precipitation is greater. In the central Arctic snow depth reaches 40 cm by the end of the first winter (Maykut and Untersteiner, 1971; Wadhams et al., 1992), but then the snow melts in summer so that the snow thickness on multi-year ice is a function of time of year rather than age of the ice.
3. In the Antarctic the snow thickness is sufficient to push the ice-snow interface below sea level in 17% of cases sampled for first-year ice (Wadhams et al., 1987; Ackley

et al., 1990) and up to 53% for second- and multi-year ice (Eicken *et al.*, 1994). This permits flooding of the ice surface and the formation of a new ice type, snow-ice or meteoric ice, from the freezing of a mixture comprising the overlying snow and penetrating sea water. It has been estimated (Eicken *et al.*, 1994) that snow and meteoric ice make up 16% of the ice mass in the Weddell Sea. In the Arctic the snow cover is very seldom thick enough to permit flooding, which only occurs in rare cases when isostatic equilibrium is breached, e.g. very close to a pressure ridge, where the random disequilibrium induced by the piling up of ice blocks can cause local depression of the ice surface and flooding.

4. In the Antarctic much of the ice has a fine-grained structure of randomly oriented crystals. It has formed from the freezing of a frazil ice suspension to form pancakes, then the freezing together of pancakes to form consolidated pancake ice, the typical first-year ice type forming in the advancing winter ice edge region (Wadhams *et al.*, 1987; Lange *et al.*, 1989). In the Arctic most ice has formed by congelation growth and so shows a crystal fabric of columnar-grained ice with horizontal c-axes, giving it highly anisotropic strength properties.

5. In the Antarctic most ridges appear to be formed by buckling and crushing of the material of the floes themselves, and so are composed of a small number of fairly thick blocks extending to modest depths — typically 6 m or less. In the Arctic ridges tend to be formed by the crushing of thin ice in refrozen leads between floes, and so are composed of a large mass of small blocks, extending to greater depths — typically 10–20 m, with significant numbers extending to 30 m or more (Wadhams, 1978a) and even to 40–50 m in extreme cases.

6. Because of the divergent nature of the ice cover in the Antarctic and the generally northward trajectory of floes, almost all of the ice is two years old or less. In the Arctic a significant fraction of the ice is multi-year, which can amount to an age of 10 years or more. Ice in the Arctic can be trapped in the Beaufort Gyre and make multiple circuits of the Canada Basin, each circuit taking 7–10 years, enabling the oldest ice to reach extreme ages. In the Antarctic there are no such orbits, and trajectories almost always bring a floe to the ice edge and thus to melting within 2 years of formation.

7. Antarctic sea ice carries a heavy load of chlorophyll and phytoplankton (Sullivan *et al.*, 1985; Garrison *et al.*, 1986; Eicken, 1992), due partly to its thin character and partly to its mode of formation, in which plankton are trapped in the originating frazil suspension. Arctic ice has much less biological material in it, and tends instead to carry a mineral burden due to turbulent flotation of sediments in shallow waters at the time of ice formation over Arctic shelves (Reimnitz *et al.*, 1987). These properties are discussed further in chapter 8.

5.4. APPLICATIONS OF THE ICE THICKNESS DISTRIBUTION

5.4.1. Sound Propagation

One area in which sea ice morphology studies find a practical application is in estimating losses in sound propagation under ice. The sound velocity profile in the Arctic Ocean is such that sound velocity increases continuously with depth, so that all rays are upward refracted and long-distance transmission necessarily involves a number of reflections from the ice-water interface.

Guoliang and Wadhams (1979) discussed two methods for estimating the reflection coefficient from a single bounce. The method of small perturbations, which assumes small roughness elements on the surface, gives a reflection coefficient of

$$R = 1 - 2\, k^2\, \sigma^2\, \sin^2 \alpha \qquad\qquad (5.21)$$

where k is the wave number, σ is the standard deviation of the surface and α is the grazing angle (a function of the sound velocities in the water and the lower part of the ice sheet, but usually estimated to lie in the range 4–12°). The tangent plane method, which also assumes small perturbations based on a Gaussian distribution of roughness elements, gives

$$R = \exp\left(- 2\, k^2\, \sigma^2\, \sin^2 \alpha\right) \qquad\qquad (5.22)$$

Both (5.21) and (5.22) demonstrate that longer-distance propagation can occur if the wave number is small, i.e. using low frequencies. The 1988–9 tomography experiment in the Greenland Sea used 250 Hz sources for propagation over 200 km, which was severely affected by the ice cover (Jin *et al.*, 1993). The 1994 Arctic ATOC (acoustic thermometry of ocean climate) experiment, involving successful trans-Arctic propagation, used a frequency of 57 Hz (Mikhalevsky *et al.*, 1999). Equations (5.21) and (5.22) also demonstrate the crucial importance of the standard deviation of the surface. Guoliang and Wadhams estimated that for the geometry of the Greenland Sea tomography array the scattering loss would be about 15 dB in typical Greenland Sea ice conditions, compared to 7 dB or less if the ice underside were smooth.

This model requires revision if large pressure ridges are present, since these must be treated as individual powerful scatterers. A number of techniques have been developed to deal with scattering from individual pressure ridges (Diachok, 1980).

Acousticians who seek to predict scattering loss from icefields with known statistical parameters would benefit from being able to simulate an under-ice surface for input into their computer models. An interesting attempt to simulate the shape of the whole pdf was made by Hughes (1991) who found that a combination of seven lognormal distributions could be made to fit. Hughes also found that he could generate synthetic ice draft data by interspersing filtered random number sequences drawn from the fitted lognormal distributions.

5.4.2. Microwave Signatures

In section 5.2.2 we discussed the relationship which has been found between SAR backscatter level and ice thickness in the case of X-band SAR, the only case in which a direct validation has so far been possible. We showed that the high positive correlation is fortuitous in terms of physical causes, since an enhanced backscatter is associated either with volume scattering (i.e. with multi-year as opposed to first-year or young ice) or with scattering from roughness elements (i.e. with deformed as opposed to undeformed ice).

It is possible that with further validation we may be able to use microwave signatures as a way of defining ice roughness. This does not necessarily mean establishing a more perfect correlation between ice thickness and SAR brightness — it is likely that the level of correlation obtained so far is about the limit of the natural correlation which really exists. It is more likely that progress will come from improving our ability to discriminate between multi-year and first-year ice in a SAR image or in passive microwave. It is recognised that multi-year ice is characterised on SAR by a greater average brightness and a more speckled texture, due to the greater surface roughness and greater variability in volume scattering. Kerman (1998) suggested that by examining the distribution of brightness *differences* between neighbouring pixels a distribution might be obtained which bears a quantitative relationship to the ice thickness distribution. Preliminary tests of such an approach with ERS-2 data have proved inconclusive (Doble and Wadhams, 1999), possibly because of the intrinsic speckle level in ERS-2 SAR. In the case of passive microwave Comiso (1995) developed an approach to a finer-scale ice classification by using a cluster analysis of data obtained from all 7 frequency-polarisation combinations in the SSM/I sensor, allowing the Arctic or Antarctic to be divided into radiometrically distinct "ice regimes". The possibility exists that each regime may correspond to a distinct mean ice thickness value, allowing this technique to be used as a proxy for mean thickness or roughness.

5.4.3. Stimulation of Internal Waves by Ice Keels

It has been found that internal waves in the Arctic pycnocline can be forced by the deeper keels of pressure ridges at times when wind stress is driving the ice cover relative to the near-surface water masses. Observations have been reviewed by Morison (1986). The simplest kind of wave is an internal wake, in which the ice keel acts much as a ship's keel does when it generates the famous "dead water" of Ekman which was the origin of work on internal waves. This kind of wake was first studied in the field by Rigby (1976). More recently, extensive measurements of the Arctic internal wave field in relation to ice topography and driving forces was carried out on the CEAREX experiment.

Internal waves may be of importance to the heat transfer between the Atlantic and polar surface water layers, and the drag exerted by internal wave generation may be of significance to ice dynamics. Since it is the deeper ridges which are of the greatest importance in stirring internal waves, we might expect the consequences of internal wave activity to be most important in those regions of the Arctic where deep ridges are most common, such as the zone north of Greenland and Ellesmere Island (Wadhams, 1978a).

5.4.4. The Under-ice Drag Coefficient

Both the air-ice and the ice-water drag coefficients are directly dependent on the statistics of ice roughness, and thus vary over the Arctic and Antarctic both geographically and with season. A number of attempts have been made through field experiments to establish an empirical relationship between drag coefficient and some parameter related to ridging frequency, on the grounds that total drag can be broken down into skin friction drag and form drag, with the form drag component being directly proportional to the total vertical area of ridge facing the wind (or current) per unit area of icefield. Naturally the drag coefficient is also dependent on the stability of the lower atmosphere.

An example of a simple parameterisation of the air-ice drag coefficient was the work of Banke *et al.* (1976, 1980). They found that sonic anemometer data on the AIDJEX camp in the Beaufort Sea was well fitted by the relationship

$$10^3 \ C_{10} = 1.16 + 0.065 \ \zeta \tag{5.23}$$

where C_{10} is the drag coefficient and ζ is the rms surface elevation at wavelengths less than 13 m, derived by integrating the high wave number end of the power spectrum

$$z^2 = \int_{k_0}^{\infty} \phi(k) \ dk \tag{5.24}$$

Here $\phi(k)$ is the power spectral density and $k_0 = 0.5$ rad m^{-1}. This is justified on the grounds that it is local scales of roughness which most affect the drag coefficient. It is likely that similar parameterisations could be carried out for the ice-water drag coefficient, but hitherto there have not been adequate field observations of drag versus underside roughness.

Clearly, if ice roughness were known synoptically over the Arctic and Antarctic it would assist the modelling of ice dynamics, since geographically varying drag coefficients could be employed.

5.4.5. Differential Melt Rates

It can be seen from figure 5.3 that as one moves southward in the Greenland Sea the slope of the ice thickness pdf at extreme ice thickness increases. This implies that thicker ice is melting (or disintegrating, since we are dealing here with ridges) at a greater rate than thinner ice. Davis and Wadhams (1995) also found that a downstream decline in mean pressure ridge thickness could be detected in the Eurasian Basin and Greenland Sea, both south and north of Fram Strait, which was greater than the rate of decline of mean ice thickness as a whole. It was concluded that ridged ice melts faster than thinner, unridged ice, and that melt occurs within the Arctic Basin as well as in seas more generally associated with ice ablation. Wadhams (1997a) confirmed these results with a further dataset and showed (Wadhams, 1997b) that at any latitude the melt rate was proportional to ice thickness.

These results are qualitatively in line with thermodynamic models (Thorndike *et al.*, 1975), which demonstrate that the thermodynamic evolution of ice cover thickness from some initial state generated by ridging events involves thin ice getting thicker and thick ice getting thinner, with all ice trying to approach an equilibrium thickness of about 3 m. The observed *rate* of melt of ridged ice, however, is much faster than predicted by a thermodynamic model, even if we take account of possible enhanced melt due to the slope angles of ridges (Flato *et al.*, 1999). Dynamic or mechanical factors must be involved. One possibility is that the hydrodynamic flow of the shear current under a keel crest involves increased velocity at the crest (the ridge acting like an aerofoil), which will cause enhanced heat transfer from the water into the ice. In the absence of an ice-water relative current another effect may prevail, in that a ridge protruding into a stratified water column sets up a layered structure associated with the differential diffusion of heat and salt from the melting interface. This is best developed in the case of a vertical interface (Huppert and Josberger, 1980) such as a melting iceberg, but recent laboratory experiments with inclined surfaces have shown that this structure can also be developed by melting ridges. The effect may be to increase the melt rate. Finally, for very deep ridges, Lewis and Perkin (1986) have drawn attention to the effect of the pressure-induced change in melting point, generating an "ice pump" which melts deep ice while causing shallow ice to grow — but this is a very slow process. The most likely mechanism in practice is that as ridges begin to melt they start to lose structural integrity and the ice blocks become "unglued" and are strewn by currents under the neighbouring undeformed ice, preferentially reducing the concentration of very thick ice.

Because of the enhanced melt rate for thick ice, the area-averaged melt rate for an icefield can be expected to be dependent upon the ice thickness distribution and the distribution of pressure ridge depths. In the case of the Greenland Sea, for instance, if these statistics vary with time at the latitude of Fram Strait, we can expect the melt rate at any latitude, and hence the fresh water flux, also to vary with time. Hence it is important to understand the relationship between melt rate and thickness or roughness.

5.4.6. Ice Scour Rates and Pipeline Burial Depths

Ice scouring is a serious problem for offshore operations, since it makes burial of pipelines or wellheads necessary near shore. Therefore it is important to estimate the return period of scours in water of different depths, to determine a safe water depth at which seabed structures may be erected, or (in conjunction with the statistics of the the incision depths of scours) a safe burial depth for a pipeline in any depth of water.

The problem is one of estimating firstly the maximum draft of a pressure ridge occurring in a given spot in a given time (for wellheads), or the maximum draft of ridge crossing a given line in a given time (for pipelines). The problem was discussed by Wadhams (1983b), and is a typical example of an ice engineering problem which depends for its solution on appropriate environmental and glaciological data. We therefore discuss it in some detail.

For the problem of extreme depth prediction at a point, Wadhams proposed three techniques and showed that they give similar results. The first technique makes use of the observed negative exponential distribution of keel drafts, given by eqns (5.3)–(5.5). If the water depth is D, and L km is the distance drifted per year by the ice cover over the location being considered, then eqn. (5.3) gives for N_D, the number of keels per year passing the point with draft exceeding D,

$$N_D = L \int_D^\infty n(h) \, dh = L \, \mu \, \exp \left[(h_0 - D)/(h - h_0) \right] \qquad (5.25)$$

N_D is then the number of grounding ridges at that site per year, i.e. the number of scouring events, and the return period for a scour at that spot is given by

$$T_D = 1 / N_D \qquad (5.26)$$

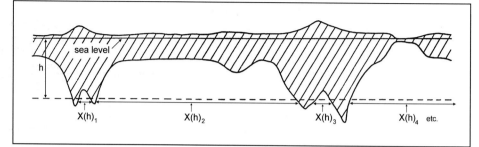

Figure 5.25. The depth crossing technique for estimating extreme keel depths.

Clearly the quantities used in (5.25) must be determined from site-specific data, e.g. moored upward sonars at the site, while L can be determined from döppler sonar or from the analysis of satellite imagery.

If we specify a return period, then rearrangement of (5.25) gives the minimum water depth at which a wellhead can be constructed with the specified degree of safety:-

$$D = h_o + (h - h_o) \ln (T_D L \mu) \tag{5.27}$$

The second method is distribution-free and uses the set of spacings between crossings of a given depth horizon by the ice bottom profile. In fig. 5.25, let X(h) be the set of distances between an upward crossing of a depth horizon h and the subsequent downward crossing of the same horizon. The return period T_D for scouring at a water depth D is then given by

$$T_D = \overline{X}(D) / L \tag{5.28}$$

If D exceeds the maximum depth to which keels are observed in any finite sample, the curve of X(h) against h can be extrapolated to greater values. As fig. 5.26 shows, for the case of the Beaufort Sea dataset analysed by Wadhams and Horne (1980), since the distribution of keel depths is found to fit a negative exponential, the function $\overline{X}(h)$ is a positive exponential which can thus be extrapolated with confidence.

A third method, also distribution-free, is probability plotting (Gumbel, 1954, 1958). The ice profile is divided into uniform sampling intervals. The deepest keel is extracted from each sampling interval and the resulting keels are ranked in order of depth. The quantity

$$\Pr(X < x) = 1 - [m / (n + 1)] \tag{5.29}$$

where m is the ranking of a given keel (m = 1 for deepest), and n is the number of keels involved, is then plotted on exponential extreme-value probability paper against the depth of the keel. In the case of the Beaufort Sea data considered so far, the result is a straight line which can be extrapolated to greater depths (fig. 5.27). This makes for a more rapid

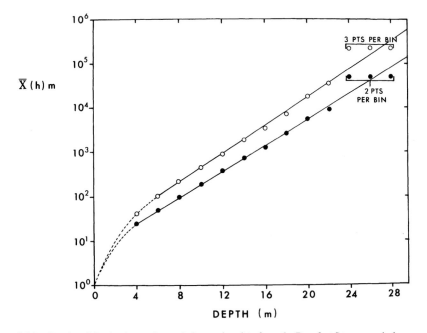

Figure 5.26. Results of the depth crossing technique using data from the Beaufort Sea: open circles represent the whole dataset of Wadhams and Horne (1990) while the closed circles represent the 100 km of data nearest the coast.

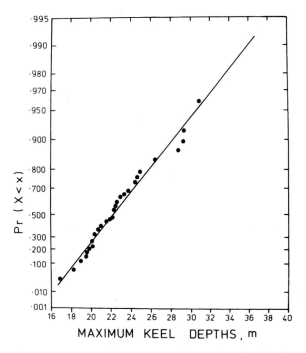

Figure 5.27. Exponential extreme-value probability plotting for maximum keel depths.

analysis, since only the deepest keel in a given section of data need be measured. Wadhams (1983b) showed that for ice in the southern Beaufort Sea these methods give return periods of 9 months for a 30 m keel, 24 years for a 40 m keel and 1010 years for a 50 m keel.

The pipeline problem (probability of ice scour along a line running at right angles to the net drift of the sea ice) is a more complex one which requires a knowledge of the along-crest statistical properties of a ridge keel. There is almost no information on this statistic, but Wadhams (1983b) proposed that along-crest ridge lengths can be deduced from the observed widths of ice scours themselves. It can be shown that the one-dimensional statistic N_D translates into a number S_D of scouring events per km of pipeline per year, where

$$S_D = N_D \, \pi \, / \, 2 \, s \qquad\qquad (5.30)$$

Here s is the mean along-crest length of keel which exceeds depth D. This assumes that keels are randomly oriented. To estimate s we assume that when a keel scours the seabed in water of depth D the width of the scour is simply the projected length of keel crest of depth exceeding D at right angles to the direction of the scouring motion. Data on s at different water depths are available from sidescan sonar statistics of scours (e.g. Lewis, 1977).

Finally, one makes use of the observed fact (Lewis, 1977; Pilkington and Marcellus, 1981) that the probability that a given scour mark has an incision depth of d or greater into the seabed is exp (– k d) where k is a site-specific parameter. This enables the required burial depth to be inferred for a pipeline of unit length so as to avoid disturbance by scouring during an interval T. The result, using (5.25) for the distribution of keel drafts, is

$$d = [\ln (L \, \mu \, T \, \pi/2 \, s) + (h_0 - D)/(h_0 - h)] \, / \, k \qquad\qquad (5.31)$$

Typical values come to 5–8 m for a proposed 76 km pipeline from an offshore wellhead to the coast.

At this point we have reached the limits of what can be achieved by statistical distributions of ridge depths and scour incision depths, since an additional factor based on the properties of the seabed material is at work. The pipeline must be buried deeper than the maximum scour depth predicted by (5.31), because the scouring ridge produces **subgouge deformation** under, and just ahead of, its actively scouring "blade", which can damage a pipeline even if the ice does not actually touch the pipe. How much extra trenching depth must we allow for the overpressure effect? The problem is an important one, because it is exceedingly expensive to bury a pipeline, and an extra metre of depth may be excessively costly or even technologically unfeasible.

Subgouge deformation has been observed, in some cases, to extend to more than twice the gouge depth (Woodworth-Lynas et al., 1996), but the extent of the deformation is very dependent on the geotechnical properties of the seabed material. Rather than calculate maximum gouge depths and add a factor for subgouge deformation, it has been proposed (Palmer, 1997) that efforts be made to directly measure the depth to which sediments have been deformed by gouging, e.g. by looking for a break in the mechanical properties of the seabed material by standard tests such as the cone penetrometer, or by using radioactive tracers to distinguish between gouged and ungouged soils. The latter method depends on the fact that radionuclides such as tritium were deposited over the sea surface in a major pulse during the last series of atmospheric nuclear tests in the early 1960s, reached the surface sediments via incorporation in the tests of foraminifera and other creatures which contribute to sedimentation, and then will have been mixed some way downwards into the sediments by repeated gouging. A discontinuity in the concentration of these tracers then marks a gouging limit (although this still requires an additional allowance for subgouge deformation that is not accompanied by reworking of sediment). We note that, in the case of icebergs, direct measurements of subgouge deformation have been made from relict scours on land dating back to the last Ice Age (Woodworth-Lynas and Guigné, 1990), in model tests (Poorooshasb *et al.*, 1989) and in centrifuge tests (Woodworth-Lynas *et al.*, 1996), and that model and centrifuge tests might achieve similar success in the case of sea ice scour.

6.   THE MARGINAL ICE ZONE

You are in the marginal ice zone of the Bering Sea in mid-winter, aboard the NOAA research ship "Discoverer", carrying out the first experiment in the MIZEX series to understand the dynamics of the MIZ ice cover. A vast expanse of thin broken ice floes stretches like a scattered jigsaw puzzle to the horizon in every direction. Over the horizon to the north of you lies St. Matthew Island, while far to the south is the ice edge. On six ice floes you have set out an array of radar transponders, which respond to the ship's radar by giving out a pulse of their own which can be tracked from the ship. This allows the ship to track the motion of all six floes simultaneously, so long as she stays within radar range of all of them. On the same floes you have other expensive equipment: ocean wave recorders to measure the response of the floes to waves, and acoustic devices to record the thinning of the ice due to melt. Soon the sky turns iron-grey and a swell starts to roll through the icefield. An intense low has spun up and is tracking across the southern Bering Sea outside the ice edge, whipping up a steep wild sea in the shallow waters of the continental shelf. Before you realise what has happened the waves have stripped away layer upon layer of the MIZ, pounding up the floes and driving the wreckage of brash ice along the ice edge towards Alaska. Suddenly you see that one of your instrumented floes has reached the ice edge and is in danger of destruction. The ship must temporarily abandon her other charges and head for the endangered floe, which is now drifting right out into the open sea. You prepare to go over the side to recover your gear, but you are prevented. It is too dangerous; only the ship's professional divers can go down, dressed in their rubber dry suits to withstand the icy water. They go over the side in the "Mummy Chair", a steel cage suspended from a crane. Waves are breaking across the floe; they have washed away all the snow, leaving the ice as slippery as a skating rink. Slithering, sliding and tumbling across the wave-lashed floe, frequently knocked down by waves and in danger of being washed off, the divers rescue most of your equipment, although some slides away into the sea and is lost for ever. When you return to port at Unalaska Island you find that two fishing trawlers disappeared in this storm, with twelve deaths.

6.1. THE STRUCTURE OF THE ICE MARGIN

The region of the Arctic or Antarctic ice cover which lies close to an open ocean boundary is called the Marginal Ice Zone (MIZ). In this region the continuous ice cover characteristic of the central basin is broken up into floes by the flexural stress of waves and swell penetrating into the ice from the ocean. The greatest amount of break-up, and thus the smallest floes, are found closest to the ice edge. The width of the marginal ice zone is defined as the depth of penetration to which waves can fracture ice to create this morphology of floes; it is a function not only of ice thickness and mechanical properties, but also of the regional wave climatology, i.e the fetch and duration of winds outside the ice edge, and whether the ice edge lies over shallow or deep water. A wave spectrum is modified in shallow water to create shorter, steeper waves, which break floes into smaller fragments but which do not penetrate so far into the ice as the longer waves of deep water. The theory of how waves break up ice into floes, what determines the floe size distribution, and how the resulting floes attenuate the waves, is discussed in section 6.3 of this chapter.

If the wind is blowing *towards* the ice edge from the open sea, it compresses the MIZ and produces a compact icefield with a sharp outer boundary of small cakes and brash ice (fragments of broken floes), against which shorter waves break (fig. 6.1(a)). This is a region of high **ambient noise** generation, which is discussed further in section 6.6. Atmospherically, an on-ice wind is often associated with stratus cloud, because of warm moist air from the open sea being cooled over the ice, so compact ice margins are often difficult to observe from the air or from visible-range satellite sensors such as AVHRR; their structure is best revealed by SAR. The temperature change causes modifications to the atmospheric boundary layer (ABL) (Kantha and Mellor, 1989; Glendening, 1994) resulting in turbulent surface heat fluxes which extend for about 100 km into the ice, i.e. roughly the width of the MIZ itself. Thus the MIZ width is not only a buffer zone which protects the stable morphology of the polar ice cover, but is also a width scale required for the ABL to acquire a stable polar form. The incoming warm air brings heat into the MIZ, so if this is also a zone in which the ice itself is moving towards the edge (e.g. Antarctic), then it will be a zone of melt as well as break-up.

A wind blowing roughly *parallel* to the ice edge can create some complex upper ocean effects, since the differential drag of wind on water and ice creates the equivalent of a longshore current along the ice edge. The effects have been reviewed by Guest et al. (1994), and depend on whether the wind is "parallel-right" (i.e. ice lies to the right looking downwind) or "parallel-left". In the Northern Hemisphere a parallel-left wind creates an Ekman transport in the OBL that is directly away from the ice edge, resulting in an upwelling that can transport heat from deeper layers into the ice edge zone. This effect was predicted by Gammelsrød et al. (1975) and observed by Buckley et al. (1979) at the ice edge north of Svalbard. A parallel-right wind in the Northern Hemisphere interacts with the local **ice breeze** (equivalent to a land breeze at the seaside), which is produced by the horizontal temperature gradient between warm air outside the ice edge and cooled air lying over the ice edge. The interaction can produce a surface front outside the ice edge — an example was a front with a temperature jump of 6°C in 25 km, observed 100 km off the ice edge in the Barents Sea (Shapiro et al., 1989).

When the wind is blowing *away from* the edge, most of the MIZ becomes diffuse (fig. 6.1(c)), but the outermost edge, instead of just becoming an open icefield fading away into the open sea, takes on a new organisation of its own, with a series of compact **ice edge bands** forming, separated by completely open water and lying with their long axes roughly perpendicular to the wind. The dynamics of formation of ice edge bands are discussed in section 6.4. In winter this opening up of the ice edge permits new ice to form in the open water areas, leading to enhanced brine rejection in the MIZ.

The effects of the various wind regimes are summarised in table 6.1 (after Guest et al., 1994), which shows the range of wind directions relative to the ice edge which will result in distinctly different effects on the MIZ. Here the wind direction given is the geostrophic wind direction relative to the ice edge, with 0° being parallel-right, 90° directly on-ice and so on.

In some marginal ice zones, such as that of East Greenland, the ice edge region also corresponds to an oceanic surface front between polar surface water on the iceward side

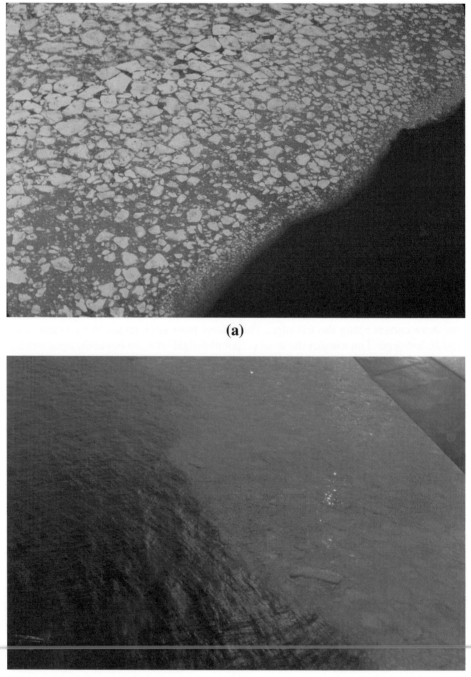

(a)

(b)

(c)

Figure 6.1. Some typical marginal ice zone scenes. (a), (b) Aerial views of compact ice edges, with on-ice wind and waves and floe size sorting with the smallest floes nearest the edge [(a) is in Greenland Sea, (b) in Labrador Sea, both winter]. (c) Interior view of an open MIZ in summer, Greenland Sea, showing ice cover of melting floes.

and warmer water in the open sea. There may also be a velocity shear across this front between the two water masses. Such a shear, or other factors such as irregular bottom topography, can cause the front to be unstable and to break up into meanders and **eddies**, which display their presence through the patterns produced by the entrainment of MIZ floes over the cold parts of the eddy. Off East Greenland the eddies are usually cyclonic (anticlockwise in circulation) with a warm water core. Eddies are discussed further in

TABLE 6.1. Effects of different wind regimes on an Arctic marginal ice zone (after Guest *et al.*, 1994). WD = geostrophic wind direction, with WD = 0° being parallel-right and WD = 90° being directly on-ice.

WD°	Regime	Ice motion	Effects on MIZ
10–50	parallel-right	on-ice	Thermal and frictional convergence, strong fronts
50–130	on-ice	on-ice	Warm, ice compaction
130–170	on-ice	off-ice	Warm, ice destruction
170–185	parallel-left	off-ice	Thermal convergence, weak fronts
185–325	off-ice	off-ice	Cold, ice divergence
325–10	off-ice	on-ice	Cold, brine rejection

section 6.5. There can also be internal wave activity near the front, since it is a wedge-shaped feature which slopes away under the ice at a gentle angle of about 1 in 1000, producing an almost horizontal shallow boundary between two water masses of different densities.

A final very important property of the Greenland Sea MIZ in winter is that local ice production in the range of latitudes (72–75°N) where polar water flows eastwards into the open sea is a triggering factor for deep ocean convection. The mechanism for this is discussed in section 6.7.

6.2. MARGINAL ICE ZONES OF THE WORLD

Wherever an ice edge abuts onto the open sea it is possible to have a marginal ice zone. Nevertheless in many cases the open sea area is of limited extent, giving a small fetch for wave generation, and so the ice edge region does not differ significantly in properties from the rest of the ice cover. Examples are the seas north of Russia in summer (Kara, Laptev, East Siberian, Chukchi), the ice-free summer channel north of Alaska, the ice-free areas of the Canadian Beaufort Sea and Archipelago in summer, and the Gulf of Bothnia. A true MIZ has its character permanently determined by abutting on to a rough ocean with a climate of long, high waves. In this sense there are only four true MIZ regions in the world:- the Greenland (and to a lesser extent Barents) Sea system, the Labrador Sea — Baffin Bay system, the Bering Sea and the circumpolar Antarctic ice edge.

6.2.1. Greenland Sea

The cold, fresh polar surface water that occupies the upper layers of the Arctic Ocean flows out along the east coast of Greenland, forming the upper part of the East Greenland Current. The eastern edge of the current corresponds to the position of the ice edge in winter, giving rise to interactions which result in ice edge eddies and other phenomena, but in summer the ice retreats westward and northward. As fig. 1.16 shows, in winter of an average year the ice reaches Kap Farvel. During December–April a bulge appears in the region 72–75°N; this is an effect of the 10-year averaging, since in any given year the feature usually takes the form of a tongue, known as Odden (section 6.7). It is a product of local ice formation in a cold surface water mass of the Jan Mayen Polar Current, an eastward diversion from the East Greenland Current which occurs just north of Jan Mayen Island. In summer the ice edge retreats to about 74°N on average, although there is a large interannual variability. In September 1996, for instance, there was a period of a month in which no ice occurred south of Fram Strait. Fig. 6.2 shows the magnitude of the 10-year variability (1966–75) for a winter and a summer month. It can be seen from fig. 1.16 and fig. 6.2 that the East Greenland Current and Barents Sea together offer the longest stretch of marginal ice zone in the Arctic, facing onto the Norwegian-Greenland Sea which is well known for its storminess.

The East Greenland Current is also of interest because it is the means by which fresh water and (negative) heat are transported out of the Arctic Basin and transferred to the

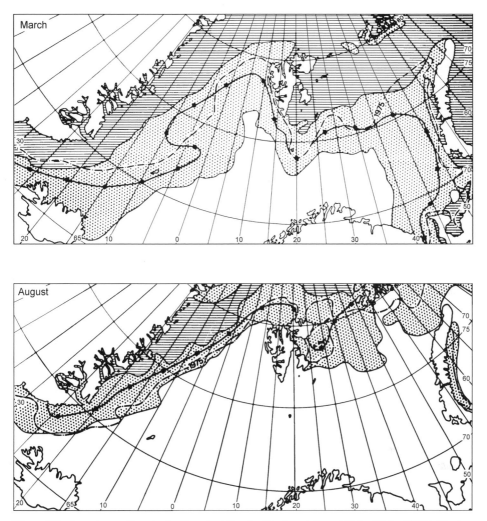

Figure 6.2.   Ice edge positions in the Greenland Sea for March and August (After Vinje, 1977b). The two types of shading mark the extreme maximum and minimum limits for 3/8 ice cover during the 10-year period 1966–75. The thick black line is the median position and dotted line is 1975. Comparison wih fig. 1.14 (1978–87) shows a general retreat in the 1980s.

Greenland Sea. The region was studied intensively during the 1980s in the MIZEX (Marginal Ice Zone Experiment) project, which included major field programmes in 1983, 1984 and 1987 (as well as a winter Bering Sea programme in 1983). Two special issues of *Journal of Geophysical Research*, produced before and after MIZEX, show how much progress was achieved (JGR, 1983, 1987; see also Further Reading section). More recently the larger-scale energy exchanges in a box bounded by Fram Strait, Denmark Strait and Mohns Ridge were studied in two other intensive series of international cruises called the Greenland Sea Project (1986–93) and the European Subpolar Ocean Programme

(1993–9). A series of summer cruises took place during the late 1980s and intensive winter measurement programmes took place during 1993–4, 1997 and 2000.

The Greenland Sea and its MIZ are also of great interest to the defence community. Russian submarines traverse the Greenland Sea to reach the North Atlantic, and an attractive cover for submarine noise is offered by the high ambient noise levels in the MIZ. Measurements show that when the wind is blowing towards the ice to create a compact ice edge, the edge acts as a line source of noise which gives levels some 20 dB louder than the open ocean, while with an off-ice wind and a diffuse ice edge where floes do not collide, the rise in noise level is only a few dB. However, measurements on the Anglo-Norwegian SIZEX experiment in 1992 showed that intense noise-generating centres are also found within ice edge eddies. The mechanism for noise generation here is unknown — perhaps shear between floes. In detecting submarines from outside or inside the ice, it is necessary to understand sound propagation across the ice edge front, and thus the effect of internal waves, eddies and convective plumes upon the sound-speed structure of the ocean. Hence much Greenland Sea MIZ research has been jointly sponsored by civilian and military agencies, since there is a common interest in understanding the processes occurring there. Further discussion is given in section 6.6.

6.2.2. Labrador Sea

The East Greenland Current continues around Kap Farvel at the southern tip of Greenland, then flows up the west coast of Greenland as the West Greenland Current. The main part of the current turns at the northern end of Baffin Bay, in Melville Bay, and becomes a southward-flowing boundary current for the west side of Baffin Bay known as the Baffin Island Current. This flows out of Baffin Bay and Davis Strait, and, still hugging the coast, continues down the coast of Labrador. In winter it carries an ice cover with it, composed of ice formed in Baffin Bay itself and small quantities of ice from East Greenland and from the Arctic Ocean transported through the narrow Nares Strait at the northern end of Baffin Bay. In spring, ice which breaks out of the fast ice regime of the Parry Channel also exits in this way. The current as it passes down the Labrador coast is called the Labrador Current. It ends by flowing over the Grand Banks of Newfoundland in filaments and forming the famous "cold wall" where it meets the much warmer Gulf Stream on the E and SE flanks of the Grand Banks. The outer edge of the Labrador Current is subject to eddies in the form of "backward-breaking waves" (Legeckis, 1978), characteristic of a velocity shear as discussed in section 6.6.

As fig. 1.16 shows, there is a classic MIZ in the Labrador Sea, resembling that off East Greenland, and lasting from December until May of a typical year. The ice retreats up the Labrador coast in June, and leaves the coast ice-free from July to November. At the same time, it can be seen that there is also an ice front with the open ocean in Davis Strait, from November/December to June/July. This ice front is rather sheltered, as it is open to wave action from only a limited range of bearings, so it does not develop full MIZ characteristics; the ice thickness in the region has been profiled by submarine (Wadhams *et al.*, 1985). Most of the ice in the Labrador Current and Baffin Bay is first-year. The Labrador Sea MIZ itself has been studied intensively in a series of collaborative remote sensing and shipborne experiments called LIMEX (Labrador Ice Margin

Experiment), of which the first was in 1987 (e.g. Liu *et al.*, 1992). An interesting phenomenon noted by Gloersen *et al.* (1992) is that ice extent anomalies in the Labrador and Greenland Seas appear to occur in antiphase, a light ice year in Greenland corresponding to a heavy year in Labrador and vice versa.

As well as sea ice, the Labrador Current carries huge numbers of Arctic icebergs, most of which were formed by calving of glaciers on the west coast of Greenland, and which end their lives by running aground on the Grand Banks. Bergs of shallow draft cross the Banks (or else grounded bergs refloat after partially melting) and reach the "cold wall", posing a danger to North Atlantic shipping. This has necessitated the presence and work of the International Ice Patrol. Icebergs in this region are discussed in more detail in chapter 7.

6.2.3. Bering Sea

The Bering Sea occupies a large area of some 2.2×10^6 km^2. About half of this is shallow continental shelf, less than 200 m deep. The rest, the Aleutian Basin, can be more than 4000 m deep. As shown in fig. 1.16, in November the ice edge advances southward through Bering Strait and develops an MIZ in the Bering Sea which lasts until May. The furthest limit of ice advance roughly corresponds to the edge of the continental shelf, where a one-layer water structure gives low ocean heat fluxes and facilitates ice growth. The ice is quite thin, only 1 m or so thick, and the floes near the ice edge are in consequence quite small. Fields of pancake ice are often seen at the ice edge.

A feature of the Bering Sea is the development of coastal polynyas downwind of St. Lawrence Island and in Norton Sound. The dynamics of these polynyas have been discussed in chapter 4.

When the ice retreats in summer into the Chukchi Sea, there continues to be an ice front, but because it is so sheltered the ice edge zone does not develop MIZ characteristics. Fig. 1.16 shows that the seasonal ice cover in the Sea of Okhotsk also can have a front with the open sea, but here the Kurile Islands protect the ice front from heavy wave action, and so again the ice cover does not tend to develop a fully-formed MIZ.

6.2.4. Antarctic Ocean

The circumpolar Antarctic ice edge is the longest and widest MIZ in the world. It faces, in winter, the stormiest ocean on the planet, with wave heights that have been mapped by Geosat satellite altimetry (Campbell *et al.*, 1994) as averaging more than 4 m throughout the year, and more than 6 m in the Pacific sector. Fig. 1.18 shows that from about April to December there is a complete circumpolar ice edge facing the open ocean, which can develop an MIZ of total length some 20,000 km.

As we have discussed in chapter 2, the early winter phase of ice edge advance (April to about September) is characterised by an ice margin at which ice is forming all the time. This accounts for much of the rate of advance of the ice edge, the rest being due to ice which forms within the pack or in coastal polynyas and moves with a northward component to its generally eastward trajectory. The ice forming in this very stormy ice edge region starts as frazil and pancake ice, and the 1986 "Polarstern" cruise showed that a belt of width 270 km is required to damp down waves to a point where the pancakes can freeze together to create consolidated pancake ice. We may speak, therefore, of a distinctive and very wide winter MIZ.

As soon as ice edge retreat begins, the ice edge becomes a melting front, since the net motion of ice is still northwards while the edge is retreating southwards. No new ice formation now occurs, and the consolidated ice becomes broken up by waves in the classic MIZ manner to create floes. In many cases the floes are made up of consolidated pancakes, but by now these are strong enough that the ice usually does not revert to the original pancakes but cracks along new lines.

6.3. WAVE-ICE INTERACTION AND THE FLOE SIZE DISTRIBUTION

6.3.1. A Review of Observations

The global geophysical role of wave-ice interaction has two aspects: the physical effects of waves on an ice cover; and the use of waves as a diagnostic tool in ice mechanics. The physical effects include the ability of waves to break up ice sheets into floes and to herd these floes into patterns which determine the morphology of the marginal ice zone (MIZ); the break-up of tabular icebergs; the calving of ice tongues; and the generation of ambient noise. The diagnostic role stems from the long-range propagation of wave energy in the form of flexural-gravity waves through ice sheets, and the information which the dispersion relation and attenuation rates can give us about ice properties and mechanics.

In this chapter we give most attention to wave propagation across the fringe of ice floes which separates the interior polar pack from the open ocean. As ice approaches the ice edge from the far interior it encounters wave energy of gradually increasing energy and gradually decreasing peak period (since short waves are attenuated more rapidly than long waves), giving a steadily increasing degree of flexure to the vast ice sheets. Eventually the flexure causes the ice to break up into fragments which themselves break again nearer the edge until a distribution of floe sizes is established with the smallest floes in the steepest wave field at the extreme ice edge. The floes thus created, which constitute the MIZ, act as a shield for the interior pack, selectively damping out the shorter waves. If the icefield is an open one so that the floes do not collide, the wave attenuation process can best be described by a scattering model. As soon as the pack becomes closer (e.g. with an on-ice wind) floe collisions occur, generating high noise levels, while a very compact pack behaves hydrodynamically either as a collection of very large floes or as a single entity. Since the floes can no longer surge in response to the waves, energy attenuation may be occurring more significantly in the form of viscous losses from the boundary layer under the ice.

Further into the MIZ, where very large floes exist, as well as in the interior pack itself, wave energy propagates as flexural-gravity waves, with losses occurring either as reflections from leads and floe edges or from pressure ridges, or as creep losses due to flexure of the ice sheet. New wave energy can be created within such a continuous cover by a sufficiently strong wind, but the amplitudes involved are very small and can only be detected by sensitive instruments.

Beginning with MIZ attenuation, the scientific problem is to understand how waves propagate in such an icefield, how they are scattered and attenuated, how the waves cause the floe size distribution itself to become modified by flexural break-up, how waves contribute to ice margin dynamics, and how attenuation is balanced by wave generation when the icefield is diffuse. Included in this theme is the problem of ambient noise generation in the marginal ice zone due to wave-induced floe collisions, and the relationship between the noise spectrum, sea state and ice conditions.

Field observations of wave decay in MIZ regions have been carried out using a variety of techniques, including shipborne wave recorders (Robin, 1963), upward sonar from a submerged hovering submarine (Wadhams, 1972, 1978b), and airborne laser profilometry (Wadhams, 1975). More common has been the use of a directional wave recording buoy in open water and diffuse ice, or a combination of accelerometers, strainmeters and tiltmeters on ice floes which perform an equivalent function to a directional wave buoy. The most extensive measurements have been described by Wadhams *et al.* (1986, 1988). Recently the extraction of ocean wave spectra from satellite SAR (synthetic aperture radar) imagery has proved to be a fruitful method of detecting and measuring waves in marginal ice zones (Wadhams and Holt, 1991).

The main conclusions from the observations are as follows:

1. The attenuation of waves with distance into the pack takes a negative exponential form, with an attenuation coefficient which decreases with increasing wave period over most of the spectral range. In heavy compact ice (e.g. East Greenland) the energy attenuation coefficient typically varies from 2×10^{-4} m^{-1} for the longest swell to 8×10^{-4} m^{-1} for 8–9 s waves, corresponding to e-folding distances of 5–1.2 km.

2. There is some evidence of a "roll-over" at the shortest periods (less than 6–8 s), where the decay rate may actually start to diminish as the wave period shortens.

3. The directional spectrum increases in spread inside the icefield until it is essentially isotropic within a few km of the edge.

4. Some wave energy is reflected from the outer edge of the icefield, but only a few percent, even when the edge is compact.

6.3.2. Models of Wave Attenuation

Scattering models

A simple scattering model (described in full in Wadhams, 1986) was developed in the early 1970s to explain these findings. It treats each floe as an elastic floating raft, solving for velocity potentials at the leading and trailing edges of the raft. Within the raft itself energy propagates as a flexural-gravity wave with an altered dispersion relation. The treatment yields an energy reflection coefficient from which an attenuation rate can be derived. Floe diameter appears to be a more critical parameter than ice thickness. The model agrees well with experimental results at normal and long wave periods, but does not always predict a "roll-over" which is often observed at short periods. Other defects of the original model are that it does not predict the observed increase in directional spread as waves pass into the ice, and that a complete solution for the matching of velocity

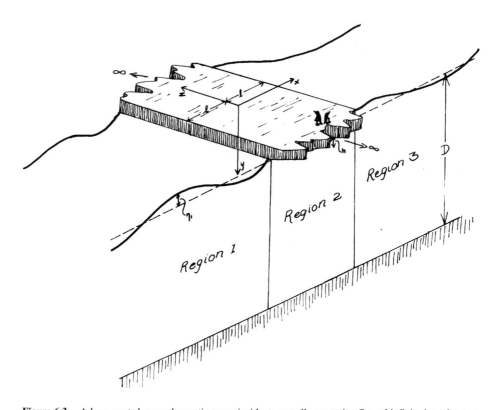

Figure 6.3. A long-crested monochromatic wave incident normally on an ice floe of infinite lateral extent.

potentials across the ice floe edge was not achieved. This has since been accomplished (Fox and Squire, 1990, 1991).

   The basis for the model is that wave scattering is due to a mismatch between the mode of propagation under an elastic raft and under an open water surface. In the simplest two-dimensional geometry (fig. 6.3), where a unidirectional wave of radian frequency ω is incident normally on the edge of a floe in water of depth D, the problem becomes one of calculating the reflection coefficient. The relevant equations for velocity potential ϕ_1 in the open water region in front of the floe are the familiar

$$\nabla^2 \phi_1 = 0 \qquad\qquad \text{(continuity)} \qquad\qquad (6.1)$$

$$\eta_1 = - g^{-1} [\partial \phi_1 / \partial t]_{y=0} \qquad\qquad \text{(linearised Bernoulli)} \qquad\qquad (6.2)$$

$$- \partial \eta_1 / \partial t = [\partial \phi_1 / \partial y]_{y=0} \qquad\qquad \text{(surface boundary condition)} \qquad\qquad (6.3)$$

$$\partial \phi_1 / \partial y = 0_{y=D} \qquad\qquad \text{(bottom boundary condition)} \qquad\qquad (6.4)$$

where η_1 is the instantaneous elevation of the water surface. These equations are satisfied by a ϕ_1 of form

$$\phi_1 = [Be^{ikx} + Re^{-ikx}]\, e^{-ky}\, e^{-i\omega t} \tag{6.5}$$

where $k = \omega^2/g$ if $D \gg 2\pi / k$, i.e. water depth is large compared to a wavelength (we assume that this is the case from here on). $|B|$ is the potential amplitude of the incident wave and $|R|$ of the wave reflected normally from the ice edge.

Under the ice, with similar notation and with $p(x,t)$ being the pressure just below the water-ice interface, we have

$$p(x,t) - L\, \partial^4\eta_2/\partial x^4 = \rho_i\, h\, \partial^2\eta_2/\partial t^2 \qquad \text{(equation of motion)} \tag{6.6}$$

$$p(x,t) = -\rho_w[g\, \eta_2 + \partial\phi_2/\partial t\, |_{y=0}] \qquad \text{(Bernoulli equation)} \tag{6.7}$$

$$-\partial\eta_2/\partial t = \partial\phi_2/\partial y\, |_{y=0} \qquad \text{(boundary condition)} \tag{6.8}$$

Here L is the flexural rigidity of the ice, given by

$$L = E\, h^3 / 12\, (1 - v^2) \tag{6.9}$$

where E is Young's modulus, h the thickness and v Poisson's ratio. A solution for ϕ_2 is a sum of potentials of form

$$\phi_2 = \sum_n [A_n\, e^{ik_n x} + B_n\, e^{-ik_n x}]\, e^{-k_n y}\, e^{-i\omega t} \tag{6.10}$$

A velocity potential of this form satisfies the boundary conditions provided

$$L\, k_n^5 + (\rho_w\, g - \rho_i\, h\, \omega^2)\, k_n - \rho_w\, \omega^2 = 0 \tag{6.11}$$

Equation (6.11) is basic to an understanding of wave interaction with sea ice in all its aspects. There are three physically feasible roots for k_n, of which the real root k_0 represents a conservative wave propagating through the ice sheet with a dispersion relation which differs from that of a wave in open water. This new type of wave, called a **flexural-gravity wave,** was first investigated theoretically by Greenhill (1887). Greenhill went on (1916) to propose the first practical application of this theory, to the problem of the waves created by skaters and the conditions under which these can cause fracture of thin ice.

A flexural-gravity wave of a given period has a wavelength which is greater than that of an open water wave of the same period, although the difference narrows for long swell (fig. 6.4). This implies that the longest swell does not "notice" the presence of ice and propagates through it unchanged. The dispersion relation of flexural-gravity waves (fig. 6.5) involves a minimum value of phase velocity c or group velocity U at periods which for typical ice thicknesses correspond to normal ocean wave or swell periods. In continuous ice sheets flexural-gravity waves can be generated by the moving pressure fluctuations of a wind blowing over the ice, but this can only occur if the wind velocity exceeds

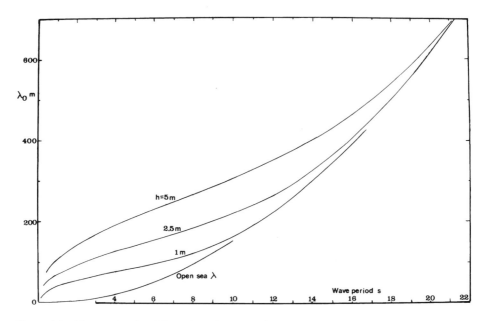

Figure 6.4. The wavelengths of flexural-gravity waves, for three ice thicknesses, compared with waves in the open sea.

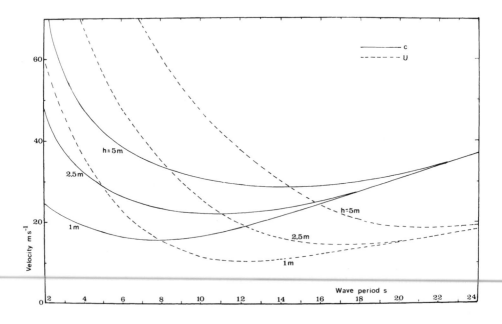

Figure 6.5. The dispersion relation for the phase (c) and group (U) velocities of flexural-gravity waves at three ice thicknesses.

the group velocity minimum (typically 10–20 m s^{-1} from fig. 6.5). Another source of such waves are moving vehicles, which can cause fracture if critical weights and speeds are exceeded, as in the case of sea ice runways. In a geophysical context such waves, if of long enough period to avoid creep losses (Wadhams, 1973b), can penetrate for thousands of km through ice-covered polar basins, once they have passed through the wave-filtering screen of the MIZ. The properties and role of flexural-gravity waves are discussed further in section 6.3.3, and in Wadhams (1986) and Squire et al. (1995, 1996).

In the present context of scattering by floes k_n defines a set of wave numbers which give the mismatch from open water propagation which is responsible for reflection at the leading edge of the floe. At the trailing edge of the floe ($x = + l$, fig. 6.3) a further mismatch occurs as the wave passes out again into open water. In region 3 the relevant velocity potential is

$$\phi_3 = T e^{ikx} e^{-ky} e^{-i\omega t} \tag{6.12}$$

so that $|T|/|B|$ is the amplitude transmission coefficient. If we set $|B|$ to unity then, with no dissipation,

$$|R|^2 + |T|^2 = 1 \tag{6.13}$$

and the following transition conditions between regions must be fulfilled at all depths:-

$$\partial^2 \eta_2 / \partial x^2 = \partial^3 \eta_2 / \partial x^3 = 0, \ x = \pm l \tag{6.14}$$

$$\phi_1 = \phi_2, \ \partial \phi_1 / \partial x = \partial \phi_2 / \partial x, \ x = - l \tag{6.15}$$

$$\phi_2 = \phi_3, \ \partial \phi_2 / \partial x = \partial \phi_3 / \partial x, \ x = l \tag{6.16}$$

From examination of the y-dependence of the forms for ϕ_n given in (6.5), (6.11) and (6.12), it is clear that further types of velocity potential must exist in order for matching to occur. An approximate solution was proposed by Hendrickson et al. (1962), which was corrected by Wadhams (1986), involving fitting potentials only at the sea surface where energy is greatest. An improvement was proposed by Wadhams (1973), in which fitting was accomplished at the surface and at a depth of $\lambda/4$, using a new set of potentials originally proposed by Ursell (1947), of form

$$\phi_a = C e^{-Kx} (K \cos Ky - k \sin Ky) e^{-i\omega t} \tag{6.17}$$

where K takes any real value. A complete solution has been derived by Fox and Squire (1990, 1991).

This simple reflection model produces an exponential decay rate with distance. If floes of diameter d ($= 2l$) in the direction of the wave vector occupy a fraction p of the sea surface, and each has an energy reflection coefficient r ($= |R|^2$) computed from the above equations, then we have an energy decay rate of form

$$- \partial E / \partial x = 2 \alpha_x E \tag{6.18}$$

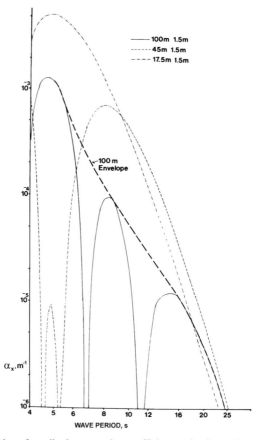

Figure 6.6. Computation of amplitude attenuation coefficient α_x for floes of different diameters but same thickness. The "envelope" shown for the 100 m case is the expected result for a real icefield made up of floes of diameters close to, but not exactly, 100 m.

where

$$\alpha_x = p\,r\,/\,2\,d + o(r^2) \qquad (6.19)$$

α_x is an **amplitude attenuation coefficient**.

The model easily copes with a distribution of floe sizes, such that floes of diameter d_i occupy a fraction p_i of the sea surface, since

$$\alpha_x = \Sigma_i\,(p_i\,r_i\,/\,2\,d_i) + o(r_i^2) \qquad (6.20)$$

It is therefore easy to test the model against observed data, so long as the experimental set of measurements of wave energy as a function of penetration into the icefield also includes a swath of aerial photographs or other imagery which gives the distribution of $(d_i,\,p_i)$ with each increment of penetration. This has been possible in most experiments (e.g. Wadhams *et al.*, 1986, 1988).

For a single floe diameter d, the model in its simplest form (surface fitting only) gives an α_x which generally decreases with increasing wave period, except for a number of resonant periods where α_x goes to a very low value (fig. 6.6). These resonant periods are critically dependent on d, therefore in a real icefield with a variety of floe diameters present, the resonances do not yield anomalously low attenuation rates at moderate wave periods, but rather are smeared out in an averaging process which produces an overall decrease of α_x with increasing wave period except at the very shortest periods. Here the model can give a fall-off of decay rate, a phenomenon which is in fact often observed in experimental data. In the case of the 100 m floe shown in the diagram, the envelope of the curve most nearly represents the attenuation in an icefield made up of floes that are all roughly, but not exactly, 100 m in diameter. Note that the attenuation of the envelope decreases with increasing wave period, except at periods less than 5 s when a "roll-over" occurs. As mentioned earlier, we expect lower attenuation rates at longer periods because the flexural-gravity wave velocity is matched more closely to the velocity of the wave in the open sea.

It has been found that the scattering model, especially in a modified form in which account is taken of multiple forward- and backward-scattering of wave vectors, gives a good fit to most of the wave measurements done in ice where floe size distributions were measured concurrently (Wadhams et al., 1988). An example is fig. 6.7, where the model is compared with data obtained from experiments in the MIZs of the Bering and Greenland Seas.

The model is inadequate, however, in the following ways:

1. Although for a single floe size the model predicts a "roll-over" at the shortest wave periods, this often disappears in the averaging process in which the effects of many floe sizes are added.
2. The model takes no account of directional spread. It was suggested by Wadhams (1978b) that the directional spread of an incident spectrum should become narrower within an icefield and more concentrated along a bearing normal to the ice edge. This is because at any distance x from the ice edge, wave components with an angle of incidence θ will have travelled a distance (x sec θ) through ice and will therefore have suffered an attenuation which increases with θ. At deep penetrations we expect an almost unidirectional spectrum propagating at right angles to the mean orientation of the local ice edge. The findings of the first directional wave measurements in ice (Wadhams et al., 1986) were that the directional spectrum widens inside the ice until it becomes essentially isotropic. This was especially true of short period waves: a 3.3-s component became isotropic within 1.2 km, 4.1 km and 0.7 km of the ice edge in three separate experiments, while the swell component took up to 17.8 km to become isotropic and did become narrower in the earliest stages. We can conclude that the effect originally proposed competes with a lateral scattering effect, whereby energy lost from the forward-going wave vector is not only backscattered but also scattered with a directional spread, so that a floe in the far interior is bathed in scattered radiation from surrounding floes, with a weakened forward-going vector. A mathematical model of two-dimensional scattering from an ice floe, which can describe the magnitude of this spread, is still required.

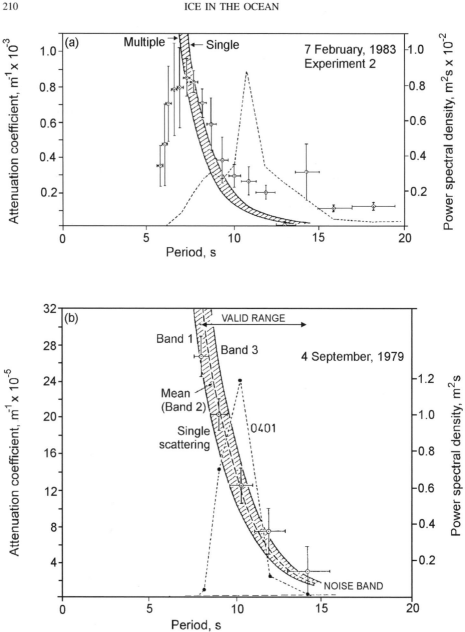

Figure 6.7. Comparison of observed attenuation rates in the MIZ of the Bering Sea (top) and Greenland Sea (bottom) with predictions from the scattering model. The dotted line is the energy spectrum (right hand scale). The data points with error bars are the measured attenuation coefficients (left hand scale). The shaded band is the range of predictions of the model.

3. When the model is used to predict wave *reflection* from the front of an icefield it is
    assumed that this reflection comes from the first row of floes encountered by the wave.
    Usually the model predicts reflection coefficients which are greater than the 2.6–
    12.7% energy reflection coefficients observed in the experiments of Wadhams *et al.*
    (1986).

The model does not predict any overall change in the dispersion relation, since it
considers only the perturbations to the forward-going wave energy vector produced by
individual floes. The implication is that in an icefield of small to moderate floes wave
energy propagates with the same phase and group velocities as in the open ocean. Support
for this idea came from the Wadhams *et al.* (1986) experiments, where no refraction was
observed across the ice edge, but more recent observations from satellite SAR images
suggest refraction. It is an open question at present whether this refraction is due to the
ice type being other than that of separate moderate floes (e.g. frazil or pancake ice, as
described below; or very large floes or floes compacted together, where flexural-gravity
dispersion might be expected to occur), or whether the scattering model can be considered
inadequate in this area too.

Viscous models

As an icefield becomes more compact the interactions between floes increase and it is
often not realistic to consider the icefield as being composed of individual floes each of
which is free to respond hydrodynamically to the incident wave. Instead floes collide,
or are held together by the stress of an on-ice wind or by freezing of brash, pancake or
frazil ice between the thicker floes. Such an icefield is approaching the condition of being
a single entity, yet does not possess the bulk property of being a uniform elastic sheet
such that flexural-gravity wave theory can be applied to propagation in it.

Under these circumstances it is tempting to ignore the detailed physics of the many
kinds of energy-consuming ice-water and ice-ice interaction processes that are occurring,
and instead model the ice cover as some kind of material with empirically-determined
properties, which covers the sea surface and attenuates waves.

The simplest such material is a collection of non-interacting mass points, producing
a load on the sea surface but possessing no strength properties of any kind. The theory
of wave propagation in this material was developed in the 1950s, and recent data suggest
that it may be applicable to frazil and pancake ice types, i.e. circumstances where the
floe diameter is negligible compared to a wavelength. This treatment is reviewed later
in this section.

A model by Weber (1987) introduced the idea of the ice cover as a viscous fluid. Liu
and Mollo-Christensen (1988) described a more physically realistic model for wave decay
which assumes that attenuation is due to the viscous boundary layer under ice. In a highly
compact ice cover it is assumed that the waves obey the dispersion relation of flexural-
gravity waves, and that the ice floes are held together by wind stress or partial freezing
of the interstitial material such that they cannot respond by surging in phase with the wave.
Thus an oscillating boundary layer develops under the ice which causes an energy loss.
Liu *et al.*(1991) developed an equation for the wave energy decay rate due to viscous
loss in this boundary layer which employed an eddy viscosity v. They state that this is

not a physical but rather a phenomenological parameter, which can only be determined as a function of the actual flow conditions. That is, it is a free parameter which is derived by matching the observed wave decay rate curve as a function of frequency with the decay rate curve predicted by the model.

The attenuation model does predict a rollover at short wave periods whereby the normal increase in wave decay rate with increasing frequency slows or even reverses. The predictions can be made to agree with the field data of Wadhams *et al.* (1988) and SAR data from the Labrador Ice Margin Experiment (LIMEX). The point at which rollover occurs depends on ice conditions, and especially on ice thickness (Liu *et al.*, 1991). The tuning parameter, the turbulent eddy viscosity which is used to measure the mixing level in the turbulent boundary layer, was found to take a wide variety of values, however. Liu *et al.* (1992) found that the two experiments in LIMEX, in 1987 and 1989, required values of 290 and 12 $cm^2 s^{-1}$ respectively to provide a good fit, while to fit data reported by Wadhams *et al.* (1988) required values ranging from 4 to 1536 $cm^2 s^{-1}$. It is clearly undesirable that a tuning parameter should take such a wide range of values, since it makes prediction of wave decay rates a very difficult business. To overcome this problem, Liu *et al.* (1992) proposed a parameterisation for v, using dimensional analysis. They suggested that v should be a linear function of (A h a / T), where A is ice concentration, h is ice thickness, a is significant wave height and T is wave period. They found a reasonable linear relationship between this parameter and the eddy viscosity which best fits the field data. However, much more development of this model is needed before it can be used for the prediction of decay rates. For instance, on this parameterisation the eddy viscosity — hence the wave decay rate — should decrease with increasing distance into the ice, since a is decreasing.

A more conventional treatment of viscous losses due to waves under ice was carried out by Wadhams (1973). He estimated energy dissipation in the boundary layer under ice by assuming realistic skin-friction and form drag coefficients and the existence of a quasi-static boundary layer, i.e. one which at each moment of the wave cycle resembles the static boundary layer which would have developed under a free stream velocity equal to the horizontal component of the wave orbital velocity. This gives an upper limit for energy losses. Even under a variety of assumptions about possible ice roughnesses, the wave amplitude decay coefficient due to viscous losses was always at least an order of magnitude lower than the decay coefficient due to the scattering model. He concluded that viscous losses are not significant as compared to scattering losses. The apparent disagreement between these results and the model of Liu *et al.* can be resolved if we consider that Liu *et al.* are forcing all forms of energy loss into the bin of viscous loss, thus requiring a high "eddy viscosity" to produce sufficient loss to match observations.

Mass loading models — waves in frazil and pancake ice

Mass loading models were the first type of model to be developed to describe wave-ice interaction. At that time (1950–3) no data existed to test the model. Ice is visualised as being a continuum composed of non-interacting mass points which exerts a pressure upon the water surface but which has no coherence or rheological properties as a material. The relevant equations for propagation of waves in the ice are thus those developed above for the wave scattering model (e.g. eqn. (6.11)) but setting $L = 0$. The theory was developed by Peters (1950), Weitz and Keller (1950), Keller and Weitz (1953) and Shapiro and Simpson (1953). It has been summarised by Wadhams (1986), who pointed out an error in the treatment by Shapiro and Simpson. This error was inadvertently repeated in a paper by Wadhams and Holt (1991). The originators of the theory envisaged it as appropriate for all kinds of icefield, but Wadhams and Holt (1991) made the case that it could be considered appropriate only for describing wave propagation in frazil and pancake ice, since these ice types have element sizes (individual cakes or ice crystals) which are negligible compared to a wavelength.

Following Wadhams (1986) we find that the wave number k_i for propagation in ice is given by

$$k_i = \rho_w \omega^2 / [\rho_w g - \rho_i h c \omega^2] \qquad (6.21)$$

where h is the thickness of the frazil slurry and c is the average ice concentration within it. This implies that a wave propagating into the ice from open water will acquire a reduced wavelength $\lambda_i = 2\pi / k_i$. This has the following implications:

1. Waves incident obliquely on the ice edge are refracted towards the normal, according to Snell's Law;
2. Since the group velocity inside the ice is also lower than in the open water, a wave entering the ice acquires a greater amplitude. This, combined with the reduced wavelength, implies a greater wave steepness;
3. There is a frequency limit implied in (6.21), given by

$$\omega_c^2 = \rho_w g / \rho_i c h \qquad (6.22)$$

Above this frequency propagation is not possible. A wave incident normally on the ice at a frequency higher than ω_c will suffer total reflection at the ice edge. Wadhams and Holt (1991) pointed out an implication for frazil ice imaging on SAR. Waves at the Bragg wavelength (30 cm for Seasat SAR considered in the paper) have frequencies only just below ω_c, so that they suffer extreme modification on passing into the ice, with a high reflection coefficient at the ice edge. There is therefore little energy present in a frazil ice slurry at these wavelengths and so frazil ice appears dark on SAR imagery. The same argument applies to pancake ice, but in this case the physical roughness of the upturned edges of the pancakes will make this type of ice appear bright to SAR.

Reflection and transmission coefficients across the ice edge were derived by Keller and Weitz (1953), and the correct forms are

$$(1 - |R|^2) = |T|^2 k^2 / k_i^2 \qquad (6.23)$$

giving an energy transmission coefficient t for normal transmission of

$$t = |T|^2 U_i / U_w = 4 k k_i / (k + k_i)^2 \qquad (6.24)$$

Figure 6.8 shows a set of reflection coefficients for normal incidence, illustrating how steeply the reflection rises as the critical period is approached.

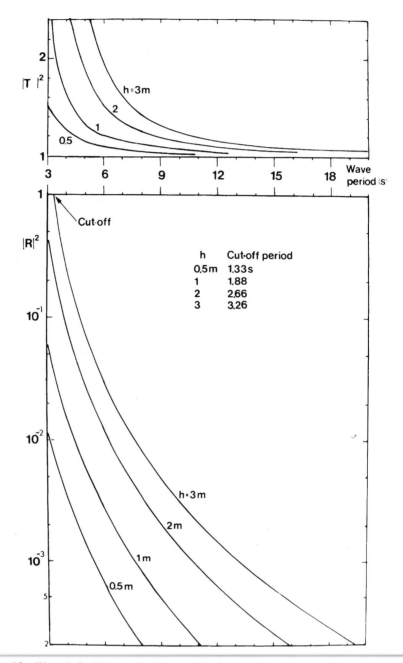

Figure 6.8. Waves in frazil ice: predicted energy reflection and transmission coefficients for waves entering a frazil icefield.

Wadhams and Holt showed that the dispersion relation given by (6.21) could explain wave refraction in frazil and pancake ice observed on Seasat SAR in the Chukchi Sea, yielding realistic (although somewhat large) values for frazil slick thickness.

Thus this model may well be taken as suitable for describing wave propagation in frazil and pancake ice. It does not yield any progressive wave decay, although it does imply wave refraction at the ice edge and the total exclusion of the shortest waves from the icefield. Clearly energy decay takes place in frazil slurries, and Martin and Kauffman (1981) observed an internal circulation within the slick in tank experiments, which would lead to viscous energy losses. Further research on this mechanism has been carried out by Wadhams *et al.* (1998), using ERS-2 SAR to observe wave refraction in pancake ice in the Greenland Sea Odden, and by Newyear and Martin (1997) in tank experiments. Results are contradictory: the SAR observations showed wavelength reductions that were greater than predicted by the model, i.e. the model when applied to the results gave ice thickness that were *greater* than physically possible values, whereas the tank experiments, at much higher frequencies, gave wavelength increases, which fitted the predictions of a viscous energy loss model developed by Keller (1997). This suggests that viscous losses may dominate the propagation at high frequencies, while at low frequencies the above model, with some modification, may be realistic.

6.3.3. Wave Decay and Floe Break-up

Thus far we have considered waves entering an icefield with a specified distribution of floe diameters. In practice there is a dynamic non-equilibrium state in the MIZ in which the incident sea state is constantly changing while random ice motion is constantly bringing large floes within range of the ice edge. This results in a continuous break-up of large floes near the ice edge into smaller floes, which then alter the attenuation rate of the waves. There is thus a tendency (never completed) towards an equilibrium situation in which the floe size at any distance into the ice does not exceed a critical value above which flexural failure would occur. Thus it is important to understand the failure criterion and flexural response of a floe which is bending in a wave field. The general problem was first addressed by Goodman *et al.* (1980) and the theory has been applied also to the breakup of tabular icebergs (Wadhams *et al.*, 1983). Lensu (1997) has described a formal model which describes mathematically how a floe size distribution evolves with time, given a specified break-up rate.

A number of mechanisms have been proposed for the way in which very large floes are broken up by waves as they approach the ice edge. The simplest approach is to consider standing waves. We have already shown (eqns. (6.12)–(6.16)) that a flexural-gravity wave incident on the trailing edge of a large floe suffers partial reflection back into the floe. The incident and reflected waves within the floe together create a standing wave, at the antinodes of which the amplitude of flexure is greater than that due to the incident flexural-gravity wave alone. Then if the incident flexural-gravity wave has an amplitude which approaches that needed to cause failure by flexure, a failure will occur as a series of parallel cracks spaced half a wavelength apart, the wavelength being defined by the wave number in the ice, i.e. k_0 in (6.11). This mechanism was discussed by Assur (1963) and Milne (1972).

A second mechanism is wind-induced tilt (Weber and Erdelyi, 1976). A wind blowing over a large ice sheet exerts a drag over the entire surface, and the combined stress causes the downwind end of the sheet to tilt downwards. Bending stresses due to the tilt can be relieved by periodic cracking of the ice. The mechanism appeared to explain tilt observations made on ice in the Beaufort Sea.

The mechanism discussed by Goodman et al. (1980) and, in more detail, by Wadhams (1986), addresses the more general and complex problem of a freely-floating large ice floe encountering a wave field. We have dealt so far with the way in which such a floe cause waves to be scattered. However, in considering the inverse problem of the effect of the waves on the floe, we have to consider that the floe has a rigid body response involving six degrees of freedom (the three rotations of pitch, roll and yaw, and three bodily translations of surge, heave and sway). These responses help to relieve the pressure produced by the wave on the floe underside, so that the flexural response of the floe is less than that of an infinite sheet of the same thickness. In general, the smaller the floe, the more it moves as a rigid body in phase with the wave, and the less it responds by flexure. A numerical treatment of this problem was offered by Squire (1983), but we here present a simplified analytical approach which brings out the main aspects of the physics.

The assumptions made are as follows:

1. The incident wave is sinusoidal, long-crested and of small amplitude.
2. Flow is irrotational and inviscid.
3. The added mass of the floe is zero. Added mass is a concept used in hydrodynamics to account for the way in which a floating body appears to have an increased mass when it is undergoing motions of translation, and a greater moment of inertia when undergoing rotations, because it is carrying some of the nearby water along with it.
4. The damping coefficient of the floe is zero. This is a damping to the floe motion due to its production of a train of progressive outgoing waves by such responses as heave and tilt.
5. The effect of wave diffraction by the body is ignored (the Froude-Krylov hypothesis).
6. The rigid body motion and the flexural response can be decoupled. We can calculate the oscillating pressure field by assuming that the body remains rigid, then use the pressure field to calculate the flexural response.
7. The effect of the floe's vertical sides can be ignored (this becomes a poor assumption when dealing with ice islands and icebergs, see chapter 7).

Using our earlier notation, we first apply Bernoulli's equation (6.7) by assuming that the pressure amplitude on the floe underside is due to the amplitude A of water particle motion, i.e. the whole wave. Hence

$$p(x,t) = - \rho_w \, g \, \eta_2 + A \, \rho_w \, g \, Re[e^{i(k \, x - \omega t)}] \qquad (6.25)$$

To allow for ice draft, A is the wave amplitude at the bottom of the floe rather than (A_0) at the water surface, where

$$A = A_0 \exp (- k \, h \, \rho_i / \rho_w) \qquad (6.26)$$

This is more important in dealing with icebergs (chapter 7).

The dynamic equation for a homogeneous elastic beam (Nevel, 1970) then yields

$$L \ (d^4X/dx^4) + (\rho_w \ g - \rho_i \ h \ \omega^2) \ X = A \ \rho_w \ g \ e^{i \ k \ x} \qquad (6.27)$$

where L is given by (6.9), and X is the profile of *elastic* flexure of the floe. This can be solved using the boundary conditions (6.14)–(6.16), that there can be no bending moment or shear at the ends of the floe. The solution for the strain ε(x,t) at the centre of the floe, i.e. at x = 0 in the geometry of fig. 6.3, is

$$\varepsilon(0,t) = - \ h \ [\partial^2h_2/\partial x^2] \ / \ 2 \ | \ _{x \ = \ 0}$$

$$= A \ \rho_w \ g \ k^2 h \cos \omega t[\Delta - \Omega \cos kl - (k/k') \sin kl \sin k'l \sinh k'l] \ / \ 2 \ L\Delta \ (k^4 + 4 \ k'^4)(6.28)$$

where

$$k' = [(\rho_w \ g - \rho_i \ h \ \omega^2) \ / \ 4 \ L]^{1/4} \qquad (6.29)$$

$$\Delta = (\sin 2 \ k \ l + \sinh 2 \ k \ l) \ / \ 2 \qquad (6.30)$$

$$\Omega = \sin \ k \ l \cosh \ k \ l + \cos \ k'l \ \sinh \ k'l \qquad (6.31)$$

Figure 6.9 shows the ratio of ε(0,t) to (A$_0$ cos ωt), i.e. the ratio of surface strain at the centre of the floe to the amplitude of the incident ocean wave, for a floe of 3 m thickness and two wave periods, 12 s and 17 s, all computed from (6.28). The ratio is plotted as a function of floe diameter. The first thing that we notice is that, as expected, the strain is very small for small floes. In the case of 17 s swell it rises to a maximum value attained at about 200 m floe diameter, and so incoming swell of this period, if of sufficient amplitude, will fracture the ice but we cannot predict the diameter of the resulting floes, except that it will be more than 200 m. For shorter waves, however, e.g. the 12 s example, there is an initial peak (B) which exceeds the stable maximum strain value. This implies that a wave of sufficient amplitude will initially break up a large ice floe into smaller floes of diameter equal to that of the peak B, i.e. about 220 m. This is a realistic value which matches observations of floe sizes in the inner part of the MIZ.

In order to see what these results mean in terms of the actual wave amplitude required to produce failure, we need some knowledge of the failure strain in sea ice. We note (Wadhams, 1979, 1986) that one case of multi-year floe breakup occurred while the author was actually monitoring the floe concerned with a strainmeter, giving a failure strain of 3×10^{-5}. The theoretical framework through which failure strains are now estimated is that of **fracture mechanics**. A full discussion of fracture mechanics as applied to sea ice goes beyond the scope of this book, but is covered fully in Sanderson (1988; see also Goodman, 1980). In simple terms, failure occurs when a crack of critical length propagates in the ice. When an ice floe is being stressed, the intensity of the stress around the tip of an existing crack is magnified by a factor proportional to the square root of the crack length. The longest existing crack is therefore the one that will yield first, and will propagate, increasing in length (and thus further increasing the stress at its tip) until it causes the entire ice sheet to fracture at that point. The concentrating effect of the stress is reduced if, as in the case of the underside of sea ice, cracks are water-filled.

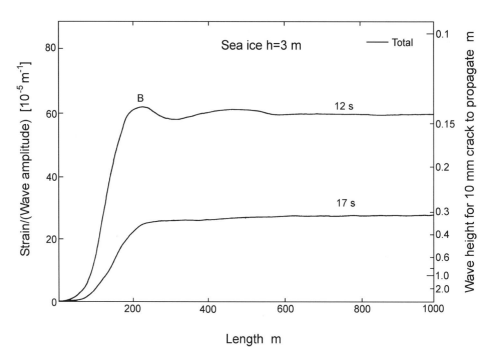

Figure 6.9. Theoretical predictions of the flexural response of a 3 m – thick ice floe to waves of period 12 s and 17 s, as a function of floe diameter. The left hand ordinate shows the ratio of surface strain at the floe centre to incident wave amplitude. The right hand ordinate shows the wave height required to allow a 10 mm-long crack at the floe centre to propagate.

It was shown in Goodman *et al.* (1980) and Wadhams (1986) that if plausible assumptions are made about the stress reduction due to water filling of cracks, and that the critical crack in an ice sheet is assumed to be 10 mm long, the failure strain becomes 4.3×10^{-5}, quite close to the single observation quoted above. In figure 6.9 we have added, as the right-hand ordinate, the wave height required to cause such a 10 mm crack to propagate, i.e. the wave height ($2 \times$ amplitude) that will produce failure in the ice if this is indeed the maximum size of pre-existing cracks.

On this basis, we can now see how well fig. 6.9 represents what is really observed in the MIZ. Consider the 12 s wave, very typical of peak wave periods in, say, the Greenland Sea. The figure shows that with a height of 14 cm this wave will start to fracture ice, typical of the situation well inside the MIZ, where much attenuation has occurred as per the model of (6.19). When the wave height reaches this value it will break large floes into floes of 220 m diameter, the critical value B. As the wave height becomes greater (i.e. as we move closer to the ice edge), the maximum surviving floe size moves down the left-hand side of the curve between 0 and B. At a wave height of 30 cm floes will have suffered further fracture down to 100 m diameter, while at 2 m wave height the maximum floe size will be 60 m. These values agree very well with what is seen in East Greenland, and the shape of the curve in fig. 6.9 demonstrates the way in which increasing wave height causes further break-up into smaller floes, giving the floe size gradation observed as one moves into the MIZ.

The results can also be used to estimate the effective width of the MIZ. If we use 12 s as the wave period, and a typical observed value of 0.8×10^{-4} m^{-1} for the amplitude attenuation coefficient, then we find that a wave of height 5 m at the ice edge (a typical rough sea) will cause break-up to a penetration of 45 km. Interestingly enough, the longer swell of 17 s period, despite requiring a higher amplitude to cause initial breakup (31 cm instead of 14 cm, fig. 6.9), decays more slowly as it penetrates the MIZ, with an attenuation coefficent of only about 0.4×10^{-4} m^{-1}. This implies that a swell of 17 s period and initial amplitude 5 m can break up ice as far as 70 km into the MIZ. The width of the MIZ therefore depends critically on the shape of the incident wave spectrum — it is the dominant wave *period* as well as incident wave *height* that determines the penetration at which fracture first occurs. It is clear that the results from these calculations are in good agreement with the observed widths of marginal ice zones facing an open sea.

6.4. ICE EDGE BANDS

When an off-ice wind blows across an ice margin, the MIZ does not just become more diffuse, with the floes moving out into the open sea. Instead, the outermost part of the icefield organises itself into a regular series of ice edge bands. These are typically 1 km wide (in the direction of the wind), 10 km long (perpendicular to the wind) and contain a dense ice cover, with open water in between the bands. A ship moving towards such an ice edge from the open sea will encounter a band of this type as the first outlier of the pack ice. The band as a whole is being moved towards the open sea and thus warmer water, and so the ice in the bands is usually melting. The melt rate limits the number of bands that can exist and the distance that they can move. Examples of ice edge bands are shown in Muench and Charnell (1977), Bauer and Martin (1980) and Martin *et al.* (1983). Fig. 6.10 shows some typical ice bands, and also shows another feature of the band structure, that the downwind edge is clean and sharp while the upwind edge often looks ragged.

Why does the ice edge organise itself in this curious way, and what determines the size and spacing of the bands? A number of ideas have been proposed, involving Langmuir circulation (Muench *et al.*, 1983), wind waves (Martin *et al.*, 1983), ice-water coupling (Häkkinen, 1986) and instabilities due to the horizontal temperature gradient between the ice interior and the open water zone (Chu, 1987). The author favours his own theory (Wadhams, 1983) which we describe here.

The mechanism depends on a phenomenon known as **wave radiation pressure**. When an ocean wave is incident on a floating body, and part of it is reflected by that body, there is a net force exerted on the body, which is given for deep water waves by (Longuet-Higgins, 1977):

$$F_r = \rho_w \, g \, (a^2 + a'^2 - b^2) / 4 \qquad (6.32)$$

Here F_r is the forward force exerted by the wave per unit width of wavefront, and a, a'

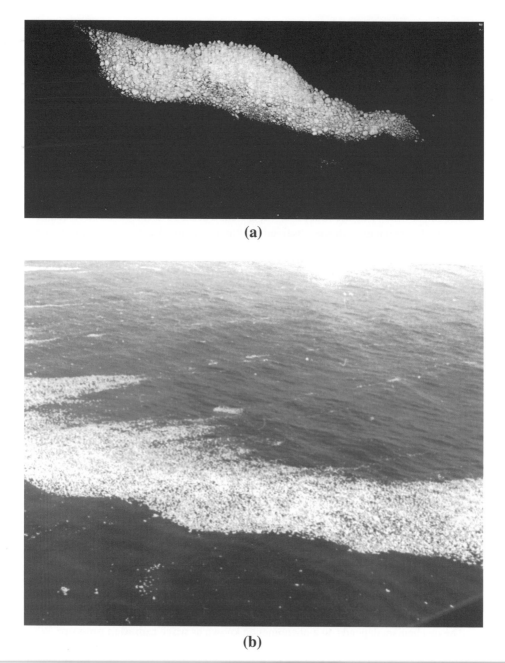

(a)

(b)

(c)

Figure 6.10.   Views of ice edge bands. (a) Aerial view of a band in the Bering Sea in March. Length 525 m, width 100 m, typical floe diameter 10 m, wind of 7 m s^{-1} perpendicular to long axis of band. (b) Close-up of band of small (1–2 m diameter) ice cakes from same region. Nearside of band is seaward side, with sharp edge and long waves incident from ocean; far side of band is iceward side, showing ragged edge and shorter waves from local wind sea coming off-ice. (c) Band systems visible on satellite SAR image.

and b are the amplitudes of the incident, reflected and transmitted waves respectively. Thus any floating object which is large enough to reflect part of the wave energy incident on it will be pushed downwave by this pressure. The magnitude of this pressure is usually less than the direct force of the wind on the object, and since the wind is generating the waves (except in the case of swell), the wave radiation pressure usually acts to slightly augment the direct wind-induced drift.

Let us now consider a single ice floe of diameter d. If its amplitude reflection coefficient to waves of a given period is R (= a′ / a), and we assume zero absorption of wave energy, then (6.32) gives, for the total wave force on the floe edge,

$$F_r \, d = \rho_w \, g \, a^2 \, R^2 \, d \, / \, 2 \qquad\qquad (6.33)$$

As an example, for a floe of 20 m diameter with perfect reflection and a 0.1 m wave height, the wave force is 1000 N. This is a substantial force in relation to wind stress, but as (6.33) shows, the force is very dependent on R. Now, as we have shown in section 6.3, R is high when the wavelength is short in relation to the floe diameter. Such short waves do not dominate the open ocean spectrum, or the spectrum of waves entering the MIZ from the ocean, but they *do* dominate the local chop that is created when a strong wind blows across a few tens of metres, or a few hundred metres, of water.

Here lies the clue as to how the wave radiation pressure creates bands. The sequence is as follows:

1. We begin with a "normal" MIZ such as in fig. 6.1(a) or (b), where wave action from the open ocean has broken up the waves into floes, with the smallest diameters nearest the ice edge. Initially the ice edge may be compact due to a moderate on-ice wind or just to the wave radiation pressure of the incoming waves and swell, a small force at the edge itself because of the long wavelengths involved.

2. A strong off-ice wind begins to blow. All floes are blown seaward, but the small floes at the extreme edge initially move faster than the larger floes at deeper penetration because of their relatively larger wind drag coefficients (see chapter 4), so initially the MIZ dilates, opening up water spaces of random sizes and shapes between the floes.

3. The wind creates a chop in these random polynyas, with the fetch-limited wave height being greater for the longer open water stretches, but the wavelength always being small — less than the diameters of the floes. An example is shown in fig. 6.10(b).

4. A sorting process now happens. The floes which are shielded from these waves by other floes continue to move only under wind stress. However, those exposed floes which have open water directly upwind are moved faster because of the extra radiation pressure of the short wind-generated waves (eqn. 6.33). They therefore move faster downwind until they bump into the next downwind floe, to which they attach themselves. They start to "herd" less favoured floes downwind, collecting up stray floes as they go. The faster motion of these wave-driven floes makes the polynya upwind of them even larger, increasing the wave force further; however, the velocity differential decreases as soon as the floe starts to push against others, as the same wave force is pushing a larger load.

5. These clumps of wave-driven floes will tend to align themselves into a band with its long axis lying cross-wind, since the forcing is downwind and the length of a collection of floes in that direction is self-limiting while there is no limit on lateral extension.

6. Once formed, a band acquires stability because of the wave force on its upwind side (which may now be distributed throughout the band as a compressive stress, once R becomes less than unity so that short waves can penetrate into the band), balanced by an opposing stress due to longer ocean waves and swell on the downwind side. Fig. 6.11 shows this final geometry of a single stable band, generated by a "starter"

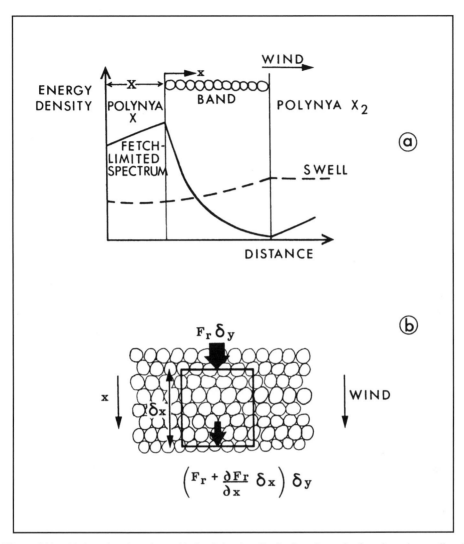

Figure 6.11. (a) A section through a stable band showing distribution of energies from incoming swell and locally-generated wind waves. (b) Geometry of internal compressive stress within band due to exponential decay of a wave.

polynya of width X and with a further "stopper" polynya of width X_2 downwind of it. Assuming that R may be less than 1, fig. 6.11(a) shows how wave and swell energy are distributed across the band, while fig. 6.11(b) shows how the case of $R < 1$ translates into an internal compressive stress. We can easily derive the magnitude of the internal compressive stress, since (6.18) and (6.19) have shown us that an initial wave amplitude a_0 diminishes in a distance x to a_x where

$$a_x = a_0 \exp(-\alpha x) \tag{6.34}$$

with

$$\alpha = p\,R^2 / 2\,d \qquad\qquad (6.35)$$

p is the fractional ice cover within the band and d the floe diameter. The compressive force per unit wavefront acting on every interior element of the icefield can be seen from fig. 6.11(b) to be $(-\partial F_r / \partial x)$, which is given, from (6.33)–(6.35), by

$$-\partial F_r / \partial x = \rho_w\,g\,p\,R^4\,a_x^2 / 2\,d \qquad\qquad (6.36)$$

Typical values might be R = 0.5, p = 1, d = 20 m, a_x = 0.1 m, giving a force of 0.2 N m^{-1}.

Summarising the above, we can say the *force originating the band* is off-ice wind drag on rough floes in the edge zone, making them move seaward slightly faster than the remainder of the pack, which dilates the MIZ and opens random polynyas. The *force consolidating the band* is the radiation pressure on floes at the downwind ends of larger polynyas due to the fetch-limited wind-wave spectrum generated in the polynya. The *forces maintaining the band* are compressive forces due to the exponential decay within the ice of the waves passing into the band from windward and leeward. The *force separating the band from the pack* is fetch-limited wave radiation pressure plus excess wind drag minus swell radiation pressure.

It is of interest to pursue this mechanism quantitatively to see if it predicts realistic band widths and spacings, and also to see if it can account for multiple bands. Firstly, let us consider whether "starter polynyas" of sufficient size can be created. Figure 6.12 is a random pattern representing an icefield dilated to 50% concentration. It is a 50×50 array representing 1 km^2 of an icefield of 20 m floes. Within just this one sample of icefield there is at least one polynya of 200 m fetch, so it is clear that randomness will throw up sufficiently large polynyas to permit wave growth. Secondly, let us consider what wave heights and periods can be produced locally. If we extend the model of fig. 6.12 to, say, a 50×100 km icefield, the largest randomly achieved fetch comes out to 460 m, so we may take 400 m as an initial X readily achievable in an MIZ. With a wind U_a of, say, 10 m s^{-1} we can estimate the dominant wave period and significant wave amplitude for a 400 m fetch from a standard wave growth model such as that of Hasselmann *et al.* (1973), obtaining 1.0 s and 0.06 m respectively. Figure 6.13, derived using the scattering model of section 6.3, shows that for a floe diameter of 20 m and various ice thicknesses, R is always close to 1 at such a low wave period. We therefore have a maximum wave force, exerted entirely on the floe or band edge first encountered. Longer waves coming in the opposite direction from the open sea will have a low R and will develop an internal compressive stress.

Next we estimate the velocities achieved by floes under wind stress alone and under wind stress plus wave radiation pressure. From chapter 4, if we equate the wind stress and water drag on an isolated floe in order to estimate a drift velocity, neglecting all other forces including Coriolis force, we obtain

$$U_i^2 = U_a^2\,\rho_a\,C_a / (\rho_w\,C_w) \qquad\qquad (6.37)$$

Figure 6.12. A random 50×50 square array with 50% occupancy, simulating 20 m floes in a 1×1 km icefield at 50% concentration, with random polynyas between them.

where U_i is the equilibrium ice drift velocity, U_a is wind velocity, ρ_a and ρ_w are air and water densities, and C_a and C_w are air and water drag coefficients. It has been found that C_a is significantly greater in the MIZ than in the central polar pack, because of the influence of raised floe edges, for instance 3.1×10^{-3} (Smith *et al.*, 1970) as compared to $1.4 - 2.1 \times 10^{-3}$ (Banke *et al.*, 1976, 1980). C_w has not been adequately measured in the MIZ; in the pack it takes typical values of 3.4×10^{-3} in winter (McPhee and Smith, 1975) and 5.5×10^{-3} in summer (McPhee, 1980). We may assume, however, that it does not increase in the MIZ as much as C_a does, because of the empirical observation by Bauer and Martin (1980) that MIZ floes start to move seaward under wind action more rapidly than interior floes. We could assume a value of, say, 7×10^{-3}. Under these circumstances, with our "standard case" of $U_a = 10$ m s^{-1}, (6.37) gives $U_i = 0.24$ m s^{-1} as the ice velocity due to wind drag alone (in good agreement with the Nansen Rule). With wave radiation pressure added under the "standard case" of a 400 m starter polynya and perfect reflection, we obtain a modified ice velocity of 0.41 m s^{-1}. Thus the exposed floes move downwind 0.17 m s^{-1} faster than floes that are masked from the waves, and the "starter polynya" widens at 10 m per minute.

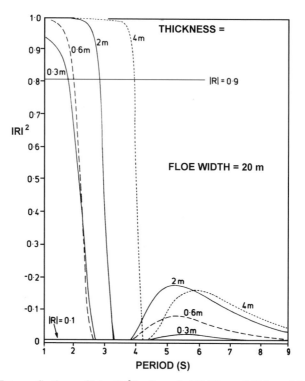

Figure 6.13.   Energy reflection coefficient |R|² for floes of width 20 m and thickness 0.3–4 m for wave periods 1–9 s, calculated from scattering model.

Such fast-moving floes soon sweep up their downwind neighbours, creating a band. But once the band has acquired a width W = n d (fig. 6.13), the wave force (6.33) is reduced by a factor (1/n), so that U_i drops towards its wind-only value, offset slightly by the increase in wave height due to the widening of polynya X. Eventually the band encounters a second polynya X_2, a "stopper polynya", which happens to be wide enough to generate its own wave field that is allowing a band 2 to be created which can just keep pace with band 1. At this point band 1 has reached its maximum size, since it is no longer sweeping up loose floes, which have all been incorporated into band 2.

Thus we can see how multiple bands are created — again the random factor produces a second polynya which is wide enough to generate its own band and which stops the further growth of the first band. It can be shown from the above equations that the final width W of band 1 is given by

$$W = d \, [X \, p - d(1 - p)] \, / \, [X_2 \, p - d(1 - p)] \tag{6.38}$$

and the final width of the "starter polynya" is

$$X(\text{final}) = X + (W - d) \, (1 - p) \, / \, p \tag{6.39}$$

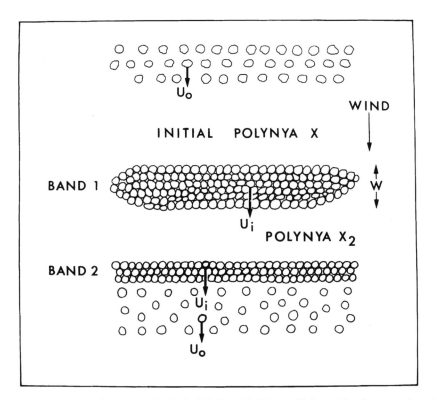

Figure 6.14. Schematic diagram of a band of width W which is in equilibrium with a "stopper polynya" of width X_2, which is just beginning to form a second band.

Assuming a stopping polynya of width 100 m, our standard case gives a final W of 95 m and a final X of 480 m for a 50% ice cover. These are certainly figures which are of the right order of magnitude in relation to observations (eg. fig. 6.10(a)).

Although this model predicts band and polynya widths successfully, and shows how multiple bands can form, the predictions depend on a knowledge of the width of the "stopper polynya". Thus, what will happen in practice is that an initial set of bands will form with a variety of widths and spacings, within a fairly limited range set by the above equations. If these bands have long lifetimes, they may then amalgamate into "mature bands", since narrow bands will tend to catch up with wider bands, whilst bands with large upwind fetches will tend to move faster than bands with small upwind fetches. It is likely, however, that in the very dynamic and rapidly-changing conditions of the MIZ, and with the bands having only a limited lifetime and range of operation because of melting, the final mature situation will never develop, so that band widths and spacings are to some extent determined by the random nature of the initial polynyas which start them off.

The above analysis is, in fact, an example of the difficulties involved in understanding the MIZ. Neither bands nor wave-induced floe size distributions normally have a chance

Figure 6.15.   Aerial photograph of a wind-driven eddy at the Antarctic ice edge north of Syowa Station (courtesy of S. Ushio, National Institute of Polar Research, Tokyo).

to achieve equilibrium states, since the dynamic variability and number of additional environmental driving forces is so much greater in the MIZ than in the interior pack.

6.5. EDDIES AND MEANDERS

As we have shown in section 6.2, the classic marginal ice zones of the northern hemisphere correspond, in their winter positions, to the locations of major fronts between cold, low salinity polar surface water and warmer, higher salinity water which cannot sustain an ice cover. Interactions can therefore occur between the ice cover and variability in the ocean structure of the front. Even if we neglect forms of interaction in which the ice plays an active role, the ice can still act as a passive tracer of upper ocean dynamics, showing the patterns of meanders, eddies and jets into which the front can break up through various forms of instability. Only in the Antarctic does the polar front lie to seaward of the furthest limit reached by the ice (although this was not true during the last ice age) so that eddies due to frontal instabilities are not traced out by the ice cover, but even here the ice shows the patterns of smaller wind-driven eddies (fig. 6.15). Such eddies have also been seen in the Sea of Okhotsk (Wakasutchi and Ohshima, 1990).

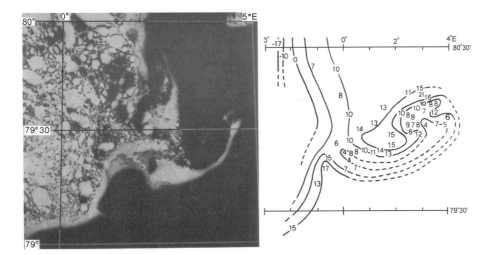

Figure 6.16. (left) Landsat image of 9 May 1976 showing ice-water eddy at edge of East Greenland Current. (right) Sea surface temperatures contoured in tenths of a degree C as observed from a Nimbus-tracked buoy, July 1976 (after Vinje, 1977a).

6.5.1. Some Eddy Observations

The most extensive observations of ice edge eddies have been made along the east coast of Greenland, and we review some of these in order to see the variety of features that occur.

There is a recurrent eddy known as the Molloy Deep eddy because it occurs in Fram Strait at about 79° 30′N in the vicinity of a 5770-m deep hole in the seabed which is surrounded by bathymetry of only 2500 m. According to some models (e.g. Smith *et al.*, 1984), this eddy is triggered by the bottom topography of Molloy Deep, while other models ascribe it simply to an instability in the polar front which develops first in Fram Strait because this is the northernmost end of the front along the East Greenland shelf. Figure 6.16, from Vinje (1977a), shows one of the earliest satellite observations of this eddy, although there are records of it going back to the time of William Scoresby (1820, vol. 1, ch. 4, sect. VIII). The photograph is a Landsat visual image showing small MIZ floes (less than 80 m across and so not individually resolvable) being drawn out into a cyclonic (i.e. anticlockwise in the northern hemisphere) eddy of diameter about 60 km. The contoured diagram alongside is a surface temperature map obtained from a satellite-tracked buoy which became entrapped in the eddy a few months later. The eddy can be seen to have a warm core, because of the melted ice and the warm (1.5°C) temperature shown by the buoy at the centre.

Fig. 6.17 shows a cross-section through an occurrence of this eddy in October 1976. It was obtained by the author using a sound velocimeter and an upward-looking echo sounder from the submarine HMS "Sovereign". The sound velocity was recorded at 85 m depth, and the lower half of the profile shows the submarine crossing the polar front at

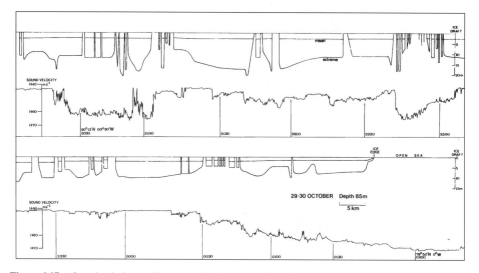

Figure 6.17. Sound velocity profile across East Greenland Polar Front at 79–80°N, October 1976, showing two warm lenses inside the zone of polar surface water. The schematic ice profile was derived from an upward-looking sonar record (after Wadhams *et al.*, 1979).

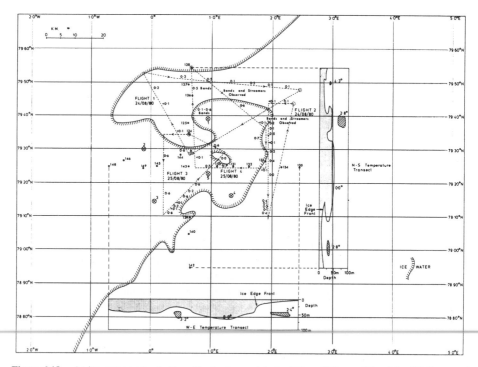

Figure 6.18. An ice-water vortex in Fram Strait, observed during "Ymer-80" expedition (after Wadhams and Squire, 1983). Ice edge location was mapped from aerial photographic transects, with flight lines shown. Horizontal and vertical temperature maps are summaries of CTD sections obtained in N–S and E–W transects of the eddy.

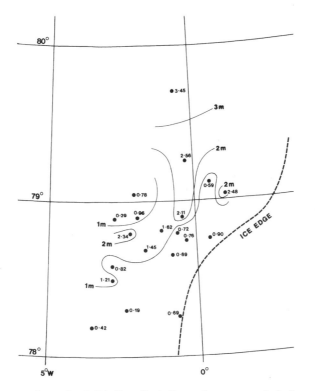

Figure 6.19. Contours of mean ice draft in Fram Strait. Data points are mean drafts from centroid positions of 50 km profiles obtained by upward-looking sonar, HMS "Superb", May 1987 (after Wadhams, 1992).

this depth *after* she has crossed the ice edge (a tracer of the polar front at the surface), indicating that the polar surface water occupies a wedge-shaped volume against the coast of Greenland. The upper profile shows two lenses of high-velocity (i.e. high temperature) water well inside the polar water zone, each with a reduced ice concentration above it. These are two limbs of the warm-core eddy; they are 20 km and 12 km wide, and are situated 80 km and 40 km from the polar front repectively.

Fig. 6.18 shows another manifestation of this recurrent eddy in August 1980. The shape of the ice edge was mapped by aerial photography from the icebreaker "Ymer", with flight lines shown, while the two temperature sections were obtained N–S and E–W across the centre of the eddy by the ship using her CTD. In each section water below 0°C, i.e. the polar surface water, is lightly shaded; it changes in thickness from 10 m under ice-free regions to 40–60 m under ice. Lenses of warm water (up to 4.3°C) are shown cross-hatched. In each case there is a warm lens, in fact the same lens, well inside the region where the polar surface layer is thick, but corresponding to the interior open part of the eddy.

A final illustration of the role of this eddy is fig. 6.19, a set of mean ice drafts obtained from 50 km sections of upward looking sonar data from the submarine HMS "Superb" in May 1987. The normal gradation of mean draft from the ice edge inwards is seen, but

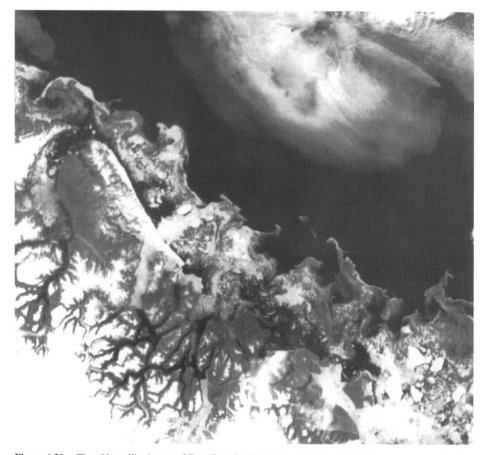

Figure 6.20. Tiros-N satellite image of East Greenland pack ice, showing meanders, eddies and a detached ice mass (probably a decaying eddy), 24 July 1979. Scoresby Sund is at left of image.

then in the west, at about 79°N 2°W, the ice becomes thinner again. Our interpretation is that in winter the ice edge lies further to the east than the eddy, such that the edge itself does not take up a vortex-like shape; instead, an interior warm region shows itself by the thinner ice above it. We presume that this is evidence of an eddy. Note that the eddies seen in all of these observations lies in slightly differing locations, as was also the case in observations in August 1984 from the icebreaker "Northwind" (Bourke *et al.*, 1987).

Although many observations have been made of the eddy in Fram Strait, it is clear that similar eddies exist all along the edge of the East Greenland Current, as can be seen in the satellite image of fig. 6.20. Many of these eddies have the "backward-breaking wave" shape that is often observed in sea surface temperature (SST) maps of eddies at the edge of the Labrador or Antarctic Circumpolar Currents. This is explained through an initially straight ice edge developing meanders through instability. As the meanders grow in amplitude they themselves become unstable, and because there is a velocity shear across the front, the wave-like shape appears to "break" backwards.

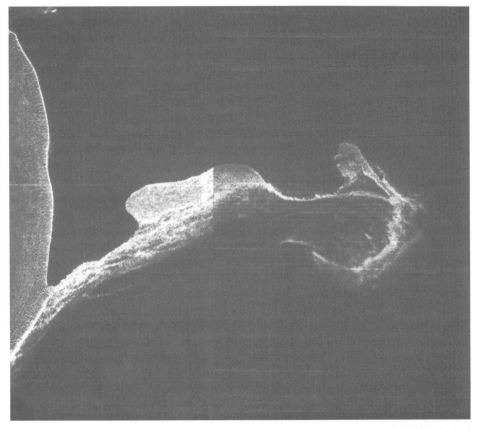

Figure 6.21. Airborne X-band SAR imagery from 78°N at East Greenland ice edge, April 1 1987, obtained during MIZEX-87 experiment (Johannessen *et al.*, 1994), showing a tongue which has evolved into a vortex pair.

Finally, we mention a entirely different range of eddy species on a smaller scale. Again, the type specimens come from east Greenland, and some observations have been described by Johannessen *et al.* (1994). They observed two types of short-lived feature emerging from the ice edge: a narrow tongue of ice, and a mushroom-shaped vortex pair (fig. 6.21). Vortex dipoles are very common features in the ocean, and have been observed frequently in SST maps. Fedorov and Ginsburg (1986, 1989) describe them as "one of the most widespread forms of non-stationary coherent motions in the ocean". In many cases (e.g. fig. 6.21) a tongue actually developed into a vortex pair, conserving overall angular momentum. It was estimated from these and other observations (Johannesen *et al.*, 1987a), that tongues and eddies are responsible for a melt of up to 300 km^3 of ice per year from the edge of the East Greenland Current, comprising about 12% of the annual transport of 2600 km^3 a^{-1} past 79°N (Vinje *et al.*, 1998). Thus **eddy melt** is a significant contributor to the freshwater budget of the Greenland Sea.

6.5.2. Generating Mechanisms

Several mechanisms have been proposed to account for eddy and tongue structures seen at the ice edge. Beginning with the small vortex pairs and tongues just mentioned, Johannessen *et al.* (1994) developed a numerical model based on the sea ice acting as a passive tracer for surface water motion. The dynamics were those of a current moving along a shelf break, where the depth increases from 400 m to 2600 m in 20 km of horizontal distance. A current is applied occurring, as observed, in a narrow jet centred on the shelf break. In such cases, depending on initial conditions, the flow develops meanders after a few days, which develop into tongues protruding from the ice edge by 30–35 km. In a case where baroclinicity is enhanced by having the jet flow only in the upper layer of the ocean, the jet develops more rapidly (in 5 days) and after 2–3 more days turns into vortex pairs as observed.

For the larger eddies, in the case of the East Greenland Current, it was first demonstrated (Jones, 1977) that a wedge-shaped boundary current of this kind is unstable if it is wider than about 25 km. Opinions differ as to the form of the instability. Wadhams and Squire (1983) considered a baroclinic instability based on the model of Pedlosky (1979), a two-layer model with vertical shear. Typical East Greenland conditions lead to a growth of unstable modes, the source of energy being the potential energy in the mean flow. The instability takes the form of waves, of wavelength about 50 km (in fact the geometric mean of the Rossby radii for the two layers), and the waves travel slowly downstream along the ice edge at a speed given by

$$c = (U_1 h_1 + U_2 h_2) / (h_1 + h_2) \qquad (6.40)$$

where U_1, U_2 are the velocities in the upper and lower layers of thickness h_1, h_2. The particular recurring eddy in Fram Strait was ascribed by Smith *et al.* (1984) to the presence of the Molloy Deep, as already mentioned. In his model, in which a 0.1 m s^{-1} current jet is imposed on the topography, a cyclonic eddy is generated with both a barotropic and a later-developing baroclinic component. The eddy remains trapped in the vicinity of the depression.

It has also been proposed (Johannessen *et al.*, 1987b) and partly confirmed by the tracking of subsurface floats (Gascard *et al.*, 1995), that some eddies in the East Greenland Current, particularly in the Fram Strait area, originate as instabilities in the West Spitsbergen Current. These instabilities produce eddies which split off and cross Fram Strait in a northwesterly direction, interacting with the East Greenland Current and appearing in satellite imagery as ice edge eddies.

6.6. AMBIENT NOISE AT THE ICE EDGE

When floes respond to wave action in the MIZ they tend to collide, since adjacent floes are moving out of phase with one another. Collisions are a major source of the high ambient noise levels found in the MIZ. Shen and Ackley (1991) proposed a model for wave-induced ice floe collisions which included drag and constant added mass effects on the floes and predicted conditions under which two floes remain closely in contact for some period during the collision, implying a possibility for freezing together during winter conditions (e.g. the consolidation of a field of pancake ice).

Gao (1992) has proposed a model which he was able to test against observational evidence from LIMEX (Labrador Ice Margin Experiment). The model itself was a simple one in which floes are assumed to have a perfect surge response to waves in amplitude and phase. In the observational test, aerial photographs were analysed to yield statistical distributions of the spacings between centres of floes and edges. He showed persuasively that lognormal distributions provide the best fit for both. He then re-expressed the collision probabilities derived for a simple case of two floes in terms of these mathematical distributions, yielding estimates of collision rates which he compared with observations made by accelerometers on the floes themselves. He found that the observed collision rates were an order of magnitude greater than the predictions.

The results of Gao, while not supportive of his theory, are important in that they are a careful experimental test, which indicates that an important physical process has been left out of consideration (the same applies to the Shen and Ackley model which is fundamentally similar). A strong candidate for the process in question is as follows. When two floes of unequal size collide due to wave-induced oscillatory motion, there will be a net momentum in the x- and y-directions imparted to each floe which is independent of the fact that the floe continues to be subject to oscillatory forcing by the wave. The momentum is damped out by frictional drag, but while it exists it will cause the floe to collide much more readily with its next neighbour, imparting momentum to that floe as well. In other words, we have a case like that of the molecules in an imperfect gas. Energy is transferred from the oscillatory motion of the floes to a motion of translation, giving each floe a "mean free path" and an equivalent kinetic energy which it shares with other floes via collisions. So some of the collisions are simple wave-induced collisions as analysed by Gao, but the majority are due to this random motion of floes which is equivalent to the energy possessed by gas molecules by virtue of their temperature. A complete analysis of the problem would deal with this question in a stochastic way. To do so requires an understanding of what happens in a single collision — is it inelastic, elastic, or something in between? — which has not yet been determined experimentally. Evidence in support of this idea was the fact that collision rates as observed by Gao's accelerometers were similar in the direction of the wave vector and at right angles to it, indicating the importance of random motion.

The problem of noise generation by floe collisions was treated by Rottier (1991), who developed a model in which the noise level is related to a collision rate which depends on ice concentration, floe sizes, and the presence or absence of brash ice between the larger floes. Rottier also acquired supporting data consisting of concurrent ambient noise and wave spectrum measurements in ice made during the MIZEX-87 and SIZEX experiments in the Greenland and Barents Seas. Clearly there is considerable promise in this kind of approach to ice edge ambient noise, since if the noise is primarily generated by collisions, and if a valid treatment of collision rates can be developed in terms of wave field and ice floe sizes, then *in-situ* data should yield the empirical factor relating collision rates to noise, giving a predictive capability.

6.7. THE ODDEN ICE TONGUE AND GREENLAND SEA CONVECTION

In the central Greenland Sea, just south of the main gyre centre, an ice tongue usually develops during winter. It grows eastwards from the main East Greenland ice edge in the vicinity of 72–74°N latitude and often curves round to the northeast until it reaches east of the prime meridian. It is called Odden, and in its curvature it embraces a bay of open water, centred on the gyre centre, which is known as Nordbukta. It is believed to form mainly by local ice production, since the tongue-shaped region corresponds to the region of influence of the Jan Mayen Polar Current (the southern part of the Greenland Sea Gyre) which maintains cold surface water. There is currently great interest in Odden, for two reasons. Firstly, the contribution of local ice production as opposed to advection of older ice from the East Greenland Current is not know for certain. Secondly, and related to this, the local ice production occurs largely in the form of frazil and pancake, since the intense wave field in the winter Greenland Sea inhibits ice sheet formation, while remote sensing imagery shows rapid changes in the size, shape and position of Odden, such as would not occur if it were composed of large floes. Again, frazil and pancake production implies high growth rates and high salt fluxes, possibly on a cyclic basis related to cold air outbreaks from Greenland. In models of winter convection in the central Greenland Sea (e.g. Rudels, 1990; Backhaus and Kämpf, 1999), this periodic salt flux is believed to play an important role in triggering narrow convective plumes.

A series of research programmes was carried out in the 1980s (Greenland Sea Project) and 1990s (European Subpolar Ocean Programme, ESOP) by European nations and the USA to understand winter convection and the role of sea ice-ocean interactions in it (Wadhams et al., 1999). Field operations have involved the direct study of convective plumes and the developing water structure in the central Gyre region associated with convection, and investigations of the ice characteristics, thickness and variability within the convection region. A 1993 campaign was described by Wadhams and Wilkinson (1999). Work began in early February using the vessels "Valdivia" and "Northern Horizon" and a BAC 1–11 research aircraft of the Royal Aerospace Establishment, Farnborough. This established that Odden was developing as a tongue of dense pancake ice, with little or no older ice present. From February to April FS "Polarstern" operated in the Greenland Sea, with intensive studies of the Odden in early April. Passive microwave and SAR images had shown that the Odden ice tongue completely separated from its root in March and became a long island of ice oriented N-S centred on the prime meridian. This island later broke into separate north and south islands, and (on passive microwave) the north island appeared to dissolve and reform late in March while the south island steadily shrank.

The ice physics programme carried out in April by "Polarstern" (Wadhams and Viehoff, 1993; Wadhams et al., 1996; Wadhams and Wilkinson, 1999) showed that the southern island was composed of a dense mass of pancake ice. The pancakes had a mean diameter of 0.96 m (sample of 2501 pancakes in an aerial photograph) with a maximum of 4.8 m. Pancakes recovered and brought on board ranged in thickness from 9 cm to 40 cm, and the larger pancakes had low salinities (2.8–5.5 psu) normally regarded as characteristic of first-year ice, implying that they had been stable components of the ice cover for some time. Frazil sampled between the pancakes had a thickness corresponding to 4–20 cm fresh water equivalent. More recent field work in Odden, in 1997 and 2000, has involved the deployment of drifting buoys to track the motion of the pancakes, and

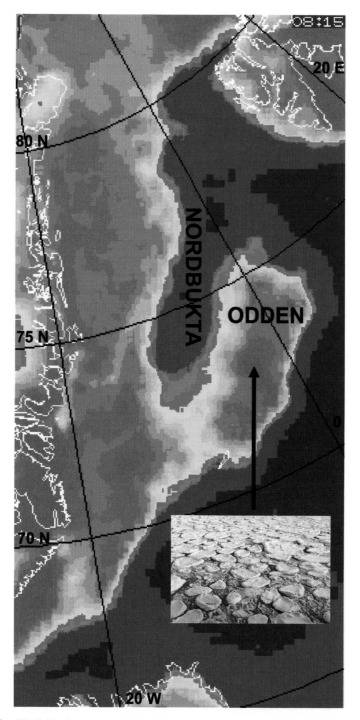

Figure 6.22. The Odden ice tongue and Nordbukta embayment as shown by passive microwave imagery in winter 1997 with (inset) the type of pancake icefield of which Odden was composed.

has found that there is a gradation of salinity (and age) in the pancakes, with the youngest being in the western part of the Odden.

A conclusion from these direct physical examinations, and from a concurrent study of SAR imagery (which distinguishes frazil from pancake by the dark return from the former and bright return from the latter) is that local ice production as frazil (turning into pancake) occurs both in the Odden and in the apparently ice-free Nordbukta region (the bay between the Odden ice tongue and the East Greenland pack ice, fig. 6.22). Cold air outbreaks from Greenland produce a cold westerly and north-westerly flow which not only facilitates the freezing process but also drives ice from Nordbukta into Odden, where it eventually melts on the seaward (SE) side of the Odden at the boundary with the warmer water of the Greenland Sea. Therefore, over a winter, there is a net positive salt flux into the ocean surface waters from ice production in Nordbukta, compensated by a net fresh water flux at the seaward edge of Odden. It is within Nordbukta, therefore, that local ice production can contribute most effectively to the increase in density of surface water, leading to overturning and convection. It is also in Nordbukta that the few direct observations of convective structures have been made, especially in the region of 75°N, 0–5°W.

The reason why the role of ice in the convection mechanism is important to assess is that if the ice were to retreat from the central Gyre region on account of climatic change it may cause deep convection to cease. Already there is evidence from tracer studies (Schlosser et al., 1991) of a severe reduction during the last two decades in the volume and depth achieved in Greenland Sea convection. If convection were to cease it would have a positive feedback effect on global warming, since the ability of the world ocean to sequestrate CO_2 through convection would be reduced. What is happening today is also important in that it may be a modern parallel to a sudden "switching-on" and "switching-off" of convection and of the North Atlantic thermohaline circulation which has been hypothesised (Broecker, 1995) as a cause of rapid climate change events during the last glacial period.

7.   ICEBERGS

You are in a sleeping bag in a tent on top of an Antarctic iceberg. It is a huge tabular berg shaped like a grand piano, 3 km long and drifting through the ocean off the South Sandwich Islands. Your tent is in a snow hole, and a clammy fog blanket lies around and above you, condensing on the tattered fabric of the tent and running down into your bag. You are down here to measure how tabular bergs flex in response to ocean swell, and to find out how big a swell is needed to break up a berg. The Navy research ship has launched a wave buoy into the sea near the berg to measure the spectrum of the waves. Then it has used its helicopters to transport you, your two assistants, equipment and six Marines onto the smooth snowy tabletop of the berg, 150 feet above the sea. Last of all the helicopters drop three 3-man survival packs. Three sheathbills alight on the snow and watch with interest as you set up an automatic meteorological station and a satellite-tracked buoy to map the future motion of the berg. The Marines dig snow pits down to a firn layer where you install strainmeters and accelerometers to measure the bending and heaving of the berg. Suddenly you notice that fog has come down. You had forgotten that icebergs at sea carry around their own weather; the

Figure 7.1. A tabular Antarctic iceberg within the pack ice zone.

cold snow surface can produce a local fog bank which sits just over the top of the berg. The helicopters cannot get in to bring you off. You must spend the night on the berg. To the Marines it is wonderful fun. They open the ancient survival packs, erect the ragged tents, roar with laughter at the cans of beans and sausages that make up the rations, manage to make a rusty Primus stove work, and soon have a piping hot meal ready. As you lie awake in your tent at night you look at the oscillating record of flexure on the strainmeter recorder and try to calculate how close the berg is to fracturing under the huge rolling swell. You feel no movement, yet your instrument tells you that this immense reassuring mass is close to breaking up. Morning comes and the berg is still intact. It did not reach its failure strain that night. The helicopter brings you back to the ship. Two days later the satellite signal from the iceberg comes to an abrupt halt.

7.1. MODE OF FORMATION

Gleaming in the sun, and sailing majestically through the pack ice or the swell of the Southern Ocean, a great tabular iceberg is the most magnificent sight in the Antarctic (fig. 7.1). A typical berg will have a diameter of several kilometres, a thickness of 200–300 metres and a freeboard of 30–50 metres, giving it a mass of around a billion tons.

Antarctic tabular icebergs originate by calving from the ice shelves which occupy about one-third of the coastline of Antarctica (fig. 7.2). The shelves are fed at their landward

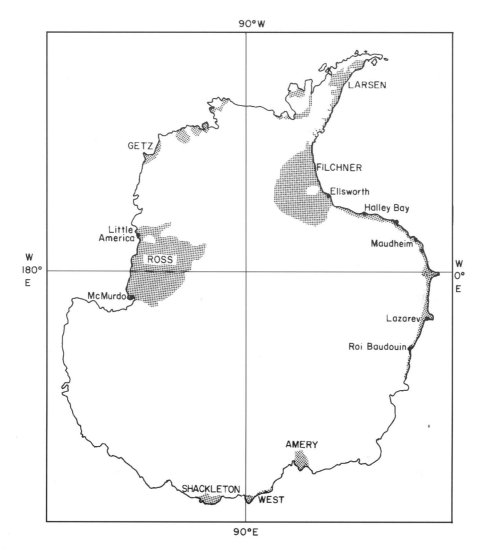

Figure 7.2. The main ice shelves surrounding the Antarctic continent (after Weeks and Mellor, 1978).

ends by ice from the Antarctic ice sheet, and their seaward ends are afloat. Under the pressure of the ice squeezing outwards from the centre of the continent, the shelves move inexorably seaward at 0.3–2.6 kilometres per year. Their floating ends are exposed to the stresses of currents, tides, ocean swell in summer and the pressure of drifting pack ice in winter. Eventually the end fractures, often along an existing line of weakness such as a crevasse, and the product is a freely floating iceberg. A special mechanism occurs in the case of **glacier tongues**, long narrow floating ice tongues which are the output of especially fast-flowing glaciers. Here the calving is thought to be caused by long ocean waves which cause the tongue to oscillate until it fails in flexure (Holdsworth and Glynn, 1978), a mechanism discussed in chapter 6 in connection with sea ice flexure.

Figure 7.3. A capsized iceberg in Davy Sund, East Greenland, showing the scalloped structure which develops at a melting ice-water interface due to small convective cells. Photo by the author.

Some apparently minor ice shelves are prolific generators of icebergs because of their rapid velocity; the Amery Ice Shelf, for instance, although small in size, drains one-eighth of the Antarctic ice sheet and produces 31 km^3 of iceberg volume per year (Budd *et al.*, 1967).

Arctic icebergs calve from the glaciers which squeeze their way through the coastal mountains fringing the Greenland ice sheet. Because the glaciers are narrow, fast-flowing and riven with crevasses the icebergs are usually smaller than Antarctic bergs and more randomly shaped. Often they remain trapped for months or years in their fjords of origin by the fact that the draft of the newly-calved berg exceeds the water depth of the sill which lies at the mouth of the fjord. An iceberg which sits in a fjord through a summer will have its vertical profile changed through the differential melt rates at different depths due to the warming of the fjord water. This change of shape may cause the iceberg to lose stability and capsize. The profile of the capsized berg is then a product of the summer temperature profile in the fjord, while its detailed surface structure of small scalloped indentations (fig. 7.3) is caused by the scale of the small convection cells involved in the melt process at the ice-water interface.

A special kind of thin tabular iceberg found in the Arctic Ocean itself is the **ice island**. These can be up to 30 km in diameter and 60 m thick. The main source is the Ward Hunt Ice Shelf on northern Ellesmere Island, which has been steadily retreating through calving of ice islands since observations began in the 1950s (Hattersley-Smith, 1963). It has now retreated so far that it is not expected to produce more ice islands. Ice islands calve into

Figure 7.4.   An ice island off Nordostrundingen, NE Greenland (photograph by author). Note annual layering. Free board 10 m.

the Beaufort Gyre and may make several revolutions within the Canada Basin before exiting from the Arctic Ocean via the Trans Polar Drift Stream and Fram Strait. The most famous ice island was T-3, which is thought to have calved in 1935 and which was discovered by Joe Fletcher in 1947 as one of three large radar targets off the north coast of Alaska. A research station was established on it in 1952 (Koenig *et al.*, 1952; Crary, 1958, 1960) and it was then manned almost continuously; in 1984 it exited from the Arctic Ocean via Fram Strait, and the last remnants of its research facilities were washed up on the west coast of Greenland a few months later. Other ice islands have become grounded along the Beaufort Sea coast of Alaska, usually breaking into fragments as in an occurrence in 1969 (LaBelle *et al.*, 1983), and more recently an ice island ("Hobson's Choice") was equipped by the Canadian Government with a research station for an expected circuit around the Arctic (Hobson *et al.*, 1989); however, its drift track took it into Nansen Sound in Ellesmere Island instead.

A secondary source of ice islands in the Arctic is the Flade Isblink, a small ice cap on Nordostrundingen, on the NE corner of Greenland, which calves thin tabular ice islands with clearly defined layering (fig. 7.4) into Fram Strait. An ice island of 35 m thickness and 400 m length shown in fig. 7.5 probably originated from this source, since it was seen in Kong Oscars Fjord further down the coast of East Greenland. A survey in 1984 showed 60 thin bergs of freeboard 12–15 m off Nordostrundingen, grounded in water of

Figure 7.5. Vertical photograph of a 400 m-long ice island of 35 m thickness in Kong Oscars Fjord, East Greenland.

37–53 m depth (Massom, 1984). An accumulation of such bergs produced a partial blockage of the western part of Fram Strait for a number of years (Vinje, 1982).

7.2. PHYSICAL CHARACTERISTICS

7.2.1. Sizes

Antarctic tabular icebergs can be extremely large. The largest reliably observed was seen off Clarence Island by the whalecatcher "Odd I" in 1927; it was 180 km square with a freeboard of 30–40 m (Wordie and Kemp, 1933). Recently much publicity has been given to giant icebergs in the Antarctic, with the implication that they are evidence of climate change. It is certainly true that the Antarctic Peninsula area has been warming, and that two ice shelves on the Peninsula, the Wordie Ice Shelf on the west side and the Larsen Ice Shelf on the east side, have been disintegrating, emitting large numbers of icebergs. Those from the Larsen Ice Shelf included a giant berg event in February 1995. However, the remainder of the Antarctic does not seem to be warming at present, and emissions of large icebergs are sporadic events which are probably not happening more frequently than in the past but are just more easily detected with the aid of satellites. A recent giant berg which approached the coast of Argentina in November 1999 was iceberg B-10A (labelling system of Joint Ice Center, Suitland), of dimensions 38×77 km and a reported freeboard of 90 m. It calved from the Thwaites Glacier in the Amundsen Sea in 1992, and spent long periods aground; in 1995 it broken into two (D. Long, Brigham Young Univ., personal commun.).

A decade earlier, two giant iceberg events occurred on opposite sides of the Antarctic. Iceberg B-9 calved from the Ross Ice Shelf in 1987 (Keys, 1990) between the Bay of Whales and King Edward VII Peninsula; in fact its calving destroyed the Bay of Whales as a feature of the Ross Ice Shelf front (Jacobs et al., 1986). At calving it was 154 km long (the second longest iceberg ever) and 35 km wide with an area of 4750 km^2. The ice shelf front was advancing at 0.84 km a^{-1}, so calving was inevitable, although the exact mechanism by which the rift developed is not known. Some 24 other large icebergs were produced in this event. The average thickness of the area of shelf that calved was 230 m, giving a volume of 1100 km^3 for B-9, equivalent to a whole average year's calving from the entire Antarctic coast, and equal to about 0.6% of the volume of the Ross Ice Shelf. The berg remained in the eastern Ross Sea region for two years, moving mainly with the depth-integrated current. In the Weddell Sea two independent calving events occurred in 1986. On the Larsen Ice Shelf a 95×95 km section broke away in January-March 1986 followed by a 55×50 km section in July. On the Filchner Ice Shelf a 90×210 km area broke out in June, carrying with it the Russian Druzhnaya research station (Ferrigno and Gould, 1987; Jacobs and Barnett, 1987). The Filchner icebergs remained in place for some time after calving (fig. 7.6), but the Larsen icebergs quickly emerged into the Southern Ocean east of Clarence Island.

In the open Southern Ocean surveys of iceberg diameters (e.g. fig. 7.7, Wadhams, 1988) show that most bergs are of moderate diameter, with a peak at 300–500 m and few exceeding 1 km. The distribution of diameters gave a good fit to a lognormal distribution. We show in section 7.5.2 how stability calculations lead to the conclusion that wave-induced flexure breaks down most larger tabular bergs to this range of sizes, while leaving a few exceptionally large bergs intact.

Arctic bergs are generally smaller than Antarctic bergs, from the time of calving onwards. The largest Arctic iceberg (excluding ice islands) on record was seen off Baffin Island in 1882. It was 13×6 km in size with a freeboard of 20 m. Most bergs, however, are of order 100–300 m in diameter. Table 7.1 is a size classification based on the usage of the International Ice Patrol and thus applicable mainly to Arctic icebergs; the terms "bergy bit" and "growler" are discussed in section 7.5.1.

7.2.2. Shapes

The main distinction in iceberg shape is between Antarctic bergs, which are frequently tabular at the time of calving, and Arctic icebergs, which are usually more random in shape from the start. Antarctic bergs evolve via further calving to tilted shapes, or by erosion of the above-water part to a variety of Arctic-like shapes. The surface may be domed or concave, depending on the local shape of the ice shelf at calving. A classification system, proposed for Antarctic bergs but applicable to Arctic bergs as well, was suggested by Keys (1986). Table 7.2 shows the variety of distinct shape categories that can be found.

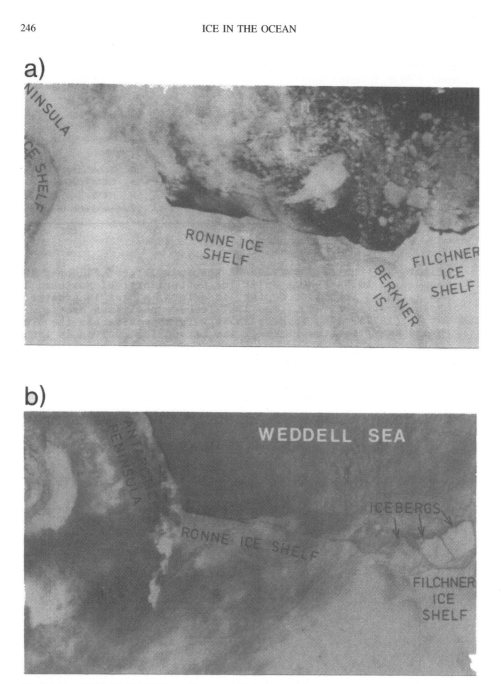

Figure 7.6.   (a). NOAA-6 AVHRR satellite image, 15 January 1986, showing Fichner-Ronne Ice Shelf front. (b) NOAA-9 AVHRR image, 21 September 1986, showing giant icebergs from Filchner calving moving into Weddell Sea (after Jacobs and Barnett, 1987).

Table 7.1. Classification of icebergs by size (adapted by Haykin *et al.* (1994) from World Meteorological Organization definitions).

Type	Height above waterline, m	Waterline length, m	Physical area above waterline, sq m	Relative size	Mass, tonnes
Iceberg	>5	>30	300	Merchant ship	180,000
Bergy bit	1–5	10–30	100–300	Small house	up to 5400
Growler	<1	<10	<100	Grand piano	up to 120

Icebergs are further classified by size as follows:
Small: Height less than 16 m, length less than 65 m
Medium: Height 16 to 48 m, length 65 to 130 m
Large: Height 48 to 70 m, length 130 to 225 m
Very large: Height greater than 70 m, length greater than 225 m

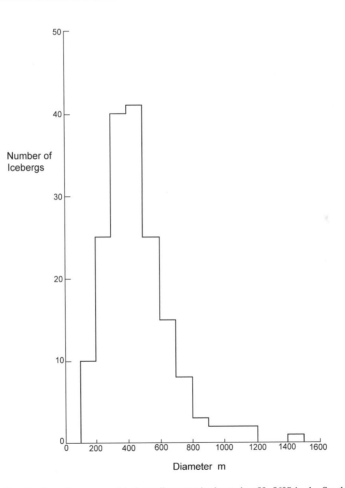

Figure 7.7. Results of a radar survey of iceberg diameters in the region 53–56°S in the South Atlantic (after Wadhams, 1988).

TABLE 7.2. Classification of icebergs by shape (as proposed by Keys, 1986).

Group	Description	Code	Shape	Remarks
1. TABULAR	recognisable tabular shape	T		original stratification normally visible on close inspection; no evidence of rollover
	horizontal	hT		top surface flat and horizontal; stratification visible and parallel to water-line; no raised or tilted waterlines visible
	uneven	uT		uneven top surface; top may be sloping or undulating and/or subaerially abated; crevasses or seracs are sometimes present; stratification is often non-planar
	domed	dT		top surface slopes down on all sides
	tilted	tT		tilting has occurred; raised waterlines often but not always present; area of original tabular surface greater or equal to half the waterline area
	blocky	kT		maximum height greater or equal to one-third width or one-fifth length at the water line

Additional description adjectives for tabular icebergs, written after shape code e.g., uT, ax

		n		iceberg with snow (and firn, and/or ice) visible
		a		iceberg with no snow (it contains firn, and/or ice - snow pack is absent
		o		no crevasses present
		v		crevasses visible only at edges
		x		crevasses traverse most of width or length of berg
		z		seracs present or heavily crevassed
		f		iceberg fractured through
2. IRREGULAR	various	I		angular or irregular features predominate or a mixture of forms is present; water-smoothed shapes are dominated by concave surfaces
	tabular remnant	mI		remnant of original tabular surface is visible together with original stratification
	pinnacled	pI		one or more pinnacles are present
	pyramidal	yI		like a pyramid
	drydock	qI		low area in middle often awash
	castellate	cI		battlement-like skyline
	jagged	jI		sawtooth or jagged skyline
	slab	bI		table-like
	blocky	kI		maximum height greater or equal to one-third width etc
	roof	eI		shaped like a roof or A-frame
	rounded	rI		generally irregular but rounded surfaces conspicuous
	combinations	eg mtrI		tabular remnant, tilted with rounded (wave shaped) platform
3. ROUNDED	rounded surfaces	R		water-smoothed, convex surfaces predominate
	subrounded	sR		some angular concave surfaces (may have complex though smoothed topography)
	rounded	rR		highest angular feature less than half maximum height
	wellrounded	wR		concave surfaces very minor; highest cliff at waterline less than one-third maximum height

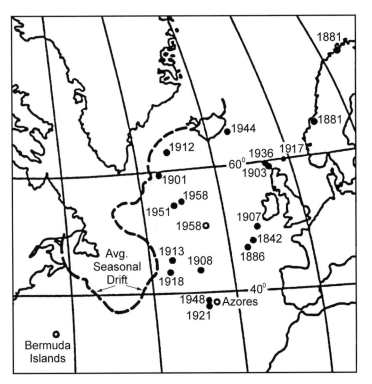

Figure 7.8.   Reports of exceptional iceberg sightings in the North Atlantic Ocean (modified from Hønsi, 1988).

7.2.3. Geographical Distribution

Antarctic icebergs can be found considerably to the north of the Antarctic sea ice limit, but are most frequent south of the Antarctic Polar Front. However, their large heat capacity and the northward trend to the wind-driven part of their trajectories (see section 7.4) can allow them to reach much lower latitudes. The record northernmost sighting was at 26° 30′S, 25° 40′W, off Brazil, while another low-latitude South Atlantic sighting was in 1828, at 35° 50′S, 18° 05′E, where clusters of bergs of about 30 m freeboard were observed (Jones, 1985).

Arctic icebergs are confined mainly to the area of influence of the East Greenland Current, West Greenland Current, Baffin Island Current and Labrador Current, as described in section 7.4. "Rogue" icebergs are those which leave this narrow conveyor belt. Figure 7.8 shows the locations of iceberg reports in the North Atlantic lying beyond the normal limits of iceberg activity (the dashed line); the southernmost sighting is a poorly documented report off Bermuda. The reasons for these exceptional sightings are various. The two reports of icebergs on the coast of Norway in spring 1881 coincided with the most extreme advance of the East Greenland sea ice ever recorded (Vinje *et al.*, 1996), so that the bergs would have been carried eastward along with the massive production and outflow of Arctic sea ice. Other sightings in the Atlantic may correspond to bergs that are either caught in cold-core eddies which move them into warmer water and

Figure 7.9.   Annual layers seen in the sidewall of a tabular Antarctic iceberg. Photograph by the author.

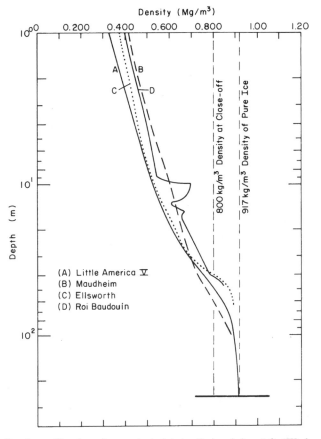

Figure 7.10. Density profiles through some typical Antarctic ice shelves (after Weeks and Mellor, 1978).

preserve them from decay, or simply to bergs that are driven far out of their normal path by prolonged storm activity — for instance, failing to round Cape Farewell but continuing S or SE instead.

7.3. GLACIOLOGICAL PROPERTIES

A new Antarctic tabular iceberg has the same structure and physical properties as its parent ice shelf. This means that the freeboard of the iceberg, the part that is above water, is not actually ice at all but rather compressed snow. The ice shelf has the same layered structure as the continental ice sheet from which it was squeezed out, with recently fallen snow on top, and older annual layers beneath. Each layer is compressed by the snow above it, so that the density increases with depth. The annual layers are clearly visible in the side of any tabular berg (fig. 7.9). Fig. 7.10 shows some typical density profiles through the parent ice shelves. At the top of the berg, where the density might be only 400 kg

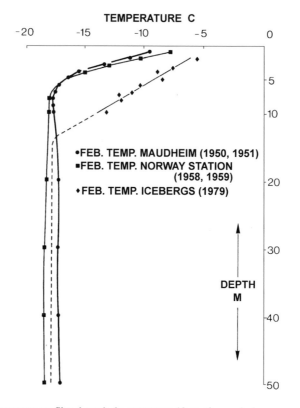

Figure 7.11. Temperature profiles through the uppermost 10 m of some icebergs near the Antarctic coast, compared with the temperatures of some nearby ice shelves (after Orheim, 1980).

m⁻³, there is free passage for air or water through the spaces between the crystal grains. Deeper down, when the density reaches 800 kg m⁻³ (pure ice has a density of 917), the air channels become closed off and turn into isolated bubbles. At this point the material can be truly called "ice", while the lower density material, when more compacted than snow, is called **firn**. The point of ice formation is not reached until 40–60 m below the surface of the iceberg, which corresponds approximately to the waterline. This corresponds to the most recent 150–200 years of snow deposition on the parent ice shelf or glacier. Further increases in density with increasing depth are associated with the compression of the air bubbles; pressures within bubbles in Greenland glacial ice have been measured at 10–15 atmospheres (Scholander and Nutt, 1960). In Greenland icebergs the air bubbles have been observed to be typically elongated, of length up to 4 mm and diameter 0.02–0.18 mm (Scholander and Nutt, 1960), while in Antarctic glaciers (and hence icebergs) they tend to be spherical or ellipsoidal, of diameter 0.33–0.49 mm (Gow, 1968; Gammon *et al.*, 1983), with mean diameter decreasing with increasing depth.

Figure 7.11 shows some typical temperature profiles through Antarctic ice shelves together with a profile through the upper layers of a recently calved berg (Orheim, 1980). Already the upper part of the berg is significantly warmer than its parent ice shelf. As soon as the berg drifts into more temperate regions, especially when it drifts free of the surrounding pack ice cover, it begins melting at its upper surface. The meltwater can percolate through the permeable uppermost 40–60 m, refreezing at depth and giving up its latent heat. By this means the above-water part of a tabular iceberg is brought relatively quickly to temperatures close to the melting point. With its warm temperature and low density, this part of the berg has little strength and is easily eroded away. The main mechanical strength of the iceberg resides in the "cold core" below sea level, where very low temperatures of –15°C to –20°C remain, and where latent heat transfer through percolation and refreezing is impossible.

An iceberg in the open sea produces squealing, popping and creaking sounds due to mechanical stresses and cracking, which can be heard underwater up to 2 km away. In summer, bergs produce a high-pitched hissing sound called "bergy seltzer", due to the release of high-pressure air bubbles from the ice as it melts in the warmer water.

Arctic icebergs often carry a top burden of dirt from the erosion of the sides of the valley down which the parent glacier ran, as well as stones and dirt on their underside, lifted from the glacier bed and later deposited out at sea. Otherwise bergs are usually white (the colour of snow or bubbly ice) or blue (the colour of glacial ice which is relatively bubble-free) (Bohren, 1983). A few deep green icebergs are seen in the Antarctic; it is believed that these are formed by the freezing of seawater rich in organic matter onto the bottoms of the ice shelves from which the icebergs later calved.

7.4. ICEBERG DYNAMICS AND DRIFT

7.4.1. Iceberg Motion

If an iceberg is viewed in an oversimplified way as equivalent to a sea ice floe which happens to be very thick (the best approximation being a large tabular berg), then the dynamical analysis developed in chapter 4 can be applied virtually unchanged. The air and water drag coefficients (4.3) and (4.5) will be greater, because of the large form drag involved in the sides of the berg, but in the case of an extremely large tabular berg the aspect ratio (diameter/thickness) can be as great as an ice floe, so that the drag coefficients are then not dissimilar. A difference is that both the atmospheric and oceanic boundary layers are severely affected by the presence of the iceberg.

It is clear, however, that a major difference lies in the Coriolis force, given by (4.6), which is proportional to the mass of the berg per unit waterline area. This is some 100 times greater for an iceberg than for sea ice. The consequences can be seen in the free drift solution. Equation (4.9), the very simplest solution for a water mass remaining at rest, gives a larger turning angle for a larger ice thickness, and in the more complete free drift solution (4.17)–(4.19), shown in fig. 4.4, the turning angle is greater and the wind factor is lower. It can be seen that the dimensionless quantity R in eqn. (4.17), which is proportional to thickness h, will be much larger for an iceberg than for sea ice, despite the possibly larger air drag coefficient. This means that (4.18) and (4.19) give a **high turning angle** and a **low wind factor**. In fact, as pointed out in section 4.2.2, as R→∞, α→1/R and θ→90°. In practice the solution breaks down at some point because the large draft of an iceberg means that simple Ekman dynamics cannot be applied.

This qualitative result helps to explain why Antarctic icebergs manage to move so far northward as compared to sea ice. The west winds which drive the sea ice and surface water in the Antarctic Circumpolar Current give the icebergs a more northerly trajectory than the sea ice; the thermal capacity of the iceberg then explains its survival as it moves northward. The result also explains why, when icebergs and loose sea ice occur together, the iceberg can usually seen to be moving relative to the sea ice and leaving a "wake" behind it. The low wind factor implies that the underlying current has a relatively greater effect on iceberg drift than the wind.

There have been many attempts to model the motion of real icebergs, taking into account their often irregular shapes. Standard approaches (e.g. Sodhi and El-Tahan, 1980) use the free drift equations with some optimised values for drag coefficients, with the additional difficulty that the underwater shape and dimensions of a given iceberg have to be guessed from a knowledge of the above-water shape. However, the fact that irregularly-shaped icebergs move in an anomalous way relative to wind and current, with their sidewalls acting as sails set in some random sense, means that trajectory prediction is often erroneous when it is most important that it should be correct, e.g. in the prediction of the track over the next few hours of an iceberg which is approaching an offshore structure. For these short-term local predictions, statistical approaches were proposed by Garrett (1984) and Gaskill and Rochester (1984). A given iceberg is tracked over a period of hours or days and the autocorrelation function of its velocity components determined. The short-term future trajectory is then based on extrapolation of the previous trajectory, with the autocorrelation providing an estimate of error, important for determining whether the tracking vessel or platform needs to take avoiding action.

7.4.2. Patterns of Iceberg Drift

Antarctic

Freshly calved Antarctic icebergs usually move westward at first, caught in the Antarctic Coastal Current. They may run aground and remain aground for years before moving on. Figure 7.12, for instance, shows the drift track of the very large iceberg "Trolltunga" which broke off the Fimbul Ice Shelf at about 0° in 1967 and then ran aground in the southern Weddell Sea for five years (Vinje, 1979). At some point the berg breaks away from the coast and come under the influence of the Antarctic Circumpolar Current or West Wind Drift, the great eastward flowing current system that circles the globe in the Southern Ocean. Driven by wind and current the iceberg's track becomes easterly, but the Coriolis force due to the Earth's rotation also gives it a northward component. Tchernia and Jeannin (1984) found that this curvature away from the coast tends to occur at four well-defined longitudes or "retroflection zones" (fig. 7.13), situated in the Weddell Sea, east of the Kerguelen Plateau at 90°E, west of the Balleny Islands at 150°E, and in the northeastern Ross Sea. These are related to the partial division of the water south of the Antarctic Circumpolar Current into gyres such as the Weddell and Ross Sea gyres, and suggests that icebergs found at low latitudes may originate from some particular sector of the Antarctic coast. Analysis of tracks of large icebergs by Swithinbank *et al.* (1977) agrees with these results.

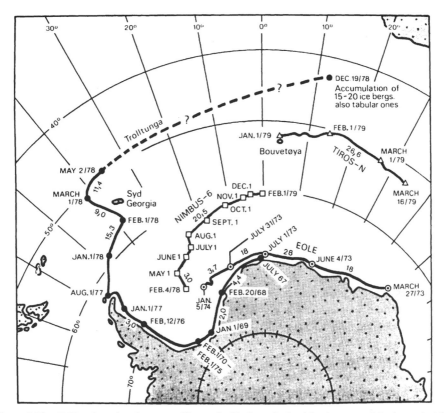

Figure 7.12. Drift trajectories of some satellite-tracked icebergs in the Atlantic sector of the Antarctic (after Vinje, 1979). The large berg "Trolltunga" was last observed on May 2 1978; a group of bergs later seen at 12°E is believed to be its fragmented remains.

Finally, after a long northeasterly voyage which may take it once round the world, the iceberg reaches a low enough latitude to break up and melt. Under extreme conditions, for instance if it is caught in a cold eddy, an iceberg may succeed in reaching an unusually low latitude — the lowest recorded was 26°S — and icebergs have been responsible for the disappearance of innumerable ships off Cape Horn.

Arctic

The highest latitude, furthest "upstream" sources of icebergs in the Arctic are islands of the Russian Arctic and of Svalbard. The annual iceberg flux from these sources is not large; Abramov (1996) estimates the total as 6.28 km³ a⁻¹, comprising 1.65 from Svalbard, 2.26 from Franz Josef Land, 2.00 from Novaya Zemlya, 0.35 from Severnaya Zemlya and 0.02 from Ushakov Island. This compares with 250–470 km³ a⁻¹ for the Arctic as a whole. Some icebergs produced by these island glaciers move directly into the shallow Barents or Kara seas where they run aground, and any shipborne transit of these seas

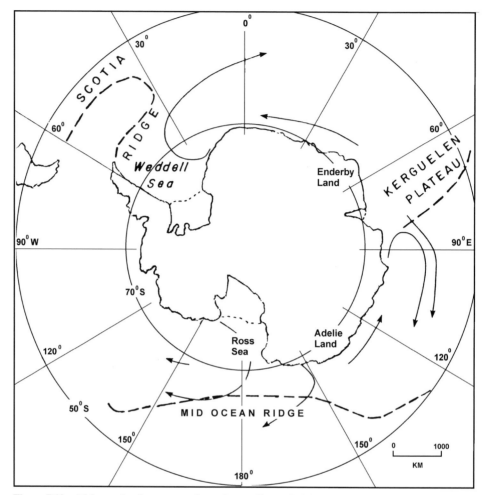

Figure 7.13. Main trends of movement shown by satellite-tracked icebergs in the Antarctic Ocean (after Tchernia and Jeannin, 1984).

reveals numbers of grounded icebergs (e.g. Overgaard *et al.*, 1983). When embedded in pack ice in winter, the grounded bergs can be detected by the looping trails of broken ice left by the inertial or tidal motion of the pack ice past the stationary bergs (Dmitriev *et al.*, 1991). Other bergs, calved from the northward-facing flanks of these islands, are injected into the Trans Polar Drift Stream and move with it into Fram Strait and thence into the East Greenland Current. Iceberg sightings in Fram Strait are rare, but fig. 7.14 shows three of a group of six small icebergs observed at the ice edge in Fram Strait at about 79° 10′N, 1°E, by the author on 23 August 1980.

These uncommon icebergs receive their first major reinforcements at the latitude of Scoresby Sund, although Davy Sund is also a minor iceberg producer (e.g. fig. 7.3). Scoresby Sund is wide enough to have an internal gyral circulation, with East Greenland

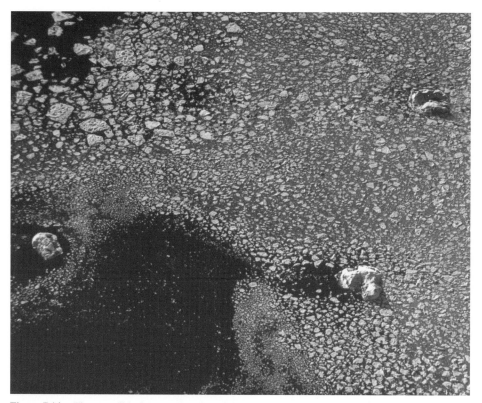

Figure 7.14. Three small icebergs at the edge of the East Greenland Current sea ice cover in Fram Strait, 79° 10′N, 1°E, 23 August 1980 (photograph by author).

Current water entering on the north side and an outward flow on the south side. This encourages the flushing of icebergs from the fjord. Narrower fjords not only offer more opportunities for icebergs to run aground, but also have an estuarine circulation, with an outward flow at the surface almost balanced by an inward flow at depth. An iceberg feels both currents because of its draft, and so does not move seaward as readily as, say, sea ice generated in the fjord.

The increased flux of icebergs reaches Cape Farewell, where most bergs turn the corner into Baffin Bay, although a few continue southwestwards directly into the Labrador Sea. Within Baffin Bay the icebergs first move northwards in the West Greenland Current, receiving enormous reinforcements from the prolific West Greenland iceberg-producing glaciers. About 10,000 icebergs per year are produced in this region. At the northern end of Baffin Bay, in Melville Bay, lies one of the most fertile iceberg-producing glacier fronts, that of Humboldt Glacier (Feazel and Kollmeyer, 1972). By this time some of the icebergs have been moved by wind or eddies across the Bay to be injected into the narrow, fast-flowing Baffin Island Current which takes them southward down the west side of the Bay. The rest make the turn in Melville Bay itself. On their southward journey their drift rates may average 12 km/day (Marko *et al.*, 1982), and they are further reinforced by calving from Baffin Island glaciers. However, because of their tendency to respond more strongly than sea ice to Coriolis force, they are constantly likely to move into the coast and run aground, where they remain for long or short periods until melt or wind and current stress causes them to start moving again. Some icebergs are affected by the gyral circulation at the entrance to Lancaster Sound, moving some way into the Sound along the northern side then out from the southern part of the entrance.

Marko *et al.* (1982) tracked icebergs which took only 8–15 months to move from Lancaster Sound to Davis Strait, but many bergs must take 3 years or more for the passage around Baffin Bay, because of groundings and inhibited motion when embedded in winter sea ice. A very variable flux of bergs then emerges from Davis Strait into the Labrador Current, where their final phase of motion occurs. Fig. 7.16, from Ebbesmeyer *et al.* (1980), shows annual average numbers of bergs crossing various latitudes within the Labrador Current. The linear decline of occurrence with latitude is mainly due to melting and break-up, or grounding followed by break-up. Finally a small number of bergs cross the 48°N parallel off northern Newfoundland and enter the zone where they are a danger to shipping in the Grand Banks area. The enormous interannual variability at this downstream limit of the iceberg drift pattern is shown in fig. 7.16, using data recorded by the International Ice Patrol. Such bergs have by now lost at least 85% of their original mass (Davidson and Denner, 1982) and are quite small and irregular. Their final fate is to melt on the Grand Banks or when they reach the "cold wall" separating the Labrador Current from the Gulf Stream at 40–44°N.

7.5. DECAY MECHANISMS

7.5.1. Erosion and Melt

We have already shown that after an Antarctic tabular iceberg emerges into the open ocean, it usually has little remaining strength in its above-water portion which has warmed nearly to the melting point through percolation and refreezing, its strength being concentrated in the cold core below the surface. Just as it has become weakened in this way, the berg becomes subject to ocean wave action. The high level of turbulence around the waterline of the berg increases the heat transport into the ice and preferentially melts out a **wave cut** which can penetrate for several metres into the berg, such that the snow and firn above it collapses. At the same time the turbulence level is enhanced around existing irregularities in the berg, such as cracks and crevasses. Waves eat their way into these, like sugar eating into a tooth cavity, and turn cracks into caves whose unsupported roofs may also collapse. Through these processes a berg in the open sea leaves behind it a trail of broken-off fragments, called **growlers** if they are small and **bergy bits** if they are large (according to the World Meteorological Organization definitions given in Table 7.1, a growler is the size of a grand piano while a bergy bit is the size of a small house). Eventually through these inroads the iceberg may evolve into a **drydock** or **pinnacled berg**, composed of apparently independent elements which are in fact joined together below the waterline. Such a berg may look like a prehistoric stone circle (fig. 7.17), but a ship would be well advised not to try to sail inside the ring of megaliths.

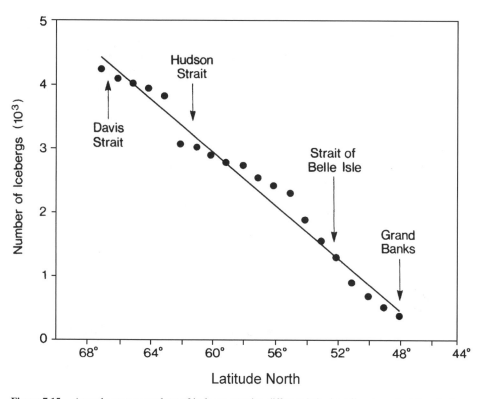

Figure 7.15. Annual average numbers of icebergs crossing different latitudes off east coast of Canada, from Davis Strait to Grand Banks, showing linear trend (after Ebbesmeyer *et al.*, 1980).

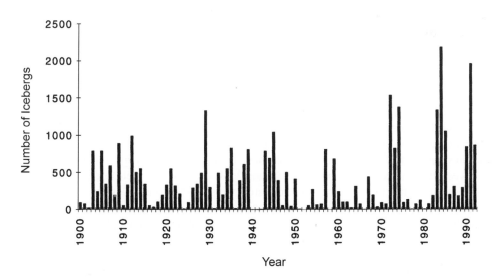

Figure 7.16. Annual numbers of icebergs crossing 48°N parallel off east coast of Canada, from 1900 to 1993 (after Haykin *et al.*, 1994).

Figure 7.17. A pair of pinnacled or drydock bergs near the South Sandwich Islands. In each case a number of independent surface-piercing blocks are connected to a single submerged main body, the result of erosion of the upper part of the berg (photograph by author).

A decaying iceberg poses a particular danger to shipping because of its trail of growlers and bergy bits. Although small in size, they have masses (up to 120 tonnes for growlers; up to 5400 tonnes for bergy bits) which are perfectly capable of damaging or sinking ships. Furthermore, as they drop into the sea they often roll over, losing their snow layers to leave a smooth surface of solid wetted ice which offers a very low radar cross-section in a heavy sea (Williams, 1975, 1979; Lewis *et al.*, 1994) and which is visually hard to discriminate against foam and whitecaps. It is often the undetected growler or bergy bit which sinks a ship while the larger parent berg has been detected and is being avoided.

Apart from loss of iceberg mass due to these mechanical erosion processes, pure melt also occurs from the sidewalls and bottom of the berg. The rate of melt depends on the salinity and temperature profile of the water column as well as the relative velocity between the berg and the near-surface water. Most estimates of melt rate are based on theoretical studies or laboratory tank measurements. Bottom melt was the subject of a theoretical study by Gade (1979) and of laboratory studies by Martin and Kauffman (1977) and Russell-Head (1980). Sidewall melting was studied by Neshyba (1977), Greisman (1979), Huppert (1980), Huppert and Josberger (1980), and Josberger (1982). An interesting variety of effects was obtained, because of the cooling and dilution that occur when ice melts into salt water. The large difference between the molecular diffusivities of salt ($\approx 10^{-7}$ m^2 s^{-1}) and heat ($\approx 10^{-5}$ m^2 s^{-1}) means that diluted water remains near the ice while the cooling diffuses further from the ice. In unstratified water (Josberger and Martin, 1981) this results in a bidirectional flow, with a horizontal flow inwards towards the wall, changing into an upwelling flow directly against the ice wall and, at greater depths, a downward flow some distance from the wall (fig. 7.18(a)). With a salinity gradient, however, a series of discrete steps can be produced, each of constant temperature and salinity, but with temperature decreasing and salinity increasing with depth (fig. 7.18(b), Huppert and Josberger, 1980). The slight inclination of the step boundaries to the horizontal results in a net upwelling as well as a flow of meltwater away from the berg. In the field, this staircase structure has been observed in measurements made close to the ice front of the Erebus Glacier Tongue (Jacobs *et al.*, 1980), but measurements close to drifting icebergs (e.g. Allison *et al.*, 1985) tend to show no such effect but simply some modification to the pycnocline depth.

In the absence of a generally accepted theoretical framework, and as a rough guide for calculating loss by melt, Weeks and Mellor (1978) produced the following rule of thumb based on observations of berg deterioration:-

$$Z = K D \qquad (7.1)$$

where

Z = loss in metres per day from the walls and bottom of the iceberg
K = a constant of order 0.12
D = mean water temperature in °C averaged over the draft of the iceberg.

This rule yields, for instance, a loss of 120 m during 100 days of drift in water at 10°C.

Finally, an eroded berg may suffer reduced stability and capsize. For an Antarctic berg complete capsize is uncommon, although tiltmeter measurements have shown that some long narrow bergs roll with a very long period, implying marginal stability (Kristensen *et al.*, 1982). In most cases an eroded Antarctic berg will settle into a new position of stability, tilted to a greater or lesser extent and thus exposing a new line of action for wave erosion (fig. 7.19).

For Arctic bergs capsize is a common event and is the defining process for the structure and properties of most Arctic icebergs seen in the open sea. Their wide range of shapes reflects properties which do not show a simple vertical gradation and horizontal uniformity as in the case of a perfect tabular berg.

7.5.2. Breakup by Flexure

Another decay mechanism for large tabular icebergs is due to the fact that they flex under the influence of long ocean swell. The theory of flexure is similar to that for an ice floe, which was considered in section 6.3.3. In large Antarctic storms an iceberg may break up into fragments through flexural failure, and this may be very important as a way of breaking up a large berg into smaller units with a bigger overall immersed surface area, producing more rapid melt.

Following the reasoning of section 3.3.3, we can apply equations (6.25)–(6.31) as before, although the approximation involved in considering the iceberg to be a thin elastic sheet is now greater. Equation (6.28) gives the strain at the top or bottom surface of the

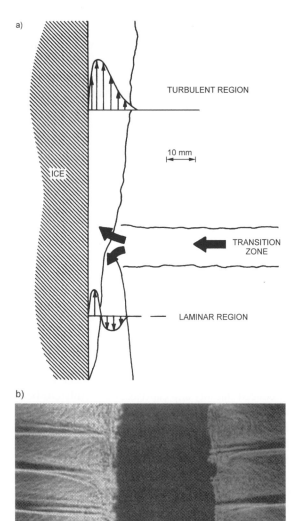

a)

TURBULENT REGION

10 mm

ICE

TRANSITION
ZONE

LAMINAR REGION

b)

Figure 7.18. (a). A sketch of the flow
observed in laboratory experiments when
a vertical ice wall melts into unstratified
salt water (after Josberger and Martin,
1980). (b) A shadowgraph of an ice column
melting into stratified salt water in a
laboratory experiment (after Huppert and
Josberger, 1980).

Figure 7.19.   A tilted Antarctic iceberg, showing a former waterline (photograph by author). The small black dots on the inclined plane are penguins.

iceberg at the centre of the berg's length. In figure 7.20 we have plotted the predictions of (6.28) for two cases, that of a 35 m-thick ice island (as shown in fig. 7.5) subjected to waves of 17 s period, and of a 100 m-thick iceberg subjected to swell of 24 s period. Let us consider the ice island case first. In considering the criteria for breakup the same arguments apply as for fig. 6.9, except that now the theory shows large positive and negative oscillations of strain as the diameter changes, instead of the single initial peak (B in fig. 6.9) followed by a uniform strain response. Thus, in fig. 7.20(a), consider a situation where the maximum crack length is 1 mm (a low value) and the wave period is 17 s. Then, from the right-hand ordinate, if a family of ice islands is exposed to a gradually increasing wave height, break-up will occur first to those islands which have a linear dimension of 600 m (point P in the figure), occurring as the wave height reaches 3 m. As the wave height increases to 10 m, those ice islands with diameters between 490 and 760 m (points Q and R) will be broken up. As the wave height exceeds this value, a second region of instability occurs, around the region S (diameter about 1500 m), while the unstable range of diameters in the first region grows wider. Finally, when the wave height reaches about 25 m (level Y), i.e. the asymptote of the strain response for very large ice islands, almost all sizes of ice island will break up including the biggest ones, although there will still be a region of stability around 1000 m (point T), a lesser one at 2000 m (U), and of course the small-diameter stable region extending part of the way along limb OP.

Consider now the 100 m iceberg (fig. 7.20(b)). Here the oscillations in strain are even more pronounced, and within the range of sizes in the computation there is no asymptote.

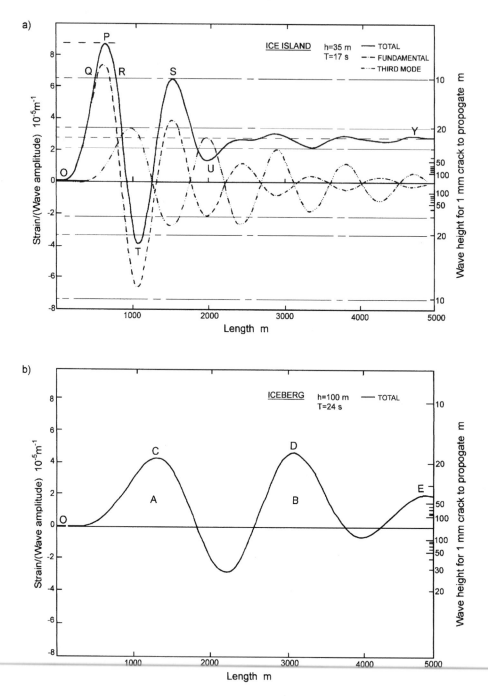

Figure 7.20. Theoretical predictions of the flexural response of a 35 m-thick ice island and a 100 m-thick iceberg to waves of period 17 s and 24 s respectively, as a function of diameter in the direction of the wave. The left hand ordinate shows the ratio of surface strain at the centre of the ice mass to incident wave amplitude. The right hand ordinate shows the wave height required to allow a 1 mm-long crack at the centre of the berg surface to propagate.

In the example shown, of a very long Southern Ocean swell, when the wave height reaches some 18 m (point D) icebergs of diameter 3000 m break up. As the wave height continues to increase, the range of unstable diameters broadens around the 3000 m value, with a second set of unstable diameters appearing at about 1200 m at some 22 m wave height (point C). Thus in a given wave field, where a range of tabular iceberg shapes and sizes are initially found, we can have a situation where some large icebergs break up, if their diameters in the direction of the waves are in these critical ranges, while others survive. In practice, of course, icebergs which have a long and a short axis will rotate so that a range of diameters is presented to the waves, and so eventual breakup is inevitable. Nevertheless, the theory does make two predictions which give a good qualitative and even quantitative fit to observations:

1. The final product of the breakup of large icebergs in a severe storm is a range of diameters extending up to only a few hundred metres (limb OC in fig. 7.20(b)).
2. It is theoretically possible for icebergs of a few km diameter to break up in a moderately severe storm, while icebergs of very large diameters (tens of km, the ultimate asymptote beyond point E) can survive the same storm.

The flexural response of Arctic ice islands was the first case to be considered theoretically and experimentally, with measurements using strainmeter arrays on an ice island in a fjord in East Greenland, by Goodman *et al.* (1980). In 1981–2 the study was extended to direct experiments on the flexure and heave of Antarctic tabular icebergs (Wadhams *et al.*, 1983). The measurements revealed significant bending even in moderate seas. Also detectable were heave and tilt oscillations at periods of 30–70 s which matched the natural bobbing and pitching frequencies of a body of the iceberg's size. When the sea state was extrapolated to that which would occur in a typical severe Southern Ocean storm, the conclusion was that most large tabular bergs would be susceptible to break-up. Support for this conclusion comes from studies of berg size distributions in the open ocean as opposed to the sheltered pack ice zone (Hult and Ostrander, 1974). The frequency of bergs increases, initially, just north of the pack ice zone while their average size decreases radically. The greatly increased number of small bergs seen in the open ocean near the ice edge is more than would be expected by pure melt of large bergs but must result from their fragmentation.

7.6. ICEBERG SCOUR

Just as sea ice can scour the seabed in shallow water, so an iceberg can plough a furrow several metres deep in the seabed when it runs aground. Such iceberg scour marks have been known in the Labrador Sea and Grand Banks areas since the early 1970s (Woodworth-Lynas *et al.*, 1984) and in 1976 the first Antarctic scour marks were discovered in 16°W off the Dronning Maud Land coast of the eastern Weddell Sea (Lien, 1981). Later observations have been made off Wilkes Land (Eittreim and Cooper, 1984) and Cape Hallett at the eastern entrance to the Ross Sea (Miller and Barnes, 1985). Labrador evidence indicates that long furrows are made when the iceberg is driven by sea ice; freely-floating bergs make only a short scour mark or a single depression. Apart from simple furrows, "washboard pattern" marks have been seen, assumed to be created when a tabular berg runs aground on a wide front and is then carried forward by tilting and ploughing on successive tides. Circular depressions have also been seen, thought to be made when an irregular iceberg touches the bottom with a small "foot" and then swings to and fro in the current like a leaf in the wind. Such movement of a berg was actually observed off Cape Hallett.

In the Arctic many of the marks, although fresh-looking, are found in water more than 400 m deep, whereas the deepest Greenland fjord sill is thought to be 220 m deep. This suggests either the existence of very deep icebergs in the past, or a very slow sedimentation rate so that old marks dating from times of lower sea level have not yet been filled in. It is also possible that an irregular berg can actually increase its draft by capsizing, although model studies suggest that the maximum gain is only a few percent, not enough to account for this discrepancy (Lewis and Bennett, 1984). A final possibility is that Greenland fjords exist with entrances of unsuspected depth; by no means all iceberg-producing fjords have been adequately surveyed bathymetrically.

7.7. CLIMATIC ROLE OF ICEBERGS

Apart from local modifications to the weather, such as the production of fog, icebergs have two potential climatic roles. One is the effect of their melt on the ocean, and the other is the effect of iceberg production on the mass balance of the parent ice sheets.

Let us consider the Antarctic first. The volume of the Antarctic ice sheet is 2.8×10^7 km^3 (Untersteiner, 1984), which represents 70% of the total stock of fresh water in the world, including groundwater sources (Skinner and Turekian, 1973). The mass balance of the ice sheet is maintained by gain from snowfall balanced by loss from melting under ice shelves and calving of icebergs. The ice sheet grows thicker by receiving snow on its top surface, the ice squeezes or **creeps** outwards, goes afloat at its outer edges in the form of ice shelves, and the shelves lose ice by melting and calving off icebergs. In a sense the two loss mechanisms are the same - if there were no melting from ice shelf bottoms the icebergs being calved from the front of the ice shelf would just be a few metres thicker, so both mechanisms can really be classed as "melt from floating ice". Additional contributions from summer runoff and from sublimation of the ice surface are negligible. It has been estimated that annual snowfall over the Antarctic continent amounts to 10^3 km^3, so if the Antarctic ice sheet is in neutral mass balance, which is believed to be approximately the case, the iceberg plus ice shelf melt flux must be close to this value. Estimates of iceberg flux do indeed start at 10^3 km^3 (Mellor, 1967), but extend up to $(1.77 \pm 0.10) \times 10^3$ km^3 (Shumskiy et al., 1964) and even as far as 2.7×10^3 km^3 (Orheim, 1984). The last figure is based on ship reports of iceberg frequencies, and if true would give the Antarctic ice sheet a negative mass balance. However there are many questions concerning the statistical validity of the various radar counts reported, and so we can cautiously discount this figure and assume that the volume flux probably lies in the range 1–2 $\times 10^3$ km^3.

These fluxes appear huge, but they are less than the mean flow rate of the Amazon River, which is 5.7×10^3 km^3 a^{-1}. It is also significant that the annual loss of ice from the Antarctic ice sheet amounts to only one ten-thousandth of its mass, so the ice sheet really is an enormous passive reservoir. The climatic implications are that global warming may lead to a greater rate of loss from the ice sheet by iceberg calving and ice shelf melting. This could make a large contribution to global sea level rise, since even at present the retreat of small glaciers in the Arctic and mountain regions is believed to contribute about 50% to the rate of rise, the rest being due to thermal expansion as the ocean warms.

In considering the second climatic effect, that of iceberg melt upon the ocean, we can do some very simple sums. Let us suppose that an annual iceberg production of 2×10^3 km^3 is melted over the area of the Southern Ocean occupied by winter sea ice, about 2×10^7 km^2 (actually it is spread over a larger area, since the icebergs can drift far north of the winter sea ice limit). The result is the addition of 0.1 m of fresh water per year to the surface water mass. This is not insignificant — it is like adding 0.1 m of extra annual rainfall — and the dilution, if averaged over a mixed layer 100–200 m deep, would amount to 0.015–0.03 psu. Melting icebergs thus make a small but not negligible contribution to maintaining the Southern Ocean pycnocline, and to keeping surface salinity in the Southern Ocean to its observed low value of 34 psu or below. If iceberg flux were to increase significantly due to global warming, this contribution would itself increase, which would tend to reduce ocean-atmosphere heat flux over the Southern Ocean, a case of a negative feedback effect.

It is interesting to note that an annual production of 2×10^3 km^3 of Antarctic iceberg ice is about one-tenth of the annual production of Antarctic sea ice (2×10^7 km^2 with a thickness of about 1 m). Icebergs therefore make up a not insignificant fraction of the total volume of floating ice in the Antarctic. Sea ice has a neutral overall effect on ocean salinity, but a very important differential effect, in that it increases ocean salinity where it forms (often near the coast), encouraging convection and bottom water formation, and decreases ocean salinity where it melts (often much further north in the Southern Ocean). Icebergs, on the other hand, always exert a stabilising effect on the water column, but only when they melt, which tends to be at lower latitudes rather than near the Antarctic coast where deep-water formation takes place.

7.8. ICEBERGS AS A FRESH WATER SOURCE

The immense amount of fresh water locked up in an iceberg — a billion tons or more in the case of a large Antarctic berg — has naturally inspired people to consider the idea of using bergs as a fresh water source. Captain James Cook and other explorers of the Southern Ocean used iceberg ice directly as a fresh water source for their ships (fig. 7.21); Cook stated that "this is the most expeditious way of Watering I ever met with". However, the earliest experiments in this direction stemmed from the use of ice as a refrigerant, before the days of industrial refrigeration. This use has a long history — the Romans cut ice from Alpine lakes and transported it in straw to Rome — but it became very important in the 19th Century in Europe and North America with their growing city populations. Perishable food such as fish had to be preserved and transported, and only natural ice was available. Ice was cut from lakes, or exported from glaciers in Norway and Alaska by ship. The first recorded instance of iceberg towing was the export of small icebergs from Laguna San Rafael in southern Chile to Valparaiso and Callao as part of a refrigerating ice supply business (Weeks and Campbell, 1973).

The idea of using iceberg ice for drinking or irrigation rather than refrigeration began with a suggestion by John D Isaacs of Scripps Institution of Oceanography in 1949 (Engel,

Figure 7.21. *"The Ice Islands of the Antarctic"*, A Voyage towards the South Pole and around the world. Performed in his Majesty's ships the Resolution and Adventure in the years 1772, 1773, 1774 and 1775, by Captain James Cook. London, W. Strahan and T. Cadell, 1977, 2 vols. (Cook Second expedition) reproduced courtesy of the Scott Polar Research Institute, Cambridge.

1961; Behrman and Isaacs, 1992). The aim was to relieve California's inherent water shortage by towing icebergs from Alaska or Antarctica. No suitable technology existed at that time, but during the 1960s and 1970s a successful iceberg towing technology was evolved for the Labrador Sea, for towing bergs away from oil drilling ships and platforms. It was found that a floating bridle around the berg, attached to several tugs, was feasible and safe. This inspired a revival of Isaacs' idea in the form of two studies, by Weeks and Campbell (1973), who considered the towing of unprotected Antarctic bergs to a Southern Hemisphere destination, and by Hult and Ostrander (1973), who considered towing trains of insulated icebergs from Antarctica to California. Much further research ensued, reported in the proceedings of two major conferences, at Ames, Iowa in 1977 (Husseiny, 1977), and in Cambridge in 1980 (Annals of Glaciology, 1980). A major stimulant was governmental interest by Saudi Arabia, which sought an alternative to desalination as a way of supplying its water needs. Further design studies ensued after 1980, but no action has yet been taken. The current view is still as summarised in a review by Wadhams (1990) and is as follows.

Any unprotected iceberg towed through warm Equatorial waters will melt, and no feasible or economically viable method of insulating an iceberg has been put forward. Therefore any scheme must involve a Southern Hemisphere destination. Even so, the melt rates implied by equation (7.1) are very restrictive. A feasible destination should have the following properties: (1) cold surface water and favourable currents along all or most of the route of the tow, to minimise total towing effort and *en route* melting; (2) deep water close inshore, to minimise the problem of transporting fragments of processed iceberg to shore-based facilities; (3) a serious local need for water, either for irrigation or drinking, so that expensive long-distance distribution systems are not needed. There are three regions which satisfy these criteria:

1. The Atacama Desert coast of northern Chile and southern Peru. Here the cold north-ward-flowing Humboldt Current assists tows from the Antarctic, the shelf is very narrow (less than 20 km in places), and irrigation could enormously increase agri-cultural productivity. Five thousand years ago the Mochica civilisation made the desert bloom through irrigation schemes, and the ruins of their cities litter the desert land-scape as a reproach to modern man, or rather to the Spanish who destroyed the canals.
2. The Namib Desert region of southern Angola, Namibia and South Africa. Here there is a zone of cold upwelling near shore, a narrow shelf, and a coastal desert, the so-called "Skeleton Coast" which is much feared by shipwrecked mariners.
3. Western Australia and the Great Australian Bight. This is not far from the limits of the Antarctic Circumpolar Current, there are places such as Rottnest Island off Perth where deep water reaches close inshore, and as well as irrigation needs there is a real need for drinking water for the expanding cities of Perth and Fremantle, now depend-ent on dwindling groundwater resources.

Figure 7.22 shows surface water temperatures in early southern Spring in relation to these three destinations, showing how short a transit through warm water is required for each.

The actual technology of finding, harnessing and towing an iceberg will involve many serious problems, but all are in principle soluble. Obviously for a given destination there will be an optimum source zone some way "upstream" in the Circumpolar Current that will involve minimal towing times. Icebergs can be seen surprisingly far north (Wadhams, 1988), but such outliers are usually in an advanced stage of melt, and younger bergs might be preferred. Even the most powerful tugs will only be modifying the iceberg's own pattern of drift by adding a knot or two of towing speed, so predictions of iceberg motion, as described in section 7.4.1 and using a model such as that of Sodhi and El-Tahan (1980) or Gaskill and Rochester (1984), will be essential in determining source area and towing strategy. Icebergs can now be easily located by satellite SAR imagery, so it will be possible to inspect a series of bergs until a suitable one is found. Towing will be by a number of powerful tugs attached to a floating bridle, as used in the Labrador Sea (Bruneau *et al.*, 1978; Benedict, 1978). Weeks and Mellor (1978) estimated that a berg of dimensions 1600 m $\times$ 400 m $\times$ 200 m thick can be towed at 1 knot with a towing force of 6.9 MN, equivalent to a bollard pull of 4760 hp, exceeding the capacity of any present wire rope and thus requiring multiple tugs. It is possible that some kind of waterline boom could be devised to reduce erosion through the wave cut effect, but this may prove technically unfeasible and considerable loss would have to be accepted. The worst case, and this may end up as a complete barrier to the use of this technique, would be if a large tabular berg fractured under swell action in a severe storm, leaving a collection of fragments within the bridle.

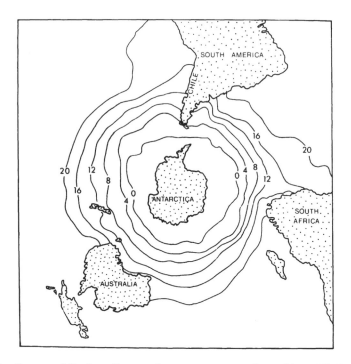

Figure 7.22. Contours of Southern Ocean surface temperatures as observed by a US Navy satellite in September 1977 (redrawn from Denner, 1978).

Assuming that these problems have been overcome and the iceberg delivered to a coastal site, it must be processed, and this must be done quite quickly as the iceberg will be melting *in situ*. The iceberg must be anchored offshore, and then reduced to pieces small enough to be towed ashore, melted and processed. Weeks and Mellor examined various blasting and cutting techniques, while Smith (1978) suggested the "hydro-wedge", whereby a artificial crack in the surface is deepened by filling it with water. Diemand (1984) found that very cold ice, such as in the lower part of an iceberg, can be fragmented successfully by the use of slow-burning explosives such as thermit, which could be implanted by drilling. The processing stage appears to be the one for which an economically attractive method has not yet been found; all techniques proposed so far are feasible but expensive.

The final stage is to make use of the iceberg fragments. Here a highly attractive idea has transfomed the potential economics. It is to make use of the latent heat of fusion of ice (33 kJ kg^{-1}) to derive energy from the iceberg as well as water. Such ideas were suggested by Heizer (1978), Roberts (1978) and DeMarle (1978), while DeMarle (1980) worked out and costed a scheme for Saldanha Bay, South Africa, where there is a natural lagoon suitable for processing iceberg fragments. The concept is called **Icetec** and works as follows (fig. 7.23). The iceberg is moored offshore and fragmented. The fragments are towed into a storage basin made out of an inner part of the bay and separated from the

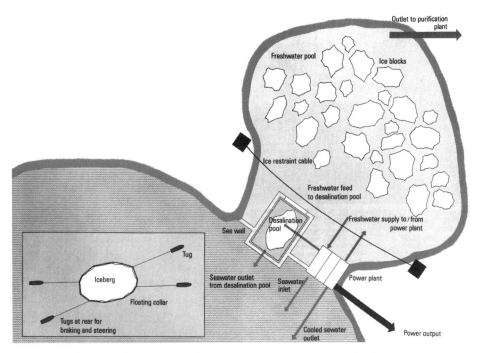

Figure 7.23. Schematic diagram of a possible towing method for icebergs and of an Icetec facility to process berg fragments for water and electricity (after Wadhams, 1982, redrawn from Heizer, 1978).

main bay by a lock. Here the fragments melt, and the melt water is fed into an ammonia heat exchanger as used in ocean thermal energy conversion (OTEC) schemes. This operates just like a steam turbine plant: the warm Saldanha Bay water (20°C) fed into the seaward side evaporates the ammonia, which drives a turbine and is then condensed by the iceberg meltwater. The warmed meltwater is fed back to the iceberg pool to enhance the melt rate of the bergs and to be cooled back to near freezing point. Some of the meltwater is drawn off for drinking or irrigation, sufficient to keep a constant level of water in the basin. DeMarle estimated that an iceberg of size 900 × 300 × 200 m would yield power worth $6.4 million at 3 cents per kWh and water worth $2.8 million at 6 cents m^{-3} (these were 1980 prices). It is noteworthy that the power is worth more than the water. Such a scheme could feed an agricultural complex with some settlement and industry attached, an attractive concept for a desert coastline. Alternatively, the water could be pumped to a more distant spot for use as drinking water, with the electricity powering the pumps. It is an attractive idea, but is very risky and requires huge initial investment, so no steps have yet been taken to implement it. Surprisingly, there are also international legal questions surrounding the free use of Antarctic icebergs (Trombetta-Panigadi, 1996).

The most recent form of iceberg exploitation dates from 1994, when the Canadian Iceberg Vodka Corporation was incorporated, to market vodka made from water harvested from growlers off Newfoundland.

8. SEA ICE, CLIMATE AND THE ENVIRONMENT

You are standing on the Beaufort Sea beside a six-foot wide dive hole that you and your colleagues have spent the morning cutting with chain saws through multi-year ice. The hole is bordered by a neat line of excavated ice blocks. On one of the blocks sits a diver, gazing into the dark water that you are keeping clear of ice by dragging a floating wooden beam across it. Will he ever pluck up the courage to slide under the ice? You are all working for the Canadian Government to investigate the way in which spilled oil behaves under ice. A helicopter ferried you, the divers, the tents and equipment and eight drums of crude oil out from Cape Parry. The blue dive tent, kept warm by a roaring propane burner, sits beside the dive hole. Here the diver struggles into his electrically heated dry suit and puts on his closed cycle breathing apparatus, so that he looks like an astronaut. He emerges with his video camera and hand-held floodlight. But he does not like the idea of sliding under four metres of ice. He sits shivering on the brink. Finally he plucks up courage. He is in the water. Time to pump oil. You rush over to the stand pipe that you had drilled into the ice, broach the first drum, insert the hand pump and pump oil down under the ice. The filthy black fluid sprays over your parka,

giving it an acrid stench. You know that you will have to throw your clothes away. The drum is empty. You roll it away and broach the second. You pump eight drums. You are exhausted and black with oil. Afterwards you watch the video monitor in the tent. There is the oil streaming away from the pipe under the ice, filmed by the diver. It runs downstream until it reaches a pressure ridge and there it gathers in a great black under-ice pool. Now you know. If an oil well blows out under ice, the ridges will act as dams to gather up the spilled oil. It was worth it.

8.1. SEA ICE AND BIOLOGY

It has been found in many field experiments that the winter sea ice cover in both the Antarctic and Arctic harbours an entire ecosystem, starting from nutrients and working up from bacteria to zooplankton. Energy pathways in the ecosystem remain to be investigated, including the role of physical parameters such as sea ice roughness (for bottom-dwelling biota); the geometry of brine drainage channels; and ice thickness (defining light levels). A critical question awaiting an answer is to what extent the sea ice ecosystem in winter is responsible for "seeding" the spring plankton bloom, which occurs with greatest intensity in the region from which the sea ice has most recently retreated (Comiso and Sullivan, 1986; Comiso *et al.*, 1993). Eicken (1992) has drawn attention to the contrasting roles of sea ice in controlling primary production under ice through reduction of irradiative fluxes by the ice and its snow cover, and in stimulating a "cryohaline" mode of life for organisms within the pores of the sea ice, controlled by salinity and ambient temperature as well as irradiance.

The ice thickness distribution and the morphology of the ice bottom have three important impacts on the ecosystem in and under sea ice. Firstly, light penetration through ice is dependent on the thickness of the ice and of the overlying snow. In the Antarctic in spring, the plankton bloom is most vigorous near the very edge of the retreating ice cover, and the question arises of whether there is sufficient light penetrating the ice to allow the bloom to begin beneath the marginal ice zone itself. This question was addressed during a cruise to the Bellingshausen Sea ice edge by RRS "James Clark Ross" in November–December 1992 (Turner *et al.*, 1995). In the light of ice conditions observed on that voyage, it appears more likely that the open nature of the retreating ice cover, which is generally divergent and is broken up into floes by wave action, is the chief agent permitting high light levels to occur under the ice cover so that plankton growth can take place. Nevertheless, in the Arctic, with its denser cover, it may well be that light penetration through the ice itself is the critical factor.

The second impact is the fact that consolidated pancake ice in the Antarctic provides a marvellous substrate for both phytoplankton and zooplankton. The mass of overlapping pancakes on the underside provides a large ice surface area per unit sea surface area, with many nooks and crannies into which zooplankton can hide from predators and consume the phytoplankton which grow attached to the ice surface.

A third factor in the potential "seeding" of the surface waters for plankton growth in spring is that chlorophyll is found throughout the depth of most ice cores analysed in the Antarctic (Eicken, 1992), because plankton are trapped in the frazil ice suspension as it

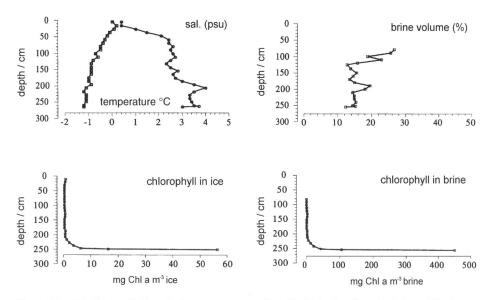

Figure 8.1. Distribution of chlorophyll-a in a summer multiyear floe in the East Greenland Current (Gradinger, 1995). The lower left and right graphs show concentrations in the ice and in the brine drainage channels respectively. Upper curves show salinity, temperature and brine volume distributions through the floe.

congeals into pancakes and ultimately into the first-year ice sheet of consolidated pancake. This material is released in spring as the ice melts, together with living plankton that have found a home within brine drainage channels. Within the Arctic the chlorophyll is concentrated much more near the bottom of the ice sheet. The main primary producers in the Arctic sea ice ecosystem are diatoms, which are adapted to start growing under very low light levels. During spring and summer they build up an in-ice algal biomass concentration which is as high as 100 milligrams of chlorophyll-a per square metre of sea surface area in fast ice areas of the Canadian Arctic, but is usually less than 10 mg Chl-a m^{-2} in the polar pack of the Transpolar Drift Stream (Gradinger, 1995). These algae are concentrated in the lowest few decimetres of the ice sheet; fig. 8.1 (after Gradinger, 1995) shows dramatically how the chlorophyll is distributed through the ice and through the liquid brine in brine drainage channels and brine cells in a summer multiyear floe.

The concentration of algae near the bottom of the ice sheet leads to the formation of a distinct sea ice bottom community of animals, the properties of which have been reviewed by Horner (1985) and Melnikov (1997). The algae are a food source for protozoans and mesozoans which are generally less than 1 mm long and which can therefore enter the brine drainage channels. They include ciliates and turbellerians in multiyear ice, and nematodes and crustaceans in shallow water areas. They also include the larvae of larger organisms such as polychaetes and molluscs, and, in the Antarctic, copepods (Kurbjeweit *et al.*, 1993).

In the region immediately below the ice sheet, another type of community is found. Firstly, the ice bottom can provide support for fronds of diatoms, specifically *Melosira*

Figure 8.2. Assemblages of the diatom *Melosira arctica* hanging from multiyear ice on the NP-22 drifting station at 81°N 138°E, July 1980 (Melnikov, 1997).

arctica, which can reach lengths of 15 m (Melnikov and Bondarchuk, 1987). Fig. 8.2 (after Melnikov, 1997), shows one such assemblage under the multi-year ice of a Russian drifting station. Less spectacular, but more common, are amphipods which can be found crawling on the ice bottom at densities of up to 60 individuals per square metre; they feed off the algal biomass and use the roughness of the ice underside as a refuge. Pelagic zooplankton such as copepods, and in the Antarctic, krill (*Euphausia superba)* temporarily ascend to the ice bottom to feed, and are themselves the prey for polar fish, thus providing a link between the ice-based primary production and the pelagic animals.

Within the water column under the ice, primary productivity is low because of low light levels. Typically the total algal biomass per unit area of water column is of about the same magnitude as the algal biomass within the ice. High phytoplankton concentrations (with overall chlorophyll-a concentrations above 40 mg m^{-2}) are confined to the marginal ice zone and to polynyas. Here ice melt leads to greater stratification, while the absence (or thinness or sparseness) of the ice cover allows higher light levels. Permanent winter-spring polynyas, such as the North East Water polynya off NE Greenland and the coastal polynyas around Antarctica, have high pelagic productivity. For this reason the MIZ and recurrent polynya sites are heavily used by birds (Brown and Nettleship, 1981) and marine mammals (Stirling, 1980), either as migration routes or, in the case of birds, as feeding grounds relative to nesting sites on neighbouring cliffs. In fact, the location of Arctic seabird rookeries is very dependent on the nearness of open water, with high latitude sites being often close to winter polynyas.

The strong link between high productivity and the existence of reduced ice cover in the MIZ or polynyas has led Gradinger (1995) to speculate on the consequences of climate change for Arctic biology. Sea ice retreat is bound to lead to a more seasonal ice regime, with more open water in winter within the pack and a winter MIZ around the Arctic ice cover where today there is none because the winter ice cover reaches the circumpolar coastlines. The consequence will be a greater biological productivity of the Arctic Ocean and thus a greater amount of biological sedimentation from shells and other body parts of marine animals (fig. 8.3). Thus the Arctic will be more efficient as a biological pump transferring CO_2 from the atmosphere via the food web to the sea floor, as predicted by Anderson et al. (1990). The Antarctic, too, will experience a retreat of the northern ice limit, so that large areas of open water replace winter sea ice. This will enhance overall biological productivity and also (as in the Arctic) will change the composition and range of many polar marine species in ways that we cannot easily predict.

8.2. SEA ICE AND ENVIRONMENTAL THREATS

8.2.1. Oil Spills and Blowouts

The role of sea ice morphology in containing and transporting oil spills received a great deal of attention during the 1970s when offshore drilling in the Canadian Beaufort Sea began (Wadhams, 1976ab, 1981b). The chief concern was that a seabed oil blowout might occur at the end of the summer drilling season (during which drillships were to be used, which had to be withdrawn to harbour in winter) such that it could not be capped off before the following summer. This would allow oil and gas to vent under the ice for a period of months.

If the blowout occurred under fast ice, calculations demonstrated that the situation is relatively straightforward. The heat carried up in the oil-gas plume by entrained water will melt a hole in the ice above the blowout. Oil will gather in this open pool and can be burned off or collected. If the ice is moving, however, the heat flux is insufficient to melt a slot in the ice above the blowout, so oil is deposited on the underside of the ice and carried downstream. Its subsequent fate, in winter, is that more ice grows beneath it, creating an "oil sandwich". The oil rises into brine drainage channels, and in spring, when these channels melt their way to the surface (which occurs in first-year ice but not necessarily in multi-year) the oil appears on the ice surface in small patches, probably at too low a concentration to be burned (fig. 8.4).

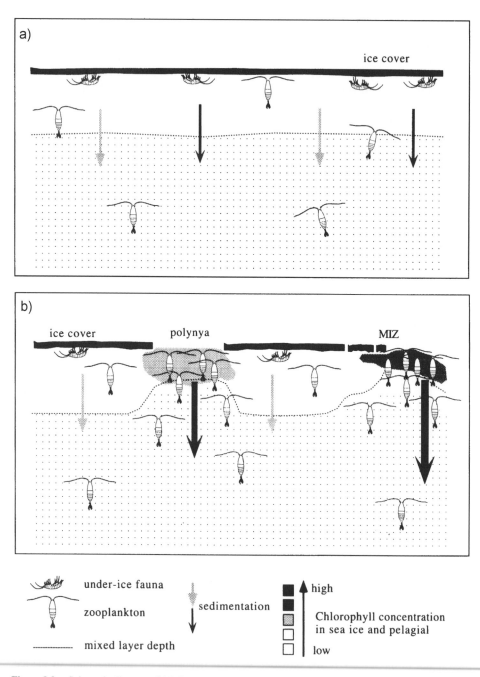

Figure 8.3. Schematic diagram of (a) the present structure of the marine ecosystem in the Arctic Ocean, (b) the likely change due to a reduction in the ice cover and a greater influence of polynyas and MIZ regions (Gradinger, 1995).

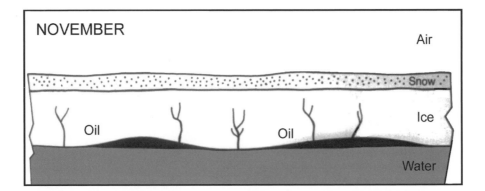

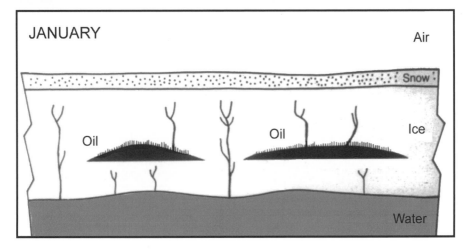

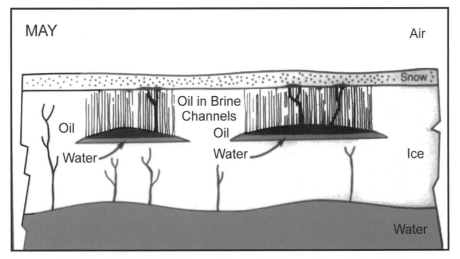

Figure 8.4. The creation of an "oil sandwich" in first-year ice by an oil spill under ice which is encapsulated by new ice growth, with opening brine drainage channels taking the oil to the surface in the following summer.

Figure 8.5. Oil from an experimental under-ice spill being moved by current and gravity in rivulets towards a pressure ridge. Multi-year ice, Beaufort Sea.

The importance of ice topography comes in considering the area of ice bottom painted by the oil from the blowout. If the ice bottom were perfectly smooth, the thickness of the oil layer is governed by surface tension and is limited to 0.5–1 cm, resulting in a very large area receiving a low concentration of pollutant. The faster the ice moves, the more episodic will be the spattering of oil patches that the underside receives from the rising plume. In rough topography, however, it is possible for the oil to flow towards high points in the underside topography, and form pools on the upstream side of pressure ridge keels, resulting in a higher concentration of pollution over a smaller area (fig. 8.5). Such flow and collection in pools was observed by divers during deliberate oil releases (NORCOR, 1977). The relevant statistic needed for the estimation of the magnitude of this effect is the along-crest variability of ice draft in a keel, since it is the minimum draft in a given length of keel crest which determines the ability of that stretch of keel to act as a dam for oil. Wadhams (1976a) was able to find only one piece of data on along-crest variability, obtained by an autonomous underwater vehicle (Francois and Nodland, 1972). To generate such data it is necessary either to steer a vehicle along a keel crest (implying use of an AUV or ROV) or else to use a sensor system which generates a genuine three-dimensional map of the ice underside. A swath-sounding sonar could provide the required information.

Offshore experimental oil spills in the Canadian Arctic are now forbidden. In the absence of new data, the earlier conclusions still stand, that an oil blowout under fast ice might be relatively benign and local in its effects, whereas a blowout under moving pack would create a trail of discontinuous oil patches on the undetectable underside of ice floes (with pools forming against pressure ridges), which would become incorporated into the ice by fresh growth underneath, and which would reach the surface in a subsequent summer, hundreds of km downstream, as small surface patches that cannot be burnt. Worse still would be the likelihood that some of the oil would end up in open summer leads, which are used by millions of migratory seabirds.

8.2.2. The Transport of Pollutants By Sea Ice

Until relatively recently the polar regions were thought to be a pristine place, perhaps the last place on Earth untouched by Man and devoid of contamination by Man's activities. That picture was wrong. Evidence has grown of high contaminant concentrations at various levels of the Arctic marine food chain: polar bears on Svalbard with high levels of PCBs (polychlorinated biphenyls, a type of pesticide) (Norheim *et al.*, 1992); sea birds with apparently fatal levels of PCB in their brains; Greenland Inuit with high mercury levels from eating seals (Hansen, 1990). Even in the Antarctic PCBs have been detected in the shells of albatross eggs, and are possibly responsible for their thinning and for consequent lower reproductive success. But it is in the Arctic that pollution is greater, clearly due to the proximity of northern hemisphere industrial centres. Does sea ice provide a way by which pollutants can be transported and concentrated?

Firstly we must ask how pollutants can enter sea ice. There are four possible sources:

(i)   the **water column** itself. Sea ice formation is a way of purifying water, in that, as we have shown in chapter 2, very few types of molecule can enter the ice crystal lattice. However, as we have also shown, liquid brine can continue to exist in sea ice in the form of brine cells, so any pollutant dissolved in water can be incorporated into the sea ice, though at a reduced concentration (by a factor of S_i/S_w, where S_i is mean ice salinity and S_w is water salinity).

(ii)   **sediment particles**. It has been found (Pfirman *et al.*, 1993) that Arctic sea ice which was formed in shallow water on the Russian shelves contains high concentrations of sediment particles distributed throughout the ice fabric. The origin of these is believed to be the **suspension freezing** process which occurs on the shelves during early autumn and which was discussed in section 1.2.4. Essentially, severe storms cause bottom sediments to become suspended in the water column through wave action, then the water freezes to form frazil and pancake ice, incorporating sediment particles as frazil nuclei or just as material adhering to the frazil mass. The subsequent freezing into a full ice sheet cements the sediment particles in place and they are subsequently carried westward in the Trans Polar Drift Stream when the ice moves off the shelf into deep water. Pollutants transported seaward in Russian rivers and then deposited off the river mouths tend to sorb onto such fine-grained sediment, and so remain part of the sediment when it becomes absorbed into sea ice.

(iii) direct pick-up of **sediment masses** by pressure ridge keels scouring the sea bed, or by a spring run-off of sediment-laden river water over fast ice, which subsequently breaks out and joins the moving pack.

(iv) deposition from the **atmosphere** of pollutants by fall-out from Arctic haze, or in snow, followed by transport by the sea ice onto which it falls.

The types of contaminant which are found in the marine environment, and hence in sea ice, have been classified by Barrie *et al.* (1992) thus:

1. Chlorinated industrial organic compounds (chlorobenzenes, CBZs; polychlorinated biphenyls, PCBs; dioxins and furans).
2. Organic pesticides (polychlorinated camphenes, PCCs; hexachlorocyclohexanes, HCHs; chlordane; chlorinated hydrocarbons such as DDT and DDE).
3. Polycyclic aromatic hydrocarbons, such as benzopyrene.
4. Heavy metals, such as mercury, cadmium and lead; and arsenic.
5. Acids such as sulphur oxides and nitrogen oxides.
6. Radionuclides, including 1950s and 1960s bomb test products; Chernobyl products such as ^{137}Cs; radioactive waste and decommissioned reactors from nuclear ships, often deliberately dumped in the offshore Russian Arctic as detailed in the Bellona reports (Nilsen and Bøhmer, 1994; Nilsen *et al.*, 1996); and potential run-off from land-based sources such as research reactors, bomb-making facilities and reprocessing plants.

Actual measurements on contaminant concentrations in sea ice are sparse, and have been reviewed by Lange and Pfirman (1998). Examples include PCB levels in Siberian seas which have been measured at 1500–2500 pg/l in sea ice, 1000 in sea water and 1500 in overlying snow (Melnikov and Vlasov, 1992). Dethleff *et al.* (1993) found 100-3000 pg/l in sea ice of the Laptev Sea. An extraordinarily high value of 15500–20300 pg/l was found at a station just north of Svalbard. Measurements on heavy metals have shown, both in the Laptev Sea and in Baffin Bay (Campbell and Yeats, 1982), that concentrations of these elements in sea ice greatly exceed their concentration in surface waters, implying transport of pollutant-laden sediments.

The main incorporation and transport processes are probably as follows. Industrial contaminants, pesticides, metals and possibly radionuclides are transported in northward-flowing Russian rivers, which drain about half of the Asian land mass including most Russian industrial centres. As shown in chapter 1, the rivers flow into the Russian shelves, where the river water is diluted by polar surface water. Dissolved pollutants can therefore spread over the whole Arctic, though diluted by mixing, and can become incorporated into sea ice via brine cells, again with further dilution. More potent is the larger mass of pollutants which are in the form of particles deposited and absorbed into the sediments near the river mouths. The suspension freezing process in early winter refloats the pollutant-laden sediment particles which become absorbed into the newly forming sea ice. A typical route for that ice is then to be transported by the Trans Polar Drift Stream through Fram Strait and into the Greenland Sea. As the ice passes through one or more summer melt seasons in the Arctic Basin the sediments become relocated within the ice sheet. Some sediment is lost to the ocean, either through the enlarged brine drainage channels or by flushing through the percolation of meltwater. Surface melt concentrates the sediments

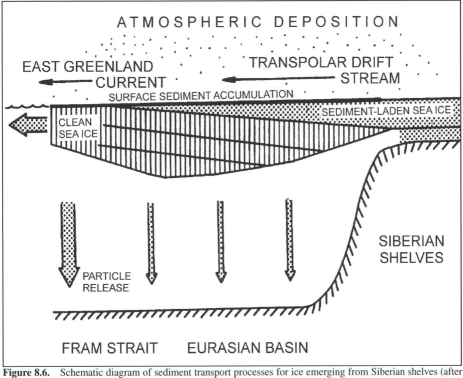

Figure 8.6. Schematic diagram of sediment transport processes for ice emerging from Siberian shelves (after Lange and Pfirman, 1998).

which are near the top of the ice into a more concentrated surface layer. If these sediments occur in clumps, the dark-coloured material will absorb more solar radiation, will warm up and melt the ice immediately around it, and will then sink into a hole of its own creation, called a **cryoconite.** In the rather similar case of wind-blown sediment on glacier and iceberg surfaces, these cryoconites can contain and nurture a self-contained ecosystem of bacteria or algae feeding off the sedimentary material and making use of the liquid water. During the following winter new clean ice grows on the ice underside, so that the sediment load becomes increasingly concentrated in the upper part of the ice sheet (fig. 8.6), where it is joined by aerial pollutants deposited in snow or directly from atmospheric fallout. The pollutants are then lost to the ocean when the ice sheet as a whole melts, i.e. either in the marginal ice zone or at the downstream limit of ice drift. The downstream ice limit in the Greenland Sea moves north and south with season, but the MIZ (e.g. in Fram Strait) moves within much narrower limits and so can be expected to be an area of high pollutant deposition into the ocean.

In the Antarctic the absence of industry, of sediment-laden rivers, and of shallow shelves means that pollution levels are far lower and are mainly confined to small amounts of air-deposited pollutants. The transport and melt mechanism is similar, with deposition in the ocean at or near the northern ice limit.

8.3. SEA ICE AND CLIMATE

8.3.1. Climate Change and the Polar Regions

We are all aware that Man is engaged on an uncontrolled experiment in modifying the climate of his planet through the increasing release of carbon dioxide and other climatically active gases such as methane into the atmosphere as a result mainly of industrial development. These gases produce the so-called **greenhouse effect,** actually a man-induced enhancement to a natural greenhouse effect caused by water vapour and naturally-occurring carbon dioxide and methane. Such gases alter the heat balance of the earth by allowing short-wave radiation from the sun through the atmosphere to warm the earth, but absorbing the long-wave radiation which the earth emits by virtue of its own temperature. This change in heat balance causes the temperature of the lower atmosphere to increase, while the temperature of the upper atmosphere decreases.

We know that the greenhouse effect is real because the **natural greenhouse effect** is what allows liquid water — and hence life — to exist on earth at all. Radiation balance calculations (Houghton, 1997) show that if the atmosphere consisted of nitrogen and oxygen only, without carbon dioxide, methane or water vapour, and thus without clouds, then the average global surface temperature would be –6°C as compared to its real value of +15°C. The **enhanced,** or **anthropogenic, greenhouse effect** is the additional effect which Man produces by virtue of adding to the natural levels of carbon dioxide and methane through fossil fuel burning, deforestation and other practices, with an accompanying change in water vapour levels (generally positive as the air temperature increases). The relative efficacy of the different components which Man is adding to the atmosphere is usually expressed in terms of their **radiative forcing** effect — that is, the net imbalance at the top of the atmosphere due to the presence of the anthropogenic gases, an imbalance which is remedied by a change in the surface temperature of the earth. Figure 8.7 shows the change in radiative forcing that is predicted during the next century due to various gases. The non-CO_2 trace gases which have the most effect are methane, tropospheric ozone (offset by stratospheric ozone, which tends to reduce the incoming radiation and so has a negative forcing) and nitrous oxide. Note the moderating effect of aerosols due to the burning of biomass and the production of sulphates from the effect of sunlight on sulphur dioxide emitted by fossil fuel burning. These bring down the total radiative forcing so that it is approximately the same as the forcing due to CO_2 alone. The scenario used for these predictions is the so-called IS92a scenario, published in the 1992 report of the Intergovernmental Panel on Climate Change (Leggett *et al.*, 1992), and regarded as the most likely development of Man's emissions in the absence of serious efforts to reduce them — it is known as the "business-as-usual" scenario. The CO_2 doubling time is about 70 years and is associated with a radiative forcing of about 4 W m^{-2}.

In recent years enormous efforts have been devoted to modelling the likely effect of this radiative forcing on global climate. The result has been a series of **General Circulation Models** (GCMs), developed on large computers by a number of university groups and meteorological institutes around the world. The models have had to take account of **feedbacks** within the climate system, which change the magnitude, rate and geographical location of the warming. The four most important feedbacks are:

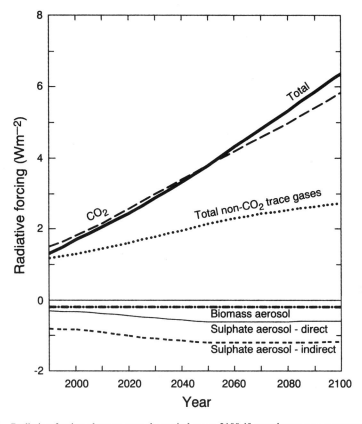

Figure 8.7. Radiative forcing changes over the period up to 2100 if greenhouse gas concentrations follow the IS92a emission scenario. The "total non-CO_2 trace gases" comprise methane, nitrous oxide, tropospheric ozone and halocarbons, offset by the negative effect of stratospheric ozone. Note the negative effect of aerosols (from Houghton *et al.*, 1996).

(i)   **Water vapour feedback,** whereby more evaporation occurs into a warmer atmosphere from ocean and wet land surfaces, giving a higher water vapour content and thus greater warming, since water vapour is a powerful greenhouse gas;

(ii) **Cloud-radiation feedback,** which can be positive (high, icy clouds which act as blankets to thermal radiation from the earth more effectively than they act as screens to incoming solar radiation) or negative (low, thick, watery clouds which act in the opposite sense);

(iii) **Ocean circulation feedbacks,** whereby the oceans' heat capacity causes them to warm more slowly than the atmosphere, and the action of ocean dynamics (which may be changed by global warming) as a dominant factor in poleward heat transport;

(iv) **Ice-albedo feedback,** the critical positive feedback for the polar regions, whereby a retreat of the seasonal snowline, or of the sea ice limit, causes high-albedo snow or ice surfaces to be replaced by low-albedo land or water surfaces. On its own this has been estimated to increase the globally averaged rate of warming by 20%.

Two types of model have emerged from these efforts. **Equilibrium models** (e.g. Mitchell *et al.*, 1990) consider the radiative effect of doubling the CO_2 concentration and then allowing the climate to reach equilibrium. They predict a surface warming at high latitudes, north and south, which is greater than the global average in winter but smaller in summer. They also predict an increase in precipitation, with a larger increase where there is larger warming, and a decrease in the area of sea ice and of seasonal snow cover. These models are unrealistic in the sense that they do not allow for time-dependent feedbacks such as (iii) above, i.e. the slowing of the rate of warming in oceanic regions due to the heat capacity of the ocean. **Time-dependent models** (e.g. Bretherton *et al.*, 1990), which have largely replaced equilibrium models, add CO_2 at a realistic rate and include rate-dependent responses. Their predictions are qualitatively similar to equilibrium models, except that at high southern latitudes the warming is predicted to be very small because of the overwhelming influence of the Southern Ocean.

As an example of the zonally and seasonally averaged predictions of a time-dependent model, we may quote the following predictions from Stouffer *et al.* (1989) for northern latitudes:

Years ahead	Warming at Equator	at 60°N	at 90°N
20	0.5°C	0.6°C	0.8°C
40	1.1	1.6	2.0
60	1.6	2.8	3.6
80	2.4	4.8	6.5

The steadily increasing enhancement at high northern latitudes is apparent. The same model predicts warming of only about 1.2°C in 80 years in the zone 60–70°S where Antarctic sea ice is found.

The main reasons why an Arctic enhancement occurs in models is that the Arctic troposphere is stable and thin, so that warming is concentrated over a shorter atmospheric column, and that the seasonal snowline on land in the Northern Hemisphere will show a rapid response to warming, retreating and thus replacing a high-albedo (about 0.8) surface by one of low albedo (about 0.15). This albedo feedback acts strongly to enhance the warming rate. A further minor feedback comes from thinning of sea ice, which allows a greater flux of heat through the ice from the ocean in winter (especially if ice concentration is reduced as well).

The regional nature of the Arctic enhancement has been explored in the most recent GCM developments. The caveat is that "confidence in predictions on a regional basis using global coupled models is still, as yet, relatively low" (Cattle and Crossley, 1995). However, since the latest GCMs are successful in hindcasting the very variable rate of global warming over the past century, and the latest Hadley Centre model (Wood *et al.*, 1999) can function without flux correction terms (which were needed to stop model predictions from diverging too far from physical reality) there is increased confidence that climate modelling, regional as well as global, is becoming more accurate and reliable. On this basis we show in fig. 8.8 the predictions of a version of the Hadley Centre model for temperature change over the Arctic in winter and summer over the CO_2 doubling period (Cattle and Crossley, 1995). It can be seen that there is relatively little change over the

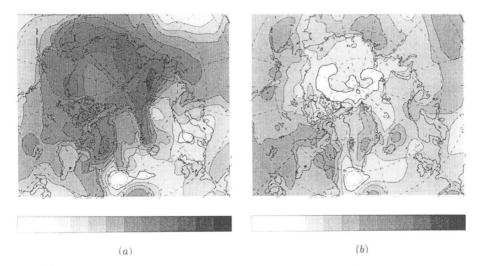

(a)   (b)

Figure 8.8.   Temperature change over the Arctic predicted for the decade of doubling of CO_2 levels using the Hadley Centre GCM (Cattle and Crossley, 1995). (a) is winter (December–February). (b) is summer (June–August).

Arctic Ocean in summer, since the air temperature is constrained by melting ice to be at or near 0°C. In winter, however, there is a large warming over many parts of the Arctic, specifically the marginal ice zones of the Greenland and Barents seas (where we may confidently expect ice retreat), the Canadian Arctic Archipelago, NE Siberia and its shelf seas. The pattern of future warming corresponds quite closely to the observed pattern of recent warming over the last 18 years, shown in fig. 8.9.

Since the albedo feedback mechanism acts more readily on the thin seasonal snow cover than on the thicker seasonal sea ice cover, we might also expect to be seeing evidence of snowline retreat in the North. Indeed, evidence is available from satellite imagery (Folland *et al.*, 1990) of a rapid decrease since 1980; over Eurasia the snow extent decrease during autumn and spring is 13% and 9% respectively for the 1980s relative to the 1970s. An additional effect which is well known and documented is the net retreat of northern hemisphere mountain glaciers since the late 19th Century, which has made a substantial contribution to global sea level rise (e.g. Meier, 1984). A more subtle question is what the effects will be upon the sea ice cover, specifically the Arctic sea ice cover since changes in the Antarctic are predicted to be quite minor.

8.3.2. Climate Change and Sea Ice Extent

The Arctic Ocean sea ice cover at present has an extent of some 16 million km^2 in winter and 9 million km^2 in summer, as we have seen in chapter 1. This is sufficient to fill the Arctic Basin in winter, as well as the Bering Sea as far as the shelf break, part of the Sea of Okhotsk, the channels and bays of the Canadian Archipelago, the Labrador coast, the East Greenland coast as far as Cape Farewell in bad years, and the shelf seas north

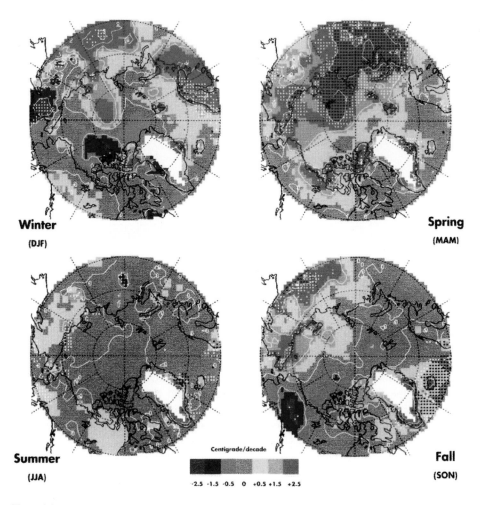

Figure 8.9.   Observed temperature changes over the Arctic during the period 1979–1997 (after I. Rigor, *Journal of Climate*, in press). Trends in grid cells that are significant at 95% level are marked with white dots; at 99% level with black dots.

of Russia. In summer the ice cover retreats to the Arctic Basin proper and part of the East Greenland coast. In most summers the ice withdraws from the Asian and American coasts of the Arctic Basin, permitting the transit of cargo vessels under icebreaker escort. This has allowed the development of a summer trade route, the Northern Sea Route, across the north of Russia (Brigham, 1991) with a seasonal length of 140–150 days. The Northwest Passage across Arctic Canada, although navigable to ice-strengthened ships, has not yet developed as a regular route for transit trade; in fact up to 1990 it had been traversed by only 50 ships in all (Pullen and Swithinbank, 1991). The summer retreat of the Arctic pack facilitates oil exploration in the coastal seas of the Arctic, since in summer it can be carried out from drill ships in ice-free waters. Thus the question of whether global warming will have a significant impact on the extent of Arctic sea ice is of importance both to economists and climatologists, as well as to the peoples of the Arctic. Improved navigability of the Arctic Ocean would enable it to fulfill a commercial role as a short cut between Europe and the Far East as well as assisting hydrocarbon development, while the climatic role of sea ice as an insulator of the sea surface and as a high-albedo reflective surface for incoming radiation suggests that any change in ice extent may have unexpected feedbacks for the world climate system as a whole.

If we adopted a simple thermodynamic view of sea ice, we might expect a retreat of sea ice limits under global warming in much the same way as a retreat of snowlines, and having much the same effect — an enhancement of warming through ice-albedo feedback. However, we have already shown how, in the Arctic, the distribution and thickness of sea ice are driven by dynamics rather than thermodynamics — for instance, the fact that the thickest ice is not found in the centre of the Arctic Ocean. Therefore the response of sea ice limits to global warming may well be a complex one.

Fortunately, much empirical data are available on the recent history of sea ice extent in the Arctic, thanks to the use of passive microwave sensors. Analyses where datasets from the SMMR and SSM/I sensors have been combined and reconciled, show that the sea ice extent in the Arctic has declined at a decadal rate of some 2.8–3% since 1978 (Bjørgo *et al.*, 1997; Cavalieri *et al.*, 1997), with a more rapid recent decline of 4.3% between 1987 and 1994 (Johannessen *et al.*, 1995). Fig. 8.10 (from Parkinson *et al.*, 1999) shows how an apparently fairly stable annual cycle of large amplitude (fig. 8.10(a)) reveals a distinct downward trend of area (fig. 8.10(b)) when anomalies from interannual monthly means are considered. Fig. 8.10(c) shows that the downward trend occurs for every season of the year; the estimated mean annual loss of ice area is $(34{,}300 \pm 3700)$ km^2.

The steady hemispheric decline in sea ice extent masks more violent regional changes. In the Bering Sea there was a sudden downward shift of sea ice area in 1976 (Niebauer, 1998), indicating a regime shift in the wind stress field as the Aleutian Low moved its position. In the Arctic Basin a passive microwave analysis (Parkinson, 1992) of the length of the ice-covered season during 1979–86 showed a see-saw effect, with amelioration in the Russian Arctic, Greenland, Barents and Okhotsk seas and a worsening in the Labrador Sea, Hudson Bay and Beaufort Sea. A further analysis (Parkinson *et al.*, 1999) extended the coverage to 1978–96 and confirmed these results: the Kara/Barents Sea region had the highest rate of decline in area, of 10.5% per decade, followed by the seas of Okhotsk and Japan and the central Arctic Basin at 9.7 and 8.7% respectively. Lesser declines were experienced by the Greenland Sea (4.5%), Hudson Bay (1.4%) and the Canadian Arctic Archipelago (0.6%). Increases were registered in the Bering Sea (1% – the starting date being later than the 1976 collapse), Gulf of St. Lawrence (2%) and Baffin Bay/Labrador Sea (3.1%). Taking the more modern data into account it is clear that the see-saw effect discovered over 1979–86 has been largely submerged in a general retreat. It can also be seen that the regions of most rapid retreat generally agree with the regions of most rapid warming predicted in fig. 8.8.

For the Antarctic, GCMs show very large discrepancies in their predictions. Early equilibrium models, such as that of Hansen *et al.* (1988), predicted that the warming in the ice regime of the Antarctic Ocean will be almost as great as in the Arctic and much greater than that occurring at low latitudes. However, time-dependent models, with their incorporation of ocean circulation feedback effects into a region dominated by the South-

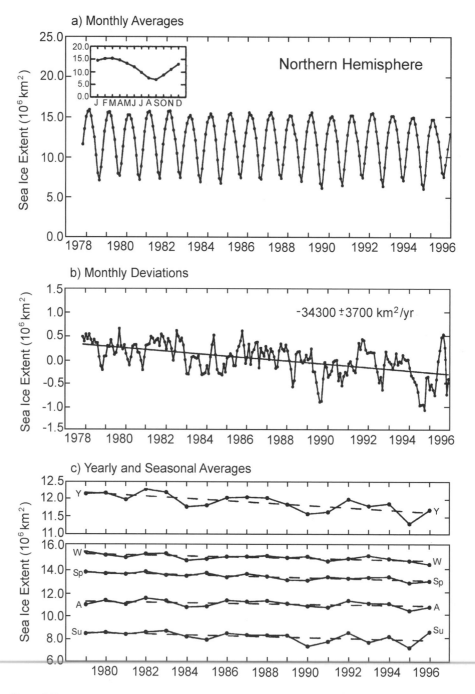

Figure 8.10. (a) Monthly averaged N Hemisphere sea ice extents from SMMR and SSM/I data, November 1978 – December 1996. Inset shows average seasonal cycle. (b) Monthly deviations of the extents from the 18-year average, with linear trend shown. (c) Yearly and seasonally averaged ice extents: W = January–March, Sp = April–June, Su = July–September, A = October–December. (Parkinson *et al.*, 1999).

ern Ocean (e.g. Stouffer *et al.*, 1989; Gates *et al.*, 1992), predict that the Antarctic sea ice zone will have the lowest temperature rise of any latitude in the globe, amounting to less than 1° in 60 years.The generation and interpretation of fresh oceanographic data, particularly in winter (e.g., Gordon and Huber, 1990), will do much to improve our understanding of the role of Antarctic sea ice in climate change, assisted by the use of models which deal with air-sea-ice interaction in the absence of ice dynamics or ocean advection and which in this way enable sensitivity studies to be carried out. An example of such a model is that of Martinson (1990), who demonstrated that a complex set of feedback mechanisms comes into play if a parameter such as air temperature is changed. The balance of lead concentration, upper ocean structure, and pycnocline depth will adjust itself to minimise the impact of changes, and he concluded that Antarctic first-year sea ice, apparently a thin and delicate skin that may easily be removed by a modest warming, is more likely to be quite resilient and resistant to the impact of warming. By contrast, some coupled sea ice — mixed layer models, such as that of Lemke *et al.* (1990), show responses of ice edge position to climate change which are essentially linear — the ice edge advances or retreats approximately in proportion to the changes in air temperature.

Observational data on Antarctic sea ice extent show no significant trend. Parkinson (1998) examined passive microwave data for the 1988–94 period. She found no evidence of an overall trend in extent, but some evidence suggesting that anomaly patterns propagate eastward, offering support for the idea of an Antarctic circumpolar wave in surface pressure, wind, temperature and sea ice extent put forward by White and Peterson (1996). During the seven years of the study, ice seasons shortened in the E Ross Sea, Amundsen Sea, W Weddell Sea, offshore eastern Weddell Sea, and east Antarctica between 40° and 80°E. Ice seasons lengthened in the W Ross Sea, Bellingshausen Sea, central Weddell Sea and the region 80°E – 135°E. Earlier evidence of a major ice retreat in the Bellingshausen Sea in the summer of 1988–91 (Parkinson, 1995) was shown to be a short-lived phenomenon. A longer-term statistical analysis of passive microwave data from 1978 onwards (Cavalieri *et al.*, 1997) gave a small and not statistically significant upward trend in overall Antarctic ice extent, of some 1.3% per decade. The conclusion is that sea ice extent and GCM predictions agree in showing no strong trend.

8.3.3. Climate Change and Fast Ice Thickness

Clearly if climatic warming occurs one might expect a thinning of the sea ice cover due to the increased radiation flux, warmer air temperatures and the reduced heat flux through the ice. However, as we have seen from our treatment of thermodynamics in chapter 3, two compensating effects may occur. Warming may increase the overall precipitation over the drainage basins of Siberian and northern North American rivers, and together with a contribution from retreating mountain glaciers this could increase the annual run-off into the Arctic Ocean. The low salinity surface layer of the Arctic Ocean preserves the ice cover, and a further reduction in its salinity would increase the steepness of the pycnocline and reduce oceanic heat flux to the ice underside — thus increasing thermodynamic ice thickness. Similarly, if warmer, moister winds prevailed over the Arctic such that snowfall were greatly increased over the present 15–30 cm (Gorshkov, 1983), then this too would increase the equilibrium ice thickness; as we showed in chapter 3, the increase must be so great that snow remains on the ice throughout the summer. So even in the case of thermodynamic ice growth it is not clear that warming would necessarily lead to a thinning of the ice cover.

There is one case where crude predictions for temperature increase can be applied directly to the ice response. Fast ice is ice which has grown thermodynamically around the coasts of the Arctic. It may be fast to the seabed, often via pinning from grounded pressure ridges as in the case of the north coast of Alaska (Reimnitz *et al.,* 1978), or it may be fast to the sides of a channel or basin of limited dimensions, as is the case in much of the Canadian Arctic Archipelago. The underlying water mass is usually only a single-layer system of polar surface water, so that oceanic heat flux is negligible. The only factors affecting thickness are then radiation fluxes, air temperature and snow cover. Under these circumstances the thickness of the ice is determined almost entirely by air temperature history (modified by the thickness of the snow cover, which alters the growth rate). The growth equation was first developed by Stefan (1890), and relationships between thickness achieved and the number of degree days of freezing since the beginning of winter have been proposed and tested against field observations (Bilello, 1961), with a further treatment for melt rates in summer (Bilello, 1980). A set of growth curves for Resolute in the Canadian Arctic was shown in fig. 3.7; it can be seen that increasing snow thickness gives a negative offset to ice thickness. We can easily see from such curves that if the average daily air temperature increases by a known amount θ, the number of degree-days of cold will decrease by 365θ, the ultimate ice thickness will diminish by an amount which can be read off the curves, and the ice-free season will lengthen. Using this technique, Wadhams (1990c) predicted that in the Northwest Passage and Northern Sea Route an air temperature rise of 8°C (equivalent to about a century of Arctic warming) will lead to a decline in the winter fast ice thickness from 1.8–2.5 m (depending on snow thickness) to 1.4–1.8 m, and an increase in the ice-free season from 41 days to 100 days. This effect would be of great value to oil exploration, production and transport, and to the extension of the navigation season in the Northern Sea Route and the Northwest Passage.

Even in this comparatively simple case, however, there are still possible negative feedbacks. If Arctic warming produces increased open water area and thus increased atmospheric water vapour content, it would lead to increased precipitation. As already mentioned, thicker snow cover decreases the growth rate of fast ice, as has been directly observed (Brown and Cole, 1992), until the thickness is increased to the point where the snow does not all melt in summer, in which case the protection that it offers the ice surface from summer melt leads to a large increase in equilibrium ice thickness.

8.3.4. Climate Change and Moving Pack Ice Thickness

The fact that the Arctic Ocean has a dynamic rather than a static ice cover makes the effect of climate change far more difficult to predict. Thermodynamic growth and decay rates no longer determine the area-averaged mean thickness of the ice cover. Wind stress acting on the ice surface causes the ice cover to open up to form leads, and later under a convergent stress refrozen leads and thicker ice elements are crushed to form pressure ridges. One result is a redistribution of ice from thinner to thicker categories, with the accompanying creation of open water areas. Another is to make the ice cover as a whole more resistant to convergent than to divergent stresses, and this causes its motion field under wind stress to differ from that of the surface water. Thus the exchanges of heat, salt and momentum are all different from those that would occur in a fast ice cover. We can only indicate a few of the interactions and how they may be changed by global warming.

Firstly, the effects of variable thickness are very important. On account of the creation of leads by the ice dynamics and the very high growth rates which occur in thin ice, it has been found that the overall area-averaged growth rate of ice is dominated (especially in autumn and early winter when much lead and ridge creation take place) by the small fraction of the sea surface occupied by ice less than 1 m thick (Hibler, 1980). In fast ice, climatic warming increases sea-air heat transfer by reducing ice growth rates. However, over open leads a warming would decrease the sea-air heat transfer, so the area-averaged change in this quantity over moving ice (and hence its feedback effect on climatic change itself) depends on the change in the rate of creation of new lead area, which is itself a function of a change in the ice dynamics, either driving forces (wind field) or response (ice rheology).

Already we see that ice dynamics are important through this thickness feedback. But they also have other effects. In the Eurasian Basin the average surface ice drift pattern is a current (the Trans Polar Drift Stream) which transports ice across the Basin, out through Fram Strait, and south via the East Greenland Current into the Greenland and Iceland Seas where it melts. This implies that a typical parcel of ice forms by freezing in the Basin, the latent heat being transferred to the atmosphere; is then transported southward (which is equivalent to a northward heat transport); and then when it melts in the Greenland or Iceland Sea it absorbs the latent heat required from the ocean. The net result is a heat transfer from the upper ocean in sub-Arctic seas into the atmosphere above the Arctic Basin. A change in area-averaged freezing rate in the Basin would thus cause a change of similar sign to the magnitude of this long-range heat transport. An identical argument applies to salt flux, which is positive into the upper ocean in ice growth areas and negative in melt areas. Thus salt is also transported northward via the southward ice drift. A relative increase in area-averaged melt would cause increased stabilisation of the upper layer of polar surface water, and hence a reduction in heat flux by mixing across the pycnocline, while a relative increase in freezing would cause destabilisation and possible overturning and convection. Deep convection already occurs in winter in the central Greenland Sea, and a mechanism has been proposed (Rudels, 1990) that assigns local freezing a dominant role in the process. If warming were to reduce the freezing rate in this region it could turn off the convection, so that ventilation of the deep Greenland Sea would cease, Greenland Sea Deep Water production would stop, the convection process could no longer play a useful role in CO_2 sequestration, and ultimately the Atlantic thermohaline circulation might slow down (see section 8.3.5).

Finally, Hibler (1989) has drawn attention to the role of ice deformation in reducing the sensitivity of ice thickness to global warming in areas of net convergence. The largest mean ice thicknesses in the Arctic — 7 m or more — occur off the Canadian Arctic Archipelago (Hibler, 1979) where the wind stress drives ice against a downstream land boundary. Here the mean ice thickness is determined by mechanical factors, largely the strength of the ice, which set a limit to the amount of deformation by crushing that can occur. In this area the thickness is likely to be insensitive to atmospheric temperature changes. The main sensitivity would be to a change in the overall pattern of winds over the Arctic.

Given the complexity of these interactions and feedbacks, it is not at all clear what the overall effect of an air temperature increase on the Arctic ice cover and upper ocean would be. Sensitivity studies using coupled ocean-ice-atmosphere models are required, but results are not yet available. At this time the best that we can do is to look for empirical evidence of trends in ice thickness. Some of these have already been discussed in chapter 5, while new evidence is discussed in the next section.

In the Antarctic where the ice cover is divergent and where land boundaries are less important, it is more reasonable to suppose that the main effect of global warming will be a simple retreat of the ice edge southward, which would be accompanied by a thinning. Data on Antarctic sea ice thickness to date are far too sparse to permit any comment on observed trends.

8.3.5. Recent Changes in the Arctic Ocean

In the preceding sections we have examined from a theoretical point of view what changes we might expect to see in sea ice under climate change. Remarkably, it seems that many of these effects are already with us. During the past five years, field measurements of ice and ocean characteristics in the Arctic have revealed some startling changes. The data have come from deep penetrations of the Arctic by the German icebreaker "Polarstern" and the Swedish "Oden" between 1987 and 1996; a complete Trans-Arctic section by the US "Polar Sea" and Canadian "Louis S. St. Laurent" in summer 1994; and under-ice cruises by US and British submarines.

The first discovery was that the Atlantic sublayer in the Arctic Ocean, which lies beneath the polar surface water and which derives from the North Atlantic Current, has warmed substantially (by 1–2°C at 200 m depth) and increased its range of influence relative to water of Pacific origin (e.g. Carmack et al., 1997; Morison et al., 1998). The front separating the two water types has now shifted from the Lomonosov to the Alpha-Mendeleyev Ridge. This warmer and shallower sublayer should increase the ocean heat flux into the bottom of the ice, and indeed this period has also corresponded to a thinning of the ice cover in the Eurasian Basin and northern Greenland Sea (Wadhams, 1990), with further thinning seen in the latest data sets (Wadhams and Davis, in press; Rothrock et al., 1999). Rothrock et al. compared US submarine data acquired during 1993–7 with data from 1958–76 and found a decrease of mean ice draft for the end of the melt season from 3.1 m to 1.8 m, with the effect being greatest in the central and eastern Arctic, i.e. the Trans Polar Drift Stream. The structure of the ice thickness distribution in datasets from the Trans Polar Drift Stream shows that the best explanation for its downstream evolution is that bottom melt now begins within the Arctic Basin, even in winter, and that the melt rate is greater for thicker ice (Wadhams, 1997b).

The next discovery was that the structure of the polar surface layer has itself changed. In the Eurasian Basin there was formerly a cold halocline layer in the 100–200 m depth range, where temperature stayed cold with increasing depth despite salinity rising. Its existence was due to riverine input from Siberia, which has recently diverted eastward due to a changed atmospheric circulation, causing a retreat of the cold halocline (Steele and Boyd, 1998) and possibly associated with a recently observed summertime retreat of sea ice in the Beaufort Sea sector (Maslanik *et al.*, 1996; McPhee *et al.*, 1998) (fig. 8.13).

Further changes relate to the Atlantic sub-Arctic seas. The area flux of sea ice through Fram Strait has been increasing since 1979 (Kwok and Rothrock, 1999; Vinje *et al.*, 1998) while the Denmark Strait overflow has warmed (Dickson *et al.*, 1999). Between these two gateways the central part of the Greenland Sea gyre, in the vicinity of 75°N 0–5°W, has seen profound changes. This is normally the site of strong wintertime convection, driven by salt fluxes from local ice growth over the cold Jan Mayen Current, which produces the tongue-like Odden feature (Wadhams, 1999). Cold off-ice winds move newly formed ice eastward within the tongue so that the net salt flux in the western part of the feature is strongly positive; this is where convection occurs. Tracer experiments have shown that deep convection has failed to reach the bottom since about 1971 (Schlosser *et al.*, 1991) and in recent years has been greatly reduced in volume and confined to the uppermost 1000 m, while ice production has also been reduced, with no Odden forming at all in 1994 and 1995 (Wadhams *et al.*, 1999), nor in 1999 and 2000.

What is the reason for these changes? We have already shown how climatic simulations by GCMs predict that the global warming effect due to increased atmospheric CO_2 should be amplified in the polar regions, particularly the Arctic, mainly through the ice-albedo feedback effect (Manabe and Stouffer, 1993; Cattle and Crossley, 1995). Recently the application of one of these models, the CM3 model of the Hadley Centre, Bracknell, to the thermohaline circulation has led to the prediction that convection in the Labrador Sea will collapse quite soon, during the period 2000–2030 (Wood *et al.*, 1999), although it makes no specific prediction about the Greenland Sea. If these driving mechanisms for the thermohaline circulation were to fail, the result would be a lesser transport of heat northwards in the Atlantic by currents. This would diminish the advantage which NW Europe currently enjoys (fig. 8.14) of having a climate significantly warmer than the zonal mean, because of the influence of the Gulf Stream. Instead, it is possible that in the long term our climate could actually cool, and this has been specifically predicted for the 21st Century in a model by Rahmstorf and Ganopolski (1999), nor in 1999 and 2000.

A more immediate cause of many of the observed changes can be identified as a changed pattern of atmospheric circulation in high latitudes. In the North Atlantic sector of the Northern Hemisphere this can be represented by the North Atlantic Oscillation (NAO) index, the wintertime difference in pressure (mb) between Iceland and Portugal, which was low or negative through most of the 1950s–1970s but which has been rising since the 1980s and which has been highly positive throughout the current decade (fig. 8.11). A high positive NAO index is associated with an anomalous low pressure centre over Iceland which involves enhanced W and NW winds over the Labrador Sea (cold winds which cause increased cooling hence increased convection), enhanced E winds over the Greenland Sea in the 72–75° latitude range (causing a reduction in local ice growth in the Odden ice tongue, and a reduced separation between growth and decay regions, hence a reduced rate of convection); enhanced NE winds in the Fram Strait area, causing an increased area flux of ice through the Strait (although the ice may have a reduced thickness, so the volume flux is not necessarily increased); and an enhanced wind-driven flow of the North Atlantic Current, allowing more warm water to enter the Atlantic layer of the Arctic Ocean.

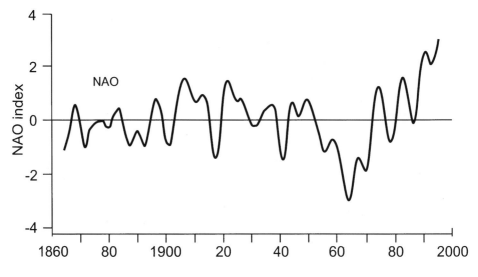

Figure 8.11. The North Atlantic Oscillation index since 1860.

Within the Arctic Basin this pattern is incorporated into a large-scale wintertime pattern (with associated index) called the Arctic Oscillation (AO) (Thompson and Wallace, 1998), which involves a seesaw of sea level pressure (SLP) between the Arctic Basin and the surrounding zonal ring. The anomaly appears to extend into the upper atmosphere, and so represents an oscillation in the strength of the whole polar vortex. Kwok and Rothrock (1999) have analysed the differences in SLP over the Arctic Ocean between high and low NAO years (corresponding also to different phases of the AO index). Fig. 8.12 shows that what had been thought of as the Arctic "norm", i.e. a high over the Beaufort Sea leading to the familiar Beaufort Gyre and Trans Polar Drift Stream as the resulting free drift (along the isobars) ice circulation pattern, is actually the result of a low NAO index (situation b). With a high NAO index (situation a) the Beaufort High is suppressed and squeezed towards the Alaskan coast, causing a reduction in the area and strength of the Beaufort Gyre, and a tendency for ice produced on the Siberian shelves to turn east and perform a longer circuit within the Basin before emerging from the Fram Strait (this applies also to the trajectory of fresh water from Siberian rivers). The weakening of the Beaufort Gyre may well explain the anomalously low summer sea ice extents observed in the Beaufort Sea in 1996–8 (fig. 8.13), since locally melting ice is not replaced by new inputs of ice from the NE. The difference field (situation c) aptly demonstrates these changes if one considers the differential ice drift vectors as occurring along the isobars shown. Proshutinsky and Johnson (1997) were the first to identify situations a and b as two distinct patterns of Arctic Ocean circulation, which they called "cyclonic" and "anticyclonic" respectively.

The key question which awaits an answer is whether the recent Arctic changes are primarily the product of a temporary change of phase of the NAO or AO (leaving aside the question of what causes these changes of phase) or whether an inexorable

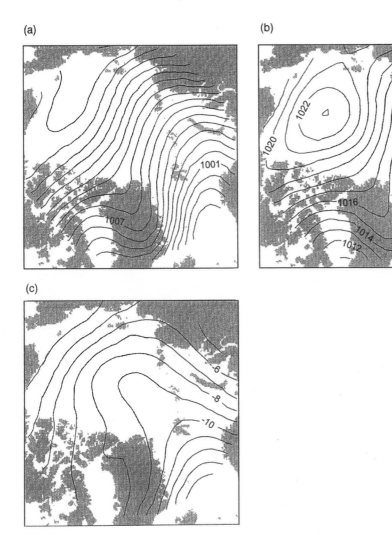

Figure 8.12. Pressure fields over the Arctic Ocean corresponding to (a) a high NAO index, (b) a low NAO index, (c) the difference field (after Kwok and Rothrock, 1999).

general warming trend is already causing irreversible effects in a zone where we already expect greenhouse amplification. If the cause is merely an atmospheric circulation anomaly, then we can expect a change back to earlier conditions at some stage, with a reversion to strong convection in the Greenland Sea, weak convection in the Labrador Sea, extended ice limits in the European Arctic, a colder Atlantic layer in the Arctic Ocean and thicker ice. If an underlying trend is at work instead (or as well), then we can never again expect to see Arctic conditions which resemble those occurring in the 1950–1960s when the Arctic and its ice cover were being intensively investigated for the first time.

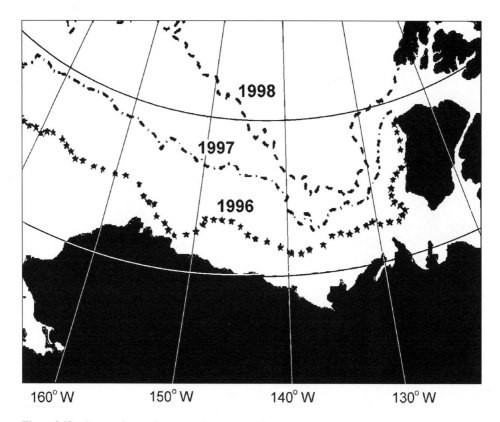

Figure 8.13.   Recent changes in summer ice edge position in the Beaufort Sea.

8.4. THE FUTURE COURSE OF RESEARCH ON SEA ICE

It is certain that the pace of research on sea ice will increase, as the realisation grows of its importance in the global climate system.

We can expect to see an increase in *buoy programmes,* since these offer a means of monitoring ice motion and of providing the basic temperature, air pressure and wind speed data needed to model ice thermodynamics and provide better synoptic pressure fields for modelling ice dynamics. In regions which are difficult or dangerous to access by ship, e.g. multi-year ice regions, "smart" buoys, which possess a large suite of ocean current, temperature and salinity sensors, can provide vital data at a fraction of the cost of a manned survey. One such buoy is the Ice-Ocean Environmental Buoy (IOEB) designed by S. Honjo at Woods Hole Oceanographic Institution. In the Arctic, a buoy programme, the International Arctic Buoy Programme, has been in existence for two decades, funded originally by the US but now with international input. In the Antarctic a variety of national buoy programmes have existed (UK, German, US, Australian, Japanese, Finnish etc), and in 1995 these were combined into the International Programme for Antarctic Buoys (IPAB) under the auspices of the World Climate Research Programme, in order to obtain

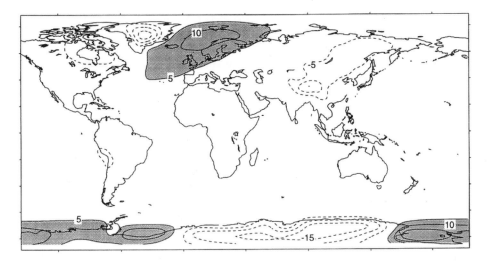

Figure 8.14. Mean average temperature anomalies relative to annual zonal means, showing warming effect of Gulf Stream upon NW Europe and the European Arctic, and anomalously warm coastline of W Antarctica.

maximum scientific benefit from the sharing of circumpolar data. Future buoy development can profit from the two-way communication offered by data links to low-orbit satellite systems such as Orbcomm. This enables not only data to be transferred from the buoy to the laboratory, but commands for switching sensors on and off, or other more complex tasks, to be transmitted from laboratory to buoy.

We can expect improved *satellite monitoring* in future years. SAR satellites with greater capabilities than the current ERS-2 (European) and Radarsat (Canadian) will have wide-beam steerable antennas and multiple frequencies. It will become routine to monitor the entire field of ice motion, ice extent, ice type and ice concentration. Improved passive microwave sensors will be available, while sea ice freeboard will be mapped from advanced radar altimeter systems (the European Cryosat) or laser altimeter systems (the GLAS system of NASA).

It is possible that large-scale *field programmes* involving drifting ice camps will become rare. Such programmes are exceedingly expensive and can only flourish under conditions of genuine international co-operation. In the Arctic, Russia has not maintained a drifting station on the central pack since 1991, and the two most recent drifting experiments were the US SIMI programme (Sea Ice Mechanics Initiative) of 1993–4 in the Beaufort Sea, and the US-Canadian SHEBA (Surface Heat Budget of the Arctic) experiment in 1997–8. In the Antarctic a camp in the western Weddell Sea was successfully deployed and recovered in 1992 in the so-called Anzone programme (LDGO, 1989) involving the US and Russia. On the other hand, the increased availability of ice-capable research ships will mean that *shipborne* field programmes will flourish. An example was the Trans Arctic Section of 1994 (Wheeler, 1997) involving US and Canadian icebreakers, which accomplished the first complete oceanographic section across the Arctic Ocean by surface ship. *Submarine* programmes have been carried out for many years by the US and UK navies, but an innovation was the US SCICEX programme of 1993–9 in which the entire submarine was turned over to civilian researchers rather than incorporating the research work into an operational military cruise. Unfortunately the decreasing availability of nuclear submarines makes it less likely that such programmes will continue.

There will be new techniques of *acoustic monitoring* of changes in the ocean and possibly the ice. Munk and Forbes (1989) carried out in 1991 a preliminary trial for a Global Acoustic Transmission Experiment, based on a source deployed off Heard Island. The aim was to monitor changes in the average temperature and structure of the upper ocean through changes in acoustic transmission time. Receiving systems on the Antarctic coastline (e.g. at Davis Station) allowed the particular properties and variability of the Antarctic Circumpolar Current to be measured. Tomography-type experiments may also be carried out, and it has been predicted (Guoliang and Wadhams, 1989; Jin *et al.*, 1993) that acoustic tomography can be used to measure mean ice thickness within a region by travel time changes. Trans Arctic acoustic propagation was accomplished in 1994 (Mikhalevsky *et al.*, 1999) between Svalbard and Alaska, and it is very likely that these acoustic techniques will be developed further.

Other forms of innovative field measurement that involve monitoring the ice and ocean by remote means include the use of coastal HF radar systems mounted on shore to map surface currents and waves in features such as Antarctic coastal polynyas (Oceanography, 1997), or the use of AUVs (autonomous underwater vehicles) to carry out oceanographic and ice sounding experiments under sea ice and ice shelves, replacing the expensive manned submarine. The use of these and other unmanned systems is bound to increase for economic as well as scientific reasons.

The biggest change in research emphasis, however, as in many other aspects of environmental science, will be the increasing dominance of *modelling* as an approach to sea ice problems. Rapidly increasing computer power makes it possible to develop ice-ocean-atmosphere models which can simulate processes with high time and space resolution. Even general circulation models are now reducing their grid scale to a few km, so that processes in sea ice which affect the global climate system can be incorporated into large-scale models with unprecedented accuracy. The only problem is that the dominance of modelling over the increasingly sparse resources devoted to field programmes may mean that models will be comparing their output with one another rather than with field observations, while unsuspected processes which happen in the real world, and which ought to be incorporated into models, will remain undiscovered.

An important aspect of all of these research strategies will be international co-operation. The sea ice zone is a vast region encompassing every longitude and of interest to every nation. International co-operation is essential for accomplishing fundamental advances in understanding. There are now wide networks for international collaboration in polar science. The most important from the point of view of sea ice and its relation to climate is probably the World Climate Research Programme's ACSYS project (Arctic Climate System Study), which is being developed further into a bipolar programme called CLIC (Climate in the Cryosphere).

Any further attempts at prediction in sea ice research will be literally out of this world, since the increasing evidence that Jupiter's satellite Europa possesses something resembling a floating ice cover means that sea ice research may end up becoming a branch of astronomy.

FURTHER READING

In this section we confine ourselves to recommending a few vital works that will take the interested reader deeper into different aspects of this subject. More specialised works are referenced in the text.

Physics of ice

The physics of ice as a material are covered in comprehensive detail in *Ice Physics* by P.V. Hobbs (Oxford University Press, 1974). A new book which takes account of more recent research is *Physics of Ice*, by V.F. Petrenko and R.W. Whitworth (Oxford University Press, 1999).

Historical

The study of sea ice is a modern field of research which really took off in North America only in the 1950s and in Europe not until the 1960s, although serious work was going on in Russia from the 1920s and Japan from the 1930s. Before this time, there are only a few works which shine out like beacons in the darkness.

Earliest of all, and still of value today, was a magnificent work of 1820 by William Scoresby Jr., a whaling captain, later a vicar and Fellow of the Royal Society. *An Account of the Artic Regions, with a History and Description of the Northern Whale-Fishery* distilled a lifetime of Arctic experience and of careful observations into the first systematic book on Arctic ice and ocean conditions. His observations on the sea ice in the Greenland Sea are still highly relevant. The two-volume work was reprinted in 1969 by David and Charles, Newton Abbot, UK, and copies of the reprint can still be found although the original edition (Constable, Edinburgh) is rare.

We move on to the greatest giant of all, Fridtjof Nansen. His *Farthest North* (1897), a classic work of world literature describing the 1893–6 Arctic drift of the "Fram", contains much information on the Arctic Ocean and ice as well as the best invocation of what it feels like to be drifting across it. Later, his many years of research with Bjørn Helland-Hansen on the oceanography of the Nordic Seas produced another classic work, *The Norwegian Sea* (Helland-Hansen and Nansen, 1909).

The first really important work on the physics of sea ice to be published in the West was Finn Malmgren's *On the properties of sea ice* in 1927 (Malmgren, 1927) based on work done aboard Amundsen's "Maud" expedition. Only a year later Malmgren was to die heroically following the crash on the Arctic ice of Nobile's airship "Italia"; he set off to get help for the survivors and died in his tracks. Meanwhile enormous efforts in sea ice research were being organised in Russia, unknown to the West. For instance, there is relatively little about sea ice in that classic work of oceanography *The Oceans* by Sverdrup, Johnson and Fleming, first published in 1942, despite the fact tht Harald

Sverdrup was himself a polar scientist of great distinction, who risked his life aboard the first submarine to attempt to reach the North Pole in 1931. The extent of the Russian effort became clear three years later in 1945 when *L'dy Arktiki* (Arctic Ice) by N.N. Zubor was published in Moscow. This is the first and greatest systematic account of sea ice physics, behaviour and distribution, summarising decades of work by a devoted group of scientists. The 1963 English translation by the U.S. Navy Hydrographic Office is still available from the National Technical Information Service, Springfield, Virginia (AD426972).

The War which delayed publication of Zubov's work gave Lauge Koch, an internee of the Germans, the opportunity to complete his major review *The East Greenland Ice* (1945), published in English in the Danish journal *Meddelelser om Grønland* (Bd. 130 no. 3), and now available mainly from second-hand bookshops in Copenhagen since the journal no longer retains back numbers. After Zubov and Koch the modern era of sea ice research begins.

Sea ice classification

A classic work in this field was *Illustrated Glossary of Snow and Ice*, edited by Terence Armstrong, Brian Roberts and Charles Swithinbank (Scott Polar Research Institute, 2nd edition, 1973). The official World Meteorological Organization terminology, *WMO Sea-Ice Nomenclature* (WMO, no. 259), first published in 1970, has undergone several revisions as new internationally agreed types of ice classification have been developed.

Sea ice distribution

NASA has produced three reports on sea ice distribution observed from passive microwave observations. They are *Antarctic Sea Ice, 1973–1976: Satellite Passive-Microwave Observations* by H. J. Zwally, J. C. Comiso; C. L. Parkinson, W. J. Campbell, F. D. Carsey and P. Gloersen (NASA SP-459, 1983); *Arctic Sea Ice, 1973–1976: Satellite Passive-Microwave Observations* by Parkinson, Comiso, Zwally, Cavalier, Gloersen and Campbell (NASA SP-489, 1987); and *Arctic and Antarctic Sea Ice, 1978–1987. Satellite Passive-Microwave Observations and Analysis* by Gloersen, Campbell, Cavalieri, Comiso, Parkinson and Zwally (NASA SP-511, 1992). The last-named is the most generally useful since it deals with both hemispheres and with results from the multi-fequency SMMR while the earlier atlases deal with the single-frequency ESMR.

Sea ice geophysics

The standard advanced text is *The Geophysics of Sea Ice* (ed. N. Untersteiner, Plenum, New York, 1986), the proceedings of a 1981 NATO Advanced Study Institute held in Maratea, Italy. Its successor is *The Physics of Ice-Covered Seas* (ed. M. Leppäranta, University of Helsinki Press, 1998), the proceedings of a 1994 Advanced Study Institute held at Savonlinna, Finland.

Marginal ice zones and polynyas

Three special issues of *Journal of Geophysical Research* were devoted to marginal ice zones, in 1983 (volume 88 no. C5), 1987 (volume 92 no. C7) and 1991 (volume 96 no.

C3), and one to leads and polynyas in 1995 (volume 100 no. C3), while volume 2 of *Cold Regions Science and Tchnology* (1980) was devoted to properties of the seasonal sea ice zone.

Icebergs

There is no single textbook on icebergs in general. However, at the time when iceberg utilization was being considered seriously, two conferences on the subject were held in which the Proceedings contain valuable general papers on iceberg properties. The first was *Iceberg Utilization* (ed. A.A. Husseiny, 1978), and the second was published as Volume 1 of *Annals of Glaciology* (International Glaciological Society, Scott Polar Research Institute, Cambridge, 1980). A newsletter *Iceberg Research* has also been produced at irregular intervals by SPRI and is available from their Library.

Sea ice engineering

The best single work in this area is *Ice Mechanics: Risk to Offshore Structures* by Tim Sanderson (Graham and Trotman, 1988) which gives an overall survey of the subject. More specialised aspects are covered in a series of monographs and reports published by the US Army Cold Regions Research and Engineering Laboratory (CRREL), Hanover, New Hampshire (a full list is available from their Library). The journal *Cold Regions Science and Technology* (Elsevier) contains many useful papers, and another useful source of information is the Proceedings of a number of regular annual or biennial conferences, notably POAC (Port and Ocean Engineering under Arctic Conditions), ISOPE (International Society of Ocean and Polar Engineers), OMAE (Offshore Mechanics and Arctic Engineering) and ITC (Ice Technology Conference). *Journal of Geophysical Reseach* had a special issue on sea ice mechanics in 1998 (volume 103 no. C10).

Sea ice, environment and climate

There is no book specifically on sea ice and, climate change, but global change issues in general are covered by the multi-volume reports of the Intergovernmental Panel on Climate Change, of which the latest is *Climate Change 2000* (Cambridge University Press). A brief, readable summary of the science of global change is Sir John Houghton's *Global Warming; the Complete Briefing* (Cambridge University Press, 2nd Edn., 1994).

Remote sensing of sea ice

A good introductory account, now a little out of date since the field is advancing so fast, is *Satellite Remote Sensing of Polar Regions* by Rob Massom (Belhaven, 1991). Systematic multi-author works on the physics of passive and active microwave are *Microwave Remote Sensing of Sea Ice* edited by Frank Carsey (American Geophysical Union, Geophysical Monograph 68, 1992); and *Remote Sensing of Sea Ice and Icebergs* edited by S. Haykin, E.O. Lewis, R.K. Raney and J.R. Rossiter (Wiley, 1994). The latter also contains much material, from Canadian workers, on detecting ice by surface radar. A recent textbook on SAR detection of sea ice is *Analysis of SAR Data of the Polar Oceans: Recent Advances* (ed. C. Tsatsoulis, R. Kwok; Springer, 1998).

Antarctic sea ice

The American Geophysical Union brought out two volumes in their Antarctic Research Series which deal specifically with Antarctic sea ice. These are *Antactic Sea Ice: Physical Processes, Interactions and Variability* (ed. Martin O. Jeffries, volume 74, 1998); and *Antarctic Sea Ice: Biological Processes, Interaction, and Variability* (ed. M. Lizotte and K. Arrigo, volume 73, 1998). A third volume, *Ocean, Ice and Atmosphere: Interactions at the Antarctic Continental Margin* (ed. Stanley S. Jacobs and Ray F. Weiss, volume 75, 1998) deals with sea ice and its interactions with the ocean and atmosphere.

REFERENCES

Aagaard, K. and E.C. Carmack (1989). The role of sea ice and other fresh water in the arctic circulation. *J. Geophys. Res.*, *94*(C10), 14485–14498.

Aagaard, K. and E.C. Carmack (1994). The Arctic Ocean and climate: a perspective. In *The Polar Oceans and their Role in Shaping the Global Environment* (ed. O.M. Johannessen, R.D. Muench, J.E. Overland), Geophys. Monograph Srs. 85, Amer. Geophys. U., Washington, 5–20.

Aagaard, K., L.K. Coachman and E.C. Carmack (1981). On the halocline of the Arctic Ocean. *Deep-Sea Res.*, *28A*, 529–545.

Aagaard, K., C.H. Pease, A.T. Roach and S.A. Salo (1990). Beaufort Sea mesoscale study. *Outer Continental Shelf Environmental Assessment Program*, *65*, 1–136.

Abramov, V. (1996). *Atlas of Arctic Icebergs.* Backbone Publ. Co., Elmwood Park, NJ, 69pp.

Ackley, S.F. (1979). Mass-balance aspects of Weddell Sea pack ice. *J. Glaciol.*, *24*(90), 391–405.

Ackley, S.F., M.A. Lange and P. Wadhams (1990). Snow cover effects on Antarctic sea ice thickness. In *Sea Ice Properties and Processes* (ed. S.F. Ackley, W.F. Weeks), CRREL Monograph 90–1, US Army Cold Regions Res. & Engng. Lab., Hanover N.H., 16–21.

Aitchison, J. and J.A.C. Brown (1957). *The Lognormal Distribution.* Cambridge Univ. Press, 176pp.

Alfultis, M.A. and S. Martin (1987). Satellite passive microwave studies of the Sea of Okhotsk ice cover and its relation to oceanic processes, 1978–1982. *J. Geophys. Res.*, *92*(C12), 13013–13028.

Alves, J., M. Bell and N. Brooks (1995). Performance review of the prototype FOAM system. Met. Office Forecasting & Res. Div., Tech. Rept. 159, Bracknell.

Anderson, D.L. (1958). A model for determining sea ice properties. In *Arctic Sea Ice*, Nat. Acad. Sci., Washington. Nat. Res. Counc. Publn. 598, 148–152.

Anderson, D.L. (1961). Growth rate of sea ice. *J. Glaciol.*, *3*(30), 1170–1172.

Anderson, L.G., D. Dyrssen and E.P. Jones (1990). An assessment of transport of atmospheric CO_2 into the Arctic Ocean. *J. Geophys. Res.*, *95*(C2), 1703–1711.

Andreas, E.L., M.A. Lange, S.F. Ackley and P. Wadhams (1993). Roughness of Weddell Sea ice and estimates of the air-ice drag coefficient. *J. Geophys. Res.*, *98*(C7), 12439–12452.

Annals of Glaciology (1980). Proceedings of the 2nd Conference on the Use of Icebergs: Scientific and Practical Feasibility, Cambridge, UK, 1–3 April 1980. *Ann. Glaciol.*, *1*, 1–136.

Assur, A. (1963). Breakup of pack ice floes. In *Ice and Snow: Properties , Processes and Applications* (W.D. Kingery, ed.). MIT Press, Cambridge, Mass., 335–347.

Backhaus, J.O. and J. Kämpf (1999). Simulations of sub-mesoscale oceanic convection and ice-ocean interactions in the Greenland Sea. *Deep-Sea Res. II*, *46*(6–7), 1427–1456.

Badgley, F.I. (1961). Heat balance at the surface of the Arctic Ocean. *Proc. Western Snow Conf.*, Spokane, Wash., 101–104.

Bagriantsev, N.V., A.L. Gordon and B.A. Huber (1989). Weddell Gyre: temperature maximum stratum. *J. Geophys. Res.*, *94*(C6), 8331–8334.

Baines, P.G. and S. Condie (1998). Observations and modelling of Antarctic downslope flows: a review. In *Ocean, Ice and Atmosphere. Interactions at the Antarctic Continental Margin* (ed. S.S. Jacobs, R.F. Weiss), Antarctic Res. Srs. 75, Amer. Geophys. U., Washington, 29–49.

Banke, E.G., S.D. Smith and R.J. Anderson (1976). Recent measurements of wind stress on Arctic sea ice. *J. Fish. Res. Bd. Canada*, *33*, 2307–2317.

Banke, E.G., S.D. Smith and R.J. Anderson (1980). Drag coefficients at AIDJEX from sonic anemometer measurements. In *Sea Ice Processes and Models* (ed. R.S. Pritchard), Univ. Washington Press, Seattle, 430–442.

Barnes, H.T. (1928). *Ice Engineering.* Renouf, Montreal. 364pp.

Barnes, P.W., D.M. Rearic and E. Reimnitz (1984). Ice gouging characteristics and processes. In *The Alaskan Beaufort Sea: Ecosystems and Environments*, Academic Press, Orlando, 185–212.

Barrie, L.A., D. Gregor, B. Hargrave, R. Lake, D. Muir, R. Shearer, B. Tracey and T. Bidleman (1992). Arctic contaminants: sources, occurrence and pathways. *Sci. of Total Envt.*, *122*, 1–74.

Bauer, J. and S. Martin (1980). Field observations of the Bering Sea ice edge properties during March 1979. *Mon. Weather Rev.*, *108*(12), 2045–2056.

Bauer, J. and S. Martin (1983). A model of grease ice growth in small leads. *J. Geophys. Res.*, *88*(C5) 2917–2925.

Behrman, D. and J.D. Isaacs (1992). *John Isaacs and His Oceans.* American Geophysical U., ICSU Press, Washington. Ch. 6, 49–54.

Benedict, C.P. (1980). A towing concept for small icebergs. In *Iceberg Utilization* (ed. A.A. Husseiny), Pergamon, New York, 334–338.

Bentham, R. (1937). The ice-foot. In *Arctic Journeys: the Story of the Oxford University Ellesmere Land Expedition 1934–5* (E. Shackleton, ed.), Appendix III, Hodder and Stoughton, London.

Bilello, M.A. (1961). Formation, growth and decay of sea ice. *Arctic*, *14*(1), 3–24.

Bilello, M.A. (1980). Decay patterns of fast sea ice in Canada and Alaska. In *Sea Ice Processes and Models* (ed. R.S. Pritchard), Univ. Washington Press, Seattle, 313–326.

Bishop, G.C. and S.E. Chellis (1990). A fractal description of ice keel small-scale surface roughness. In *Sea Ice Properties and Processes* (ed. S.F. Ackley, W.F. Weeks), CRREL Monograph 90–1, US Army Cold Regions Res. & Engng. Lab., Hanover N.H., 141–145.

Bjørgo, E., O.M. Johannessen and M.W. Miles (1997). Analysis of merged SMMR-SSMI time series of Arctic and Antarctic sea ice parameters 1978–1995. *Geophys. Res. Lett.*, *24*(4), 413–416.

Björk, G. (1989). A one-dimensional time-dependent model for the vertical stratification of the upper Arctic Ocean. *J. Phys. Oceanogr.*, *19*(1), 52–67.

Bohren, C. (1983). Colours of snow, frozen waterfalls and icebergs. *J. Opt. Soc. Amer.*, *73*, 1646.

Bourke, R.H. and R.P. Garrett (1987). Sea ice thickness distribution in the Arctic Ocean. *Cold Regions Sci. Technol.*, *13*, 259–280.

Bourke, R.H. and A.S. McLaren (1992). Contour mapping of Arctic Basin ice draft and roughness parameters. *J. Geophys. Res.*, *97*(C11), 17715–17728.

Bourke, R.H., M.D. Tunnicliffe, J.L. Newton, R.G. Paquette and T.O. Manley (1987). Eddy near the Molloy Deep revisited. *J. Geophys. Res.*, *92*(C7), 6773–6776.

Bratchie, I. (1984). Modelling sea ice floe fields. PhD thesis, Univ. Cambridge, Scott Polar Res. Inst.

Bretherton, F.P., K. Bryan and J.D. Woods (1990). Time-dependent greenhouse-gas-induced climate change. In *Climate Change. The IPCC Scientific Assessment.* (ed. J.T. Houghton, G.J. Jenkins and J.J. Ephraums). Cambridge Univ. Press, 173–194.

Brigham, L.W. (ed.) (1991). *The Soviet Maritime Arctic.* Belhaven Press, London, 336pp.

Brigham, L.W. (1996). Sea ice and ocean processes in the Laptev Sea. MPhil thesis, Univ. Cambridge, Scott Polar Res. Inst. 93pp.

Broecker, W.S. (1995). Chaotic climate. *Scientific American*, Nov. 1995, 45–50.

Bromwich, D. H., and D.D. Kurtz (1984). Katabatic wind forcing of the Terra Nova Bay polynya. *J. Geophys. Res.*, *89*(C3), 3561–3572.

Bromwich, D. H., T.R. Parish and C.A. Zorman, (1990). The confluence zone of the intense katabatic winds at Terra Nova Bay, Antarctica, as derived from airborne sastrugi surveys and mesoscale numerical modeling. *J. Geophys. Res.*, *95*(D5), 5495–5509.

Brown, R.A. (1980). Planetary boundary layer modeling for AIDJEX. In *Sea Ice Processes and Models* (ed. R.S. Pritchard), Univ. Washington Press, Seattle, 387–401.

Brown, R.D. and P. Cole (1992). Interannual variability of landfast ice thickness in the Canadian High Arctic, 1950–1989. *Arctic*, *45*(3), 273–284.

Brown, R.G. and D.N. Nettleship (1981). The biological significance of polynyas to Arctic colonial seabirds. In *Polynyas in the Canadian Arctic* (ed. I. Stirling, H. Cleator), Can. Wildlife Service, Ottawa, 59–66.

Bruneau, A.A., R.T. Dempster and G.R Peters (1978). Iceberg towing for rig avoidance. In *Iceberg Utilization* (ed. A.A. Husseiny), Pergamon, New York, 379–388.

Buckley, J.R., T. Gammelsrød, J.A. Johannessen, O.M. Johannessen and L.-P. RØed (1979). Upwelling: oceanic structure at the edge of the arctic icepack in winter. *Science*, *203*, 165–167.

Budd, W., I.L. Smith and E. Wishart (1967). The Amery Ice Shelf. In *Physics of Snow and Ice* (ed. H. Ôura), Inst. Low Temp. Sci., Hokkaido Univ., Sapporo, *1*, 447–467.

Budyko, M.I. (1974). *Climate and Life.* Intl. Geophys. Srs. 18, Academic Press, New York. 508pp.

Burns, B.A., D.J. Cavalieri, M.R. Keller, W.J. Campbell, T.C. Grenfell, G.A. Maykut and P. Gloersen (1987). Multisensor comparison of ice concentration estimates in the marginal ice zone. *J. Geophys. Res.*, *92*(C7), 6843–6856.

Calkins, D.J. (1986). Hydrologic aspects of ice jams. Cold Regions Hydrology Symp., Amer. Water Resources Assoc., Fairbanks, Alaska, 603–609.

Campbell, J.A. and P.A. Yeats (1982). The distribution of manganese, iron, nickel, copper and cadmium in the waters of Baffin Bay and the Canadian Arctic Archipelago. *Oceanol. Acta*, *4*(2), 161–168.

Campbell, R.C. (1974). *Statistics for Biologists*. 2nd Ed. Cambridge Univ. Press, 385pp.

Campbell, W.J. (1965). The wind-driven circulation of ice and water in a polar ocean. *J. Geophys. Res.*, *70*(14), 3279–3301.

Campbell, W.J., E.G. Josberger and N.M. Mognard (1994). Southern Ocean wave fields during the austral winters, 1985–1988, by Geosat radar altimeter. In *The Polar Oceans and their Role in Shaping the Global Environment* (ed. O.M. Johannessen, R.D. Muench, J.E. Overland), Geophys. Monograph 85, Amer. Geophys. U., Washington, 421–434.

Carmack, E.C., K. Aagaard, J.H. Swift, R.W. Macdonald, F.A. McLaughlin, E.P. Jones, R.G. Perkin, J.N. Smith, K.M. Ellis and L.R. Kilius (1997). Changes in temperature and tracer distributions within the Arctic Ocean: results from the 1994 Arctic Ocean section. *Deep-Sea Res. II*, *44*, 1487–1502.

Carsey, F.D. (1980). Microwave observation of the Weddell Polynya. *Mon. Wea. Rev.*, *108*(12), 2032–2044.

Carsey, F.D. (ed.) (1992). *Microwave Remote Sensing of Sea Ice*. Geophys. Monograph 68, Amer. Geophys. U., Washington, 462pp.

Cattle, H. and J. Crossley (1995). Modelling Arctic climate change. *Phil. Trans. Roy. Soc., London*, *A352*, 197–385.

Cavalieri, D.J. and S. Martin (1985). A passive microwave study of polynyas along the Antarctic Wilkes Land coast. In *Oceanology of the Antarctic Continental Shelf* (ed. S.S. Jacobs), Antartic Res. Srs. 43, Amer. Geophys. U., Washington, 227–252.

Cavalieri, D. J. and S. Martin (1994). The contribution of Alaskan, Siberian, and Canadian coastal polynyas to the cold halocline layer of the Arctic Ocean. *J. Geophys. Res.*, *99*(C9), 18343–18362.

Cavalieri, D.J., P. Gloersen, C.L. Parkinson, J.C. Comiso and H.J. Zwally (1997). Observed hemispheric asymmetry in global sea ice changes. *Science*, *278*, 1104–1106.

Cherepanov, N.V. (1964). Structure of very thick sea ice. *Trudy Arkt. Antarkt. N.I. Inst..*, *267*, 13–18.

Cherepanov, N.V. (1971). Spatial arrangement of sea ice crystal structure. *Probl. Arkt. Antarkt.*, *38*, 176–181 (Eng. trans. Amerind Publ. Co., New Delhi, 1973).

Cherniyovskii, N.T. (1966). Radiational properties of the Central Arctic ice cover. In *Soviet Data on the Arctic Heat Budget and Its Climate Influence* (ed. J.U. Fletcher, B. Veller, S.M. Ulenicoff), RAND Corp., Santa Monica, Rept. RM-5003-PR, 151–174.

Chu, P.C. (1987). An instability theory of ice-air interaction for the formation of ice edge bands. *J. Geophys. Res.*, *92*(C7), 6966–6970.

Colony, R.L., I. Rigor and K. Runciman-Moore (1991). A summary of observed ice motion and analyzed atmospheric pressure in the Arctic Basin, 1979–1990. Appl. Phys. Lab., Univ. Washington, Seattle, Tech. Rept., APL-UW TR9112, 106pp.

Comiso, J.C. and A.L. Gordon (1987). Recurring polynyas over the Cosmonaut Sea and the Maud Rise. *J. Geophys. Res.*, *92*(C3), 2819–2833.

Comiso, J.C. and A.L. Gordon (1996). Cosmonaut Polynya in the Southern Ocean: structure and variability. *J. Geophys. Res.*, *101*(C8), 18297–18313.

Comiso, J.C. and C.W. Sullivan (1986). Satellite microwave and in situ observations of the Weddell Sea ice cover and its marginal ice zone. *J. Geophys. Res.*, *91*(C8), 9663–9681.

Comiso, J.C. and H.J. Zwally (1989). Polar microwave brightness temperatures from Nimbus 7 SMMR: time series of daily and monthly maps 1978–1987. NASA Reference Publn. RP1223, NASA, Washington D.C., 102 pp.

Comiso, J.C., P. Wadhams, W.B. Krabill, R.N. Swift, J.P. Crawford, and W.B. Tucker III (1991). Top/bottom multisensor remote sensing of Arctic sea ice. *J. Geophys. Res.*, *96*(C2), 2693–2709.

Comiso, J.C., C.R. McClain, C.W. Sullivan, J.P. Ryanand C.L. Leonard (1993). Coastal Zone Color Scanner pigment concentrations in the Southern Ocean and relationships to geophysical surface features. *J. Geophys. Res.*, *98*(C2), 2419–2451.

Connors, D.N., E.R. Levine and R.R. Shell (1990). A small-scale under-ice morphology study in the high arctic. In *Sea Ice Properties and Processes* (ed. S.F. Ackley, W.F. Weeks), CRREL Monograph 90–1, US Army Cold Regions Res. & Engng. Lab., Hanover, N.H.. 145–151.

Coon, M.D. (1980). A review of AIDJEX modeling. In *Sea Ice Processes and Models* (ed. R.S. Pritchard), Univ. Washington Press, Seattle, 12–27.

Cooper, P.F. (1974). Landfast ice in the southeastern part of the Beaufort Sea. In *The Coast and Shelf of the Beaufort Sea* (ed. J.C. Reed and J.F. Sater), Arctic Inst. N. Amer., 235–242.

Cottier, F., H. Eicken and P. Wadhams (1999). Linkages between salinity and brine channel distribution in young sea ice. *J. Geophys. Res.*, *104*(C7), 15859–15871.

Cox, D.R. and E.J. Snell (1981). *Applied Statistics: Principles and Examples*. Chapman and Hall, London, 143–147.

Cox, G.F.N. and W.F. Weeks (1974). Salinity variations in sea ice. *J. Glaciol.*, *13*(67), 109–120.

Crary, A.P. (1958). Arctic ice island and shelf ice studies. Part 1. *Arctic*, *11*(1), 3–42.

Crary, A.P. (1960). Arctic ice island and shelf ice studies. Part 2. *Arctic*, *13*(1), 32–50.

Curry, J.A., J.L. Schramm and E.E. Ebert (1995). Sea ice – albedo climate feedback mechanism. *J. Climate*, *8*, 240–247.

Davidson, L.W. and W.W. Denner (1982). Sea ice and iceberg conditions on the Grand Banks affecting hydrocarbon production and transportation. *Proc. IEEE Oceans'82*, 1236–1241.

Davis N.R. and P. Wadhams (1995). A statistical analysis of Arctic pressure ridge morphology. *J. Geophys. Res., 100*(C6), 10915–10925.

Dayton, P.K. and S. Martin (1971). Observations of ice stalactites in McMurdo Sound, Antarctica. *J. Geophys. Res.*, *76*(C6), 1595–1599.

De La Mare, W.K. (1997). Abrupt mid twentieth-century decline in Antarctic sea ice extent from whaling records. *Nature*, *389*, 6646.

Deacon, G. (1984). *The Antarctic Circumpolar Ocean*. Cambridge Univ. Press, Cambridge. 180pp.

Defant, A. (1924). Die Gezeiten des Atlantischen Ozeans und des Arctischen Meeres. *Ann. Hydr. Mar. Met.*, *52*(8–9), 153–166, 177–184.

DeMarle, D.J. (1979). Ice cooled ocean thermal energy conversion plants. *Desalination*, *29*, 153–163.

DeMarle, D.J. (1980). Design parameters for a South African iceberg power and water project. *Ann. Glaciol.*, *1*, 129–133.

Den Hartog, G., S.D. Smith, R.J. Anderson, D.R. Topham and R.G. Perkin (1983). An investigation of a polynya in the Canadian archipelago. 3, Surface heat flux. *J. Geophys. Res.*, *88*(C5), 2911–2916.

Denner, W.W. (1978). Environmental factors along an iceberg tow route in the Indian Ocean. In *Iceberg Utilization* (ed. A.A. Husseiny), Pergamon, New York, 389–416.

Dethleff, D., D. Nürnberg, E. Reimnitz, M. Saarso and Y.P. Savchenko (1993). East Siberian Arctic region expedition '92: its significance for Arctic sea ice formation and transpolar sediment flux. *Berichte zur Polarforschung*, *120*, 1–44.

Diachok, O. (1980). Arctic hydroacoustics. *Cold Regions Sci. & Technol.*, *2*, 185–201.

Dickins D.F. and V.F. Wetzel (1981). Multiyear pressure ridge study, Queen Elizabeth Islands. *POAC 81*, Proceedings of the Sixth International Conference on Port and Ocean Engineering under Arctic Conditions, II, 765–775.

Dickson, R.R., J. Meincke, I.M. Vassie, J. Jungclaus and S. Osterhuis (1999). Possible predictability in overflow from the Denmark Strait. *Nature*, *397*, 243–246.

Diemand, D. (1984). Iceberg fragmentation by thermal shock. *Iceberg Res.*, Scott Polar Res. Inst., Cambridge, *8*, 8–10.

Dietrich, G., K. Kalle, W. Krauss and G. Siedler (1980). *General Oceanography, An Introduction*. Wiley, New York.

Dmitriev, N.E., A.Y. Proshutinsky, T.B. Loyning and T. Vinje (1991). Tidal ice dynamics in the area of Svalbard and Franz Josef Land. *Polar Res.*, *9*(2), 193–205.

Doble, M. and P. Wadhams (1999). Analysis of concurrent SAR images and submarine ice draft profiles in the Arctic Ocean. *Proc. POAC'89, Port & Ocean Engng under Arctic Condns., Helsinki, Aug.1989.* Helsinki Univ. Technol.

Doronin, Yu.P. (1963). On the heat balance of the central Arctic. *Trudy Arkt. Antarkt. Nauch. Issle. Inst.*, *253*, 178–184.

Doronin, Yu. P. and D.E. Kheisin (1977). *Sea Ice*. Gidrometeoizdat, Leningrad. Eng. trans. Amerind Publ. Co., New Delhi, 323pp.

Ebbesmeyer, C.C., A. Okubo and H.J.M. Helset (1980). Description of iceberg probability between Baffin Bay and the Grand Banks using a stochastic model. *Deep-Sea Res.*, *27A*, 975–986.

Ebert, E.E. and J.A. Curry (1993). An intermediate one-dimensional thermodynamic sea ice model for investigating ice-atmosphere interactions. *J. Geophys. Res.*, *98*(C6), 10085–10109.

Eicken, H. (1992). The role of sea ice in structuring Antarctic ecosystems. *Polar Biol.*, *12*(1), 3–13.

Eicken, H. and M.A. Lange (1989). Sea ice thickness: the many vs the few. *Geophys. Res. Lett.*, *16*(6), 495–498.

Eicken, H., M.A. Lange, H.-W. Hubberten and P. Wadhams (1994). Characteristics and distribution patterns of snow and meteoric ice in the Weddell Sea and their contribution to the mass balance of sea ice. *Ann. Geophysicae*, *12*(1), 80–93.

Eide, L.I. and S. Martin (1975). The formation of brine drainage features in young sea ice. *J. Glaciol.*, *14*(70), 137–154.

Eittreim, S.L. and A.K. Cooper (1984). Marine geological and geophysical investigations of the Antarctic continental margin, 1984. US Geological Survey Circular 935, Washington.

Ekman, V.W. (1902). Om jordrotationens inverkan på vindströmmar i hafvet. *Nyt. Mag. Naturvid.*, *40*.

Ekman, V.W. (1905). On the influence of the Earth's rotation on ocean currents. *Arch. Math. Astron. Phys.*, 2, no. 11.

Engel, L. (1961). *The Sea*. Time Inc., New York.

Ezraty, R. and A. Cavanie (1999). Intercomparison of backscatter maps over Arctic sea ice from NSCAT and the ERS scatterometer. *J. Geophys. Res.*, *104*(C5), 11471–11483.

Feazel, C.T. and R.C. Kollmeyer (1972). Major iceberg-producing glaciers of West Greenland. In *Sea Ice* (ed. T. Karlsson), Nat. Res. Counc. of Iceland, Reykjavik, 140–145.

Fedorov, K.N. and A.I. Ginsburg (1986). "Mushroom-like" currents (vortex dipoles) in the ocean and in a laboratory tank. *Ann. Geophysicae*, *4*(B), 5, 507–515.

Fedorov, K.N. and A.I. Ginsburg (1989). Mushroom-like currents (vortex dipoles): One of the most widespread forms of non-stationary coherent motions in the oceans. In *Mesoscale/Synoptic Coherent Structures in Geophysical Turbulence* (ed. J.C.J. Nihoul, B.M. Jamart), Elsevier Oceanography Srs. 50, Elsevier, Amsterdam, 1–14.

Ferrigno, G.J. and W.G. Gould (1987). Substantial changes in coastline of Antarctica revealed by satellite imagery. *Polar Rec.*, *23*(146), 577–583.

Feyling-Hanssen, R.W. (1953). Brief account of the ice-foot. *Saetr. Nor. Geogr. Tidsskr.*, *14*, 1–4, 45–52.

Flato, G.M. and W.D. Hibler III (1990). On a simple sea-ice dynamics model for climate studies. *Ann. Glaciol.*, *14*, 72–77.

Flato, G.M. and W.D. Hibler III (1995). Ridging and strength in modelling the thickness distribution of Arctic sea ice. *J. Geophys. Res.*, *100*(C9), 18611–18626.

Fletcher, J.O. (1965). The Heat Budget of the Arctic Basin and its Relation to Climate. RAND Corp., Santa Monica, Rept. R-444–PR, 179pp.

Fofonoff, N.P. (1956). Some properties of sea water influencing the formation of Antarctic bottom water. *Deep-Sea Res.*, *4*, 32–35.

Folland, C.K., T.R. Karl and K.Ya. Vinnikov (1990). Observed climate variations and change. In *Climate Change. The IPCC Scientific Assessment* (ed. J.T. Houghton, G.J. Jenkins and J.J. Ephraums). Cambridge Univ. Press, 201–238.

Folland, C.K., T.R. Karl, N. Nicholls, B.S. Nyenzi, D.E. Parker and K.Ya. Vinnikov (1992). Observed climate variability and change. In *Climate Change 1992. The Supplementary Report to the IPCC Scientific Assessment* (ed. J.T. Houghton, B.A. Callander and S.K. Varney). Cambridge Univ. Press, 135–170.

Fox, C. and V.A. Squire (1990). Reflection and transmission characteristics at the edge of shore fast sea ice. *J. Geophys. Res.*, *95*(C7), 11629–11639.

Fox, C. and V.A. Squire (1991). Strain in shore fast ice due to incoming ocean waves and swell. *J. Geophys. Res.*, *96*(C3), 4531–4547.

Francois, R.E. and W.M. Nodland (1972). Unmanned Arctic Research Submersible (UARS) system development and test report. Tech. Rept. APL-UW-7219, Applied Physics Lab., Univ. Washington, Seattle, 88pp.

Gade, H.G. (1979). Melting of ice in sea water: a primitive model with application to the Antarctic ice shelf and icebergs. *J. Phys. Oceanogr.*, *9*(1), 189–198.

Gammelsrød, R., M. Mork and L.-P. Røed (1975). Upwelling possibilities at an ice edge: homogeneous model. *Mar. Sci. Communs.*, *1*, 115–145.

Gammon, P.H., R.E. Gagnon, W. Bobby and W.E. Russell (1983). Physical and mechanical properties of icebergs. *15th Ann. Offshore Tech. Conf. 4459*, Houston, 143–149.

Gao, L. (1992). Ice floe collisions under wave action in the marginal ice zone. MEng thesis, Memorial Univ. of Newfoundland, St. John's.

Garrett, C. (1984). Statistical prediction of iceberg trajectories. *Iceberg Res.*, 7, 3–8. Scott Polar Res. Inst., Cambridge.

Garrison, D.L., C.W. Sullivan and S.F. Ackley (1986). Sea ice microbial community studies in the Antarctic. *Bioscience*, 36, 243–250.

Gascard, J-C., C. Richez and C. Rouault (1995). New insights on large-scale oceanography in Fram Strait: the West Spitsbergen Current. In *Arctic Oceanography: Marginal Ice Zones and Continental Shelves* (ed. W.O. Smith and J.M. Grebmeier), Amer. Geophys. U., Washington, 131–182.

Gaskill, H.S. and J. Rochester (1984). A new technique for iceberg drift prediction. *Cold Reg. Sci. Technol.*, 8, 223–234.

Gates, W.L., J.F.B. Mitchell, G.J. Boer, U. Cubasch and V.P. Meleshko (1992). Climate modelling, climate prediction and model validation. In *Climate Change 1992. The Supplementary Report to the IPCC Scientific Assessment* (ed. J.T. Houghton, B.A. Callander and S.K. Varney). Cambridge Univ. Press, 97–134.

Gill, A.E. (1973). Circulation and bottom water production in the Weddell Sea. *Deep-Sea Res.*, 20, 111–140.

Gill, A.E. (1982). *Atmosphere-Ocean Dynamics*. Academic Press, New York. 662pp.

Glendening, J. (1994). Dependence of boundary-layer structure near an ice-edge coastal front upon geostrophic wind direction. *J. Geophys. Res.*, 99(D3), 5569–5581.

Gloersen, P. and W.J. Campbell (1988). Variations in the Arctic, Antarctic, and global sea ice covers during 1979–1987 as observed with the Nimbus 7 Scanning Multichannel Microwave Radiometer. *J. Geophys. Res.*, 93(C9), 10666–10674.

Gloersen, P., W.J. Campbell, D.J. Cavalieri, J.C. Comiso, C.L. Parkinson and H.J. Zwally (1992). *Arctic and Antarctic Sea Ice, 1978–1987: Satellite passive-microwave observations and analysis*. National Aeronautics and Space Administration, Rep. NASA SP-511, 290pp.

Goodman, D.J. (1980). Critical stress intensity factor (K_{Ic}) measurements at high loading rates for polycrystalline ice. Proc. IUTAM Conf. on Physics and Mechanics of Ice, Copenhagen (P. Tryde, ed.).

Goodman, D.J., P. Wadhams and V.A. Squire (1980). The flexural response of a tabular ice island to ocean swell. *Ann. Glaciol.*, 1, 23–27.

Gordon, A.L. (1978). Deep Antarctic convection west of Maud Rise. *J. Phys. Oceanogr.*, 8, 600–612.

Gordon, A.L. and J.C. Comiso (1988). Polynyas in the Southern Ocean. *Scientific American*, 256(6), June 1988, 90–97.

Gordon, A.L. and R.D. Goldberg (1970). Circumpolar characteristics of Antarctic waters. *Antarctic Map Folio Srs.*, Folio 13, V. Bushnell, ed., Amer. Geog. Soc., New York.

Gordon, A.L. and B.A. Huber (1990). Southern Ocean winter mixed layer. *J. Geophys. Res.*, 95(C7), 11655–11672.

Gordon, A.L. and P. Tchernia (1972). Waters of the continental margin off Adélie coast, Antarctica. In *Antarctic Oceanology II: The Australian - New Zealand Sector.* (ed. D.E. Hayes). Antarctic Res. Srs. 19, 59–69, Amer. Geophys. U., Washington.

Gorshkov, S.G. (ed.) (1983). *World Ocean Atlas. Vol. 3, Arctic Ocean*. Pergamon Press, Oxford, 184 pl.

Gow, A.J. (1968). Bubbles and bubble pressures in Antarctic glacier ice. US Army Cold Regions Res. & Engng. Lab., Hanover N.H., Res. Rept. 249.

Gow, A.J., S.F. Ackley, W.F. Weeks and J.W. Govoni (1982). Physical and structural characteristics of Antarctic sea ice. *Ann. Glaciol.*, *3*, 113–117.

Gow, A.J., S.F. Ackley, K.R. Buck and K.M. Golden (1987). Physical and structural characteristics of Weddell Sea pack ice. CRREL Rept. 87–14, US Army Cold Regions Res. & Eng. Lab., Hanover N.H.

Gradinger, R. (1995). Climate change and the biological oceanography of the Arctic Ocean. *Phil. Trans. Roy. Soc.*, *A352*, 277–286.

Grandvaux, B. and J.-F. Drogou (1989). Saga 1, une premiere etape vers les sous-marins autonomes d'intervention. In *Arctic Technology and Economy - Present Situation and Problems, Future Issues.* Bureau Veritas, Paris.

Greisman, P. (1979). On upwelling driven by the melt of ice shelves and tidewater glaciers. *Deep Sea Res.*, *26*, 1051–1065.

Greenhill, A.G. (1887). Wave motion in hydrodynamics. *Am. J. Math.*, *9*, 62–112.

Greenhill, A.G. (1916). Skating on thin ice. *Phil. Mag.*, *31*, 1–22.

Gudmandsen, P., B.B. Thomsen, L.T. Pedersen, H. Skriver and P.J. Minnett (1995). Northeast water polynya — satellite observations summer 1992 and 1993. *Int. J. Remote Sensing*, *16*(17), 3307–3324.

Guest, P.S. and K.L. Davidson (1991). The aerodynamic roughness of different types of sea ice. *J. Geophys. Res.*, *96*(C3), 4709–4721.

Guest, P.S., K.L. Davidson, J.E. Overland and P.A. Frederickson (1994). Atmosphere-ocean interactions in the marginal ice zones of the Nordic Seas. In *Arctic Oceanography: Marginal Ice Zones and Continental Shelves* (ed. W.O. Smith and J.M. Grebmeier), Amer. Geophys. U., Washington, 51–95.

Gumbel, E.J. (1954). Statistical theory of extreme values and some practical applications. US Natl. Bureau of Standards, Applied Math. Series, *33*, 51pp.

Gumbel, E.J. (1958). *Statistics of Extremes.* Columbia Univ. Press, New York, 375pp.

Guoliang, J. and P. Wadhams (1989). Travel time changes in a tomography array caused by a sea ice cover. *Prog. Oceanogr.*, *22*(3), 249–275.

Häkkinen, S. (1986). Ice banding as a response of the coupled ice-ocean system to temporally varying winds. *J. Geophys. Res.*, *91*(C4), 5047–5053.

Häkkinen, S., G.L. Mellor and L.H. Kantha (1992). Modeling deep convection in the Greenland Sea. *J. Geophys. Res.*, *97*(C4), 5389–5408.

Hansen, J., I. Fung, A. Lacis, D. Rind, S. Lebedeff, R. Ruedy and G. Russell (1988). Global climate changes as forecast by Goddard Institute for Space Studies three-dimensional model. *J. Geophys. Res.*, *93*(D8), 9341–9364.

Hansen, J.C. (1990). Human exposure to metals through consumption of marine foods: a case study of exceptionally high intake among Greenlanders. In *Heavy Metals in the Marine Environment* (ed. R.W. Furness, P.S. Rainbow), CRC Press, Cleveland, 227–243.

Hanson, A.M. (1965). Studies of the mass budget of arctic pack-ice floes. *J. Glaciol.*, *5*(41), 701–709.

Hasselmann, K. et al. (1973). Measurements of wind wave growth and swell decay during the Joint North Sea Wave Project (JONSWAP). *Herausgegeben Deutsch. Hydro. Institut.*, Reihe A.

Hattersley-Smith, G. (1963). The Ward Hunt ice shelf: recent changes of the ice front. *J. Glaciol.*, *4*(34), 415–424.

Heizer, R.T. (1978). Energy and fresh water production from icebergs. In *Iceberg Utilization* (ed. A.A. Husseiny), Pergamon, New York, 1978, 657–673.

Hendrickson, J.A., L.M. Webb and R.J. Quigley (1962). Study of natural forces acting on floating ice fields. Rept. NBy-32215, Natl. Engng. Sci. Co., Pasadena, for US Naval Civil Engng. Lab., Port Hueneme.

Hibler, W.D. III (1972). Two dimensional statistical analysis of Arctic sea ice ridges. In *Sea Ice* (ed. T. Karlsson), Nat. Res. Counc. of Iceland, Reykjavik, 261–275.

Hibler, W.D. III (1979). A dynamic thermodynamic sea ice model. *J. Phys. Oceanogr.*, 9(4), 815–846.

Hibler, W.D. III (1980). Modeling a variable thickness sea ice cover. *Mon. Weather Rev.*, 108, 1943–1973.

Hibler, W.D. III (1989). Arctic ice-ocean dynamics. In *The Arctic Seas. Climatology, Oceanography, Geology, and Biology* (Y. Herman, ed.). Van Nostrand Reinhold, New York, 47–91.

Hibler, W.D. III and K. Bryan (1987). A diagnostic ice-ocean model. *J. Phys. Oceanogr.*, 17(7), 987–1015.

Hibler, W.D. III and L.A. LeSchack (1972). Power spectrum analysis of undersea and surface sea-ice profiles. *J. Glaciol.*, 11(63), 345–356.

Hibler, W.D. III and W.B. Tucker III (1979). Some results from a linear-viscous model of the Arctic ice cover. *J. Glaciol.*, 22(87) 293–304.

Hibler, W.D. III and J.E. Walsh (1982). On modelling the seasonal and interannual fluctuations of arctic sea ice. *J. Phys. Oceanogr.*, 12, 1514–1523.

Hibler, W.D. III and J. Zhang (1994). On the effect of ocean circulation on Arctic ice-margin variations. In *The Polar Oceans and their Role in Shaping the Global Environment* (ed. O.M. Johannessen, R.D. Muench, J.E. Overland), Geophys. Monograph 85, Amer. Geophys. U., Washington, 383–395.

Hibler, W.D. III, W.F. Weeks, A. Kovacs and S.F. Ackley (1974a). Differential sea-ice drift. I: Spatial and temporal variations in sea-ice deformation. *J. Glaciol.*, 13(69), 437–455.

Hibler, W.D. III, S.F. Ackley, W.K. Crowder, H.W. McKim and D.M. Anderson (1974b). Analysis of shear zone deformation in the Beaufort Sea using satellite imagery. In:*The Coast and Shelf of the Beaufort Sea*. (ed. J.C. Reed and J.E. Sater), Arctic Inst. N. Amer., Calgary, 285–296.

Hobbs, P.V. (1974). *Ice Physics*. Oxford Univ. Press.

Hobson, G. + XVIII (1989). Ice island field station; new features of Canadian Polar Margin. *EOS. Trans. Am. Geophys. U.*, 70(37), 833, 838–839.

Holdsworth, G. and J. Glynn (1978). Iceberg calving from floating glaciers by a vibrating mechanism. *Nature*, 274, 464–466.

Holladay, J.S., J.R. Rossiter and A. Kovacs (1990). Airborne measurement of sea ice thickness using electromagnetic induction sounding. *Proc. 9th Int. Conf. Offshore Mech. and Arctic Engng.* (ed. O.A. Ayorinde, N.K. Sinha, D.S. Sodhi), Amer. Soc. Mech. Engrs., 309–315.

Holt, B., J. Crawford and F. Carsey (1990). Characteristics of sea ice during the Arctic winter using multifrequency aircraft radar imagery. In *Sea Ice Properties and Processes* (ed. S.F. Ackley and W.F. Weeks), Monograph 90–1, US Army Cold Regions Res. & Engng. Lab., Hanover, N.H., 224 (abstract).

Hønsi, I. (1988). Isfjell i Barentshavet. Norwegian Petroleum Directorate, Stavanger.

Hopkins, M.A. (1994). On the ridging of intact lead ice. *J. Geophys. Res.*, *99*(8), 16351–16360.

Hopkins, M.A., W.D. Hibler III and G.M. Flato (1991). On the numerical simulation of the sea ice ridging process. *J. Geophys. Res.*, *96* (3), 4809–4820.

Horner, R. (1985). *Sea Ice Biota*. CRC Press, Boca Raton.

Houghton, J.T. (1997). *Global Warming. The Complete Briefing*. Cambridge Univ. Press, 2nd Ed., 251pp.

Houghton, J.T., L.G. Meira Filho, B.A. Callender, N. Harris, A. Kattenberg and K. Maskell (eds.) (1996). *Climate Change 1995: The Science of Climate Change*. Cambridge Univ. Press.

Hudson, R. (1990). Annual measurement of sea-ice thickness using an upward-looking sonar. *Nature*, *344*, 135–137.

Hughes, B.A. (1991). On the use of lognormal statistics to simulate one- and two-dimensional under-ice draft profiles. *J. Geophys. Res.*, *96*(C12), 22,101–22,111.

Hult, J.L. and N.C. Ostrander (1973). Antarctic icebergs as a global fresh water resource. Rep. R-1255–NSF, 83pp, RAND Corp., Santa Monica, Calif.

Hunkins, K. (1986). Anomalous diurnal tidal currents on the Yermak Plateau. *J. Mar. Sci.*, *44*, 51–69.

Huppert, H.E. (1980). The physical processes involved in the melting of icebergs. *Ann. Glaciol.*, *1*, 97–102.

Huppert, H.E. and E.G. Josberger (1980). The melting of ice in cold stratified water. *J. Phys. Oceanogr.*, *10*(6), 953–960.

Husseiny, A.A. (ed.) (1978). *Iceberg Utilization*. Pergamon, New York.

Jacobs, S.S. and D. Barnett (1987). On the draughts of some large Antarctic icebergs. *Iceberg Res.*, *14*, 3–13. Scott Polar Res. Inst., Cambridge.

Jacobs, S.S. and J.C. Comiso (1989). Sea ice and oceanographic processes over the Ross Sea continental shelf. *J. Geophys. Res.*, *94*(C12), 18195–18211.

Jacobs, S.S., A.F. Amos and P.M. Bruchhausen (1970). Ross Sea oceanography and Antarctic Bottom Water formation. *Deep-Sea Res.*, *17*, 935–962.

Jacobs, S.S., H.E. Huppert, H. Holdsworth and D.J. Drewry (1981). Thermohaline step induced by melting of the Erebus glacier tongue. *J. Geophys. Res.*, *86*(C7), 6547–6555.

Jacobs, S.S., R.G. Fairbanks and Y. Horibe (1985). Origin and evolution of water masses near the Antarctic continental margin: evidence from $H_2^{18}O/H_2^{16}O$ ratios in sea water. In *Oceanology of the Antarctic Continental Shelf* (ed. S.S. Jacobs). Ant. Res. Srs. 43, Amer. Geophys. U., Washington, 59–85.

Jacobs, S.S., D.R. MacAyeal and J.L. Ardai (1986). The recent advance of the Ross Ice Shelf, Antarctica. *J. Glaciol.*, *32*(112), 464–474.

JGR (1983). *Journal of Geophysical Research*, vol. 88 no. C5. Special issue, marginal ice zones.

JGR (1987). *Journal of Geophysical Research*, vol. 92 no. C7. Special issue, marginal ice zone research.

Jin, Guoliang, J.F. Lynch, R. Pawlowicz, P. Wadhams and P. Worcester (1993). Effects of sea ice cover on acoustic ray travel times, with applications to the Greenland Sea Tomography Experiment. *J. Acoust. Soc. Am.*, *94*(12), 1044–1056.

Johannessen, O.M. (1970). Note on some vertical current profiles below ice floes in the Gulf of St. Lawrence and near the North Pole. *J. Geophys. Res.*, *75*(15), 2857–2861.

Johannesen, O.M., J.A. Johannessen, E. Svendsen, R.A. Shuchman, W.J. Campbell and E. Josberger (1987a). Ice-edge eddies in the Fram Strait Marginal Ice Zone. *Science*, *236*, 427–429.

Johannessen, J.A., O.M. Johannessen, E. Svendsen, R. Shuchman, T. Manley, W.J. Campbell, E.G. Josberger, S. Sandven, J.C. Gascard, T. Olaussen, K. Davidson and J. Van Leer (1987b). Mesoscale eddies in the Fram Strait marginal ice zone during the 1983 and 1984 Marginal Ice Zone Experiments. *J. Geophys. Res.*, *92*(C7), 6754–6772.

Johannessen, O.M., L. Bjørnø, G. Bienvenu, K. Hasselmann, P.M. Haugan, J. Johnsen, U. Lie, J. Papadakis, S. Sandven, P. Wadhams and M. Zakharia (1993). ATOC — Arctic: acoustic thermometry of the ocean climate in the Arctic Ocean. Nansen Envtl. and Remote Sensing Centre, Bergen, rept.

Johannessen, O.M., S. Sandven, W.P. Budgell, J.A. Johannessen and R.A. Shuchman (1994). Observation and simulation of ice tongues and vortex pairs in the marginal ice zone. In *The Polar Oceans and their Role in Shaping the Global Environment* (ed. O.M. Johannessen, R.D. Muench, J.E. Overland). Amer. Geophys. U., Geophys. Monograph 85, 109–136.

Johannessen, O.M., M. Miles and E. Bjørgo (1995). The Arctic's shrinking sea ice. *Nature*, *376*, 126–127.

Jones, A.G.E. (1985). Icebergs in the Southern Ocean. *Iceberg Res.*, *11*, 16–19. Scott Polar Res. Inst., Cambridge.

Jones, S. (1977). Instabilities and wave interactions in a rotating two-layer fluid. PhD thesis, Univ. Cambridge.

Josberger, E.G. (1982). The oceanographic impact of melting icebergs and marine ice shelves. *Iceberg Res.*, *1*, 4–9, Scott Polar Res. Inst., Cambridge.

Kantha, L.H. and G.L. Mellor (1989). A numerical model of the atmospheric boundary layer over a marginal ice zone. *J. Geophys. Res.*, *94*(C4), 4959–4970.

Keller, J.B. (1997). Gravity waves on ice-covered water. *J. Geophys. Res.*, *103*(C4), 7663–7669.

Keller, J.B. and M. Weitz (1953). Reflection and transmission coefficients for waves entering or leaving an icefield. *Communs. Pure Appl. Math.*, *6*(3), 415–417.

Kerman, B.R. (1998). On the relationship of pack ice thickness to the length of connectivity trees in SAR imagery. Proc. 14th Intl. Symp. on Ice, *Ice in Surface Waters*, vol. 2 (H.T. Shen, ed.), Clarkson Univ., Potsdam N.Y.

Ketchum, R.D. (1971). Airborne laser profiling of the Arctic pack ice. *Remote Sensing Environ.*, *2*, 41–52.

Key, J.R. and A.S. McLaren (1989). Periodicities and keel spacings in the under-ice draft distribution of the Canada Basin. *Cold Regions Sci. & Technol.*, *16*, 1–10.

Key, J. and A.S. McLaren (1991). Fractal nature of the sea ice draft profile. *Geophys. Res. Letters*, *18*(8), 1437–1440.

Keys, H.(1986). Towards a new shape classification of Antarctic icebergs. *Iceberg Res.*, *12*, 15–19.

Keys, H. (1990). Iceberg B-9 and its drift in the eastern Ross Sea, Antarctic, up to 26 January 1989. *Iceberg Res.*, *15*, 17–27. Scott Polar Res. Inst., Cambridge.

Koch, L. (1945). The east Greenland ice. *Medd. om Grønland, 130*, part 3.

Koenig, L.S., V.R. Greenaway, M. Dunbar and G. Huttersley-Smith (1952). Arctic ice islands. *Arctic*, 5(2), 67–103.

Kovacs A. (1971). On pressured sea ice. In *Sea Ice* (ed. T. Karlsson), Proceedings of International Sea Ice Conference, Reykjavik, Iceland, May 10–13 1971, 276–295.

Kovacs, A. and J.S. Holladay (1989). Development of an airborne sea ice thickness measurement system and field test results. US Army Cold Regions Res. 9 Engng. Lab., Hanover, N.H., Rept. 89–19.

Kovacs, A. and J.S. Holladay (1990). Airborne sea ice thickness sounding. In *Sea Ice Properties and Processes* (ed. S.F. Ackley, W.F. Weeks), US Army Cold Regions Res. & Engng. Lab., Hanover, N.H., Monograph 90–1, 225–229.

Kovacs, A. and M. Mellor (1974). Sea ice morphology and ice as a geological agent in the southern Beaufort Sea. In *The Coast and Shelf of the Beaufort Sea* (ed. J.C. Reed, J.A. Sater), Arctic Inst. N. Amer., Arlington, 113–161.

Kovacs, A. and R.M. Morey (1986). Electromagnetic measurements of multiyear sea ice using impulse radar. *Cold Regions Sci. Technol., 12*, 67–93.

Kovacs A. and D.S. Sodhi (1980). Shore ice pile-up and ride-up: field observations, models and theoretical analyses. *Cold Regions Sci. Technol., 2*, 209–288.

Kovacs A., W.F. Weeks, S.F. Ackley and W.D. Hibler (1973). Structure of a multiyear pressure ridge. *Arctic, 26*(1), 22–32.

Kowalik, Z. and A.Y. Proshutinsky (1993). Diurnal tides in the Arctic Ocean. *J. Geophys. Res., 98*(C9), 16449–16468.

Kowalik, Z. and A.Y. Proshutinsky (1994). The Arctic Ocean tides. In *The Polar Oceans and their Role in Shaping the Global Environment* (ed. O.M. Johannessen, R.D. Muench, J.E. Overland), Geophys. Monograph 85, Amer. Geophys. U., Washington, 137–158.

Krabill, W.B., R.N. Swift and W.B. Tucker III (1990). Recent measurements of sea ice topography in the Eastern Arctic. In *Sea Ice Properties and Processes* (ed. S.F. Ackley and W.F. Weeks), Monography 90–1, US Army Cold Regions Res. & Engng. Lab., Hanover N.H., 132–136.

Kristensen, M., V.A. Squire and S.C. Moore (1982). Tabular icebergs in ocean waves. *Nature, 297*, 669–671.

Kurbjeweit, F., R. Gradinger and J. Weissenberger (1993). The life cycle of *Stephos longipes* - an example for cryopelagic coupling in the Weddell Sea (Antarctica). *Mar. Ecol. Progr. Ser. 98*, 255–262.

Kurtz, D. D., and D.H. Bromwich (1983). Satellite observed behaviour of the Terra Nova Bay polynya. *J. Geophys. Res., 88*(C14), 9717–9722.

Kurtz, D.D. and D.H. Bromwich (1985). A recurring, atmospherically forced polynya in Terra Nova Bay. In *Oceanology of the Antarctic Continental Shelf* (S.S. Jacobs, ed.), Antarctic Res. Srs 43, American Geophysical Union, Washington, 177–201.

Kvambekk, A.S. and T. Vinje (1992). Ice draft recordings from upward looking sonars in the Fram Strait and Barents Sea in 1987/88 and 1990/91. *Norsk Polarinstitutt Rapportserie, 79*.

Kwok, R. and D.A. Rothrock (1999). Variability of Fram Strait ice flux and North Atlantic Oscillation. *J. Geophys. Res., 104*(C3), 5177–5189.

Kwok, R., G.F. Cunningham, N. LaBelle-Hamer, B. Holt and D. Rothrock (1999). Ice thickness derived from high-resolution radar imagery. *EOS, Trans. Amer. Geophys. U., 80*(42), 495, 497.

LaBelle, J.C., J.L. Wise, R.P. Voelker, R.H. Schulze and G.M. Wohl (1983). *Alaska Marine Ice Atlas.* Arctic Envtl. Info. & Data Center, Univ. Alaska, Anchorage. 302pp.

Lake, R.A. and E.L. Lewis (1970). Salt rejection by sea ice during growth. *J. Geophys. Res.*, *75*(3), 583–597.

Lange, M.A. and H. Eicken (1991). The sea ice thickness distribution in the northwestern Weddell Sea. *J. Geophys. Res.*, *96*(C3), 4821–4837.

Lange, M.A. and S.L. Pfirman (1998). Arctic sea ice contamination: major characteristics and consequences. In *Physics of Ice-covered Seas* (ed. M. Leppäranta), Helsinki Univ. Printing House, 2, 651–681.

Lange, M.A., S.F. Ackley, P. Wadhams, G.S. Dieckmann and H. Eicken (1989). Development of sea ice in the Weddell Sea Antarctica. *Ann. Glaciol.*, *12*, 92–96.

Langhorne, P. (1983). Laboratory experiments on crystal orientation in NaCl ice. *Ann. Glaciol.*, *4*, 163–169.

Langhorne, P. and W.H. Robinson (1986). Alignment of crystals in sea ice due to fluid motion. *Cold Regions Sci. & Technol.*, *12*(2), 197–214.

LDGO (1989). AnZone: International Coordination of Oceanographic Studies of the Antarctic Zone. 23–25 May 1989 Meeting report (ed. A. Gordon). Lamont-Doherty Geological Observatory, Palisades N.Y.

Lebedev, V.V. (1938). Rost l'do v arkticheskikh rekakh i moriakh v zavisimosti ot otritsatel'nykh tempertur vozdukha. *Problemy Arktiki*, *5*, 9–25.

Ledley, T.S. and S. Pfirman (1997). The impact of sediment-laden snow and sea ice in the Arctic on climate. *Climatic Change*, *37*(4), 641–664.

Legeckis, R. (1978). A survey of worldwide sea surface temperature fronts detected by environmental satellites. *J. Geophys. Res.*, *83*(C9), 4501–4522.

Leggett, J., W.J. Pepper and R.J. Swart (1992). Emissions scenarios for the IPCC: an update. In *Climate Change 1992*, *The Supplementary Report to the IPCC Scientific Assessment* (ed. J.T. Houghton, B.A. Callander and S.K. Varney), Cambridge Univ. Press, 69–95.

Lemke, P. (ed.) (1997). Report of the Third ACSYS Sea Ice – Ocean Modelling Workshop, Victoria B.C., Canada, 17–20 August 1996. WCRP Informal Rept., 51997, Kiel.

Lemke, P., W.B. Owens and W.D. Hibler III (1990). A coupled sea ice - mixed layer pycnocline model for the Weddell Sea. *J. Geophys. Res.*, *95*(C6), 9513–9525.

Lensu, M. (1997). Correlations between fragment sizes in sequential fragmentation. *J. Phys. A: Math. Gen.*, *30*, 1–7.

Leppäranta M. (1981). Statistical features of sea ice ridging in the Gulf of Bothnia. Winter Navigation Research Board, Helsinki, Tech. Rep. 32, 46pp.

Leppäranta, M. (1998). The dynamics of sea ice. In *The Physics of Ice-Covered Seas* (ed. M. Leppäranta), Univ. Helsinki Press, *1*, 305–342.

Leppäranta, M. and T. Thompson (1989). BEPERS-88 sea ice remote sensing with synthetic aperture radar in the Baltic Sea. *Eos, Trans. Am. Geophys. U.*, *70*(28), 698–699, 708–709.

LeSchack, L.A. (1980). Arctic Ocean sea ice statistics derived from the upward-looking sonar data recorded during five nuclear submarine cruises. LeSchack Associates Ltd., 116–1111 University Blvd. W., Silver Spring, Md., Tech. Rept.

Lewis, C.F.M. (1977). Bottom scour by sea ice in the southern Beaufort Sea. Dept. of Fisheries & Envt., Ottawa, Beaufort Sea Proj., Tech. Rept. 23.

Lewis, E.L. (1967). Heat flow through winter ice. In *Physics of Snow and Ice* (ed. H. Ôura), *1*, 611–631, Inst. Low Temp. Sci., Univ. Hokkaido.

Lewis, E. L. (1990). Polynyas, their formation and significance. *Comité Arctique Internationale Commentary*, 2, February 1990, 17–20.

Lewis, E.L. and R.G. Perkin (1986). Ice pumps and their rates. *J. Geophys. Res.*, *91*(C10), 11756–11762.

Lewis, E.L., D. Ponton, L. Legendre and B. Leblanc (1996). Springtime sensible heat, nutrients and phytoplankton in the Northwater Polynya, Canadian Arctic. *Continental Shelf Res.*, *16*(14), 1775–1777.

Lewis, E.O., B.W. Currie and S. Haykin (1994). Surface-based radar: noncoherent. In *Remote Sensing of Sea Ice and Icebergs* (ed. S. Haykin, E.O. Lewis, R.K. Raney and J.R. Rossiter), Wiley, 341–442.

Lewis, J.C. and G. Bennett (1984). Monte Carlo simulations of iceberg draft changes caused by roll. *Cold Regions Sci. Technol.*, *10*, 1–10.

Lien, R. (1981). Sea bed features in the Blaaenga area, Weddell Sea, Antarctica. POAC-81, Proc. 6th Intl. Conf. on Port & Ocean Engng. under Arctic Condns., Quebec, Université Laval, 706–716.

Liu, A.K. and E. Mollo-Christensen (1988). Wave propagation in a solid ice pack. *J. Phys. Oceanogr.*, *18*(11), 1702–1712.

Liu, A.K., P.W. Vachon and C.Y. Peng (1991). Observation of wave refraction at an ice edge by synthetic aperture radar. *J. Geophys. Res.*, *96*(C3), 4803–4808.

Liu, A.K., P.W. Vachon, C.Y. Peng and A.S. Bhogal (1992). Wave attenuation in the marginal ice zone during LIMEX. *Atm.-Ocean*, *30*(2), 192–206.

Livingstone, C.E. (1989). Combined active/passive microwave classification of sea ice. *Proc. IGARSS-89*, *1*, 376–380.

Lizotte, M.P. and K.R. Arrigo (eds.) (1998). *Antarctic Sea Ice: Biological Processes, Interactions and Variability.* Antarctic Res. Srs. 73, Amer. Geophys. U., Washington, 198pp.

Longuet-Higgins, M.S. (1977). The mean forces exerted by waves on floating or submerged bodies with applications to sand bars and wave power machines. *Proc. Roy. Soc.*, *A352*, 463–480.

Lowry, R.T. and P. Wadhams (1979). On the statistical distribution of pressure ridges in sea ice. *J. Geophys. Res.*, *84*(C5), 2487–2494.

Lyon, W.K. (1961). Ocean and sea-ice research in the Arctic Ocean via submarine. *Trans. N.Y. Acad. Sci.*, srs. 2, *23*, 662–674.

McCann, S.B. and R.J. Carlisle (1972). The nature of the ice-foot on the beaches of Radstock Bay, south-west Devon Island, N.W.T., Canada. Inst. British Geographers, Special Publn. 4, 175–186.

Macdonald, R.W. (in press). Arctic estuaries and ice: a positive-negative estuarine couple. In *The Arctic Ocean Fresh Water Budget*, Proc. NATO Advanced Research Workshop, Tallinn, 28 April – 2 May 1998.

Mackintosh, N.A. and H.F.P. Herdman (1940). Distribution of the pack-ice in the Southern Ocean. *Discovery Rep.*, *19*, 285–296.

McLaren, A.S. (1988). Analysis of the under-ice topography in the Arctic Basin as recorded by the USS *Nautilus* during August 1958. *Arctic*, *41*(2), 117–126.

McLaren, A.S. (1989). The under-ice thickness distribution of the Arctic basin as recorded in 1958 and 1970. *J. Geophys. Res., 94*(C4), 4971–4983.

McLaren, A.S., P. Wadhams and R. Weintraub (1984). The sea ice topography of M'Clure Strait in winter and summer of 1960 from submarine profiles. *Arctic, 37*(2), 110–120.

McPhee, M.G. (1979). The effect of the oceanic boundary layer on the mean drift of pack ice: application of a simple model. *J. Phys. Oceanogr., 9*(2), 388–400.

McPhee, M.G. (1980). An analysis of pack ice drift in summer. In *Sea Ice Processes and Models* (ed. R.S. Pritchard), Univ. Washington Press, Seattle, 62–75.

McPhee, M.G. (1982). Sea ice drag laws and simple boundary layer concepts, including application to rapid melting. Rept. 82–4, US Army Cold Regions Res. & Engng. Lab., Hanover N.H.

McPhee, M.G. (1986). The upper ocean. In *The Geophysics of Sea Ice* (ed. N. Untersteiner), Plenum, New York, 339–394.

McPhee, M.G. and J.D. Smith (1975). Measurements of the turbulent boundary layer under pack ice. *AIDJEX Bull., 29*, 49–92, Div. Marine Resources, Univ. Washington.

McPhee, M.G., T.P. Stanton, J.H. Morison and D.G. Martinson (1998). Freshening of the upper ocean in the central Arctic; is perennial sea ice disappearing? *Geophys. Res. Lett., 25*, 1729–1732.

Makshtas, A. P. (1991). The heat budget of Arctic ice in the winter. International Glaciological Society, Cambridge, U.K., 77pp.

Manabe, S. and R.J. Stouffer (1993). Century-scale effects of increased atmospheric CO_2 in the ocean-atmosphere system. *Nature, 364*, 215–218.

Mandelbrot, B.B. (1977). *Fractals. Form, Chance and Dimension.* Freeman, San Francisco.

Mandelbrot, B.B. (1982). *The Fractal Geometry of Nature.* Freeman, San Francisco.

Marko, J.R., J.R. Birch and M.A. Wilson (1982). A study of long-term satellite-tracked iceberg drifts in Baffin Bay and Davis Strait. *Arctic, 35*(1), 234–240.

Markus, T. and D.J. Cavalieri (1998). Snow depth distribution over sea ice in the Southern Ocean from satellite passive microwave data. In *Antarctic Sea Ice. Physical Processes, Interactions and Variability* (ed. M.O. Jeffries). Antarctic Res. Srs., 74, 19–40, Amer. Geophys. U., Washington.

Marshurnova, M.S. (1961). Principal characteristics of the radiation balance of the underlying surface and of the atmosphere in the Arctic. *Trudy Arkt. Antarkt. Nauch. Issle. Inst., 229* (Eng. trans. by RAND Corp., Santa Monica, rept. RM-5003–PR, 1966).

Martin, S. and P. Kauffman (1977). An experimental and theoretical study of the turbulent and laminar convection generated under a horizontal ice sheet floating on warm salty water. *J. Phys. Oceanogr., 7*(2), 272–283.

Martin, S. and P. Kauffman (1981). A field and laboratory study of wave damping by grease ice. *J. Glaciol., 27*(96), 283–313.

Martin, S., P. Kauffman and C. Parkinson (1983). The movement and decay of ice edge bands in the winter Bering Sea. *J. Geophys. Res., 88*(C5), 2803–2812.

Martinson, D.G. (1990). Evolution of the Southern Ocean winter mixed layer and sea ice: open ocean deepwater formation and ventilation. *J. Geophys. Res., 95*(C7), 11641–11654.

Martinson, D.G. and C. Wamser (1990). Ice drift and momentum exchange in winter Antarctic pack ice. *J. Geophys. Res., 95*(C2), 1741–1755.

Martinson, D.G., P.D. Killworth and A.L. Gordon (1981). A convective model for the Weddell Polynya. *J. Phys. Oceanogr.*, *11*, 466–487.

Maslanik, J.A., M.C. Serreze and R.G. Barry (1996). Recent decreases in Arctic summer ice cover and linkages to atmospheric circulation anomalies. *Geophys. Res. Lett.*, *23*, 1677–1680.

Massom, R. (1984). Tabular icebergs off Nordostrundingen, North East Greenland. *Iceberg Res.*, *8*, 11–16. Scott Polar Res. Inst., Cambridge.

Massom, R. (1991). *Satellite remote sensing of polar regions*. Belhaven, London.

Maykut, G.A. (1986). The surface heat and mass balance. In *The Geophysics of Sea Ice* (ed. N. Untersteiner), Plenum, New York, 395–464.

Maykut, G.A. and N. Untersteiner (1971). Some results from a time-dependent thermodynamic model of Arctic sea ice. *J. Geophys Res.*, *76*(6), 1550–1575.

Maykut, G.A., T.C. Grenfell and W.F. Weeks (1992). On estimating spatial and temporal variations in the properties of ice in the polar oceans. *J. Mar. Sys.*, *3*, 41–72.

Meier, M.F. (1984). Contribution of small glaciers to global sea level. *Science, 226*, 1418–1421.

Melling, H. and D.A. Riedel (1995). The underside topography of sea ice over the continental shelf of the Beaufort Sea in the winter of 1990. *J. Geophys. Res.*, *100*(C7), 13641–13653.

Mellor, M. (1967). Antarctic ice budget (and Pleistocene variations of ice volume). In *The Encycopaedia of Atmospheric Sciences and Astrogeology* (ed. R.W. Fairbridge), 16–19, Reinhold, New York.

Melnikov, I.A. (1997). *The Arctic Sea Ice Ecosystem*. Gordon & Breach, Amsterdam. 204pp.

Melnikov, I.A. and L.L. Bondarchuk (1987). Ecology of mass accumulations of colonial diatom algae under drifting Arctic ice. *Oceanology, 27*, 233–236.

Melnikov, S.A. and S.V. Vlasov (1992). A summary report on contaminant levels in the compartments of the marine environment of the Arctic seas in 1990. Rept., Arctic & Antarctic Res. Inst., St. Petersburg.

Mikhalevsky, P.N., A.N. Gavrilov and A.B. Baggeroer (1999). The Trans-Arctic Acoustic Propagation Experiment and climate monitoring in the Arctic. *IEEE J. Oceanic Engng.*, *24*(2), 183–201.

Miller, R.G. and P.W. Barnes (1985). Formation of iceberg keel marks on the Antarctic sea floor. *Iceberg Res.* (Scott Polar Res. Inst., Cambridge), *11*, 10–12.

Milne, A.R. (1972). Thermal tension cracking in sea ice: a source of under-ice noise. *J. Geophys. Res.*, *77*(12), 2177–2192.

Minnett, P. J. (1995). Measurements of the summer surface heat budget of the northeast water polynya in 1992. *J. Geophys. Res.*, *100*(C3), 4309–4322.

Mitchell, J.F.B., S. Manabe, V. Meleshko and T. Tokioka (1990). Equilibrium climate change – and its implications for the future. In *Climate Change. The IPCC Scientific Assessment* (J.T. Houghton, G.J. Jenkins and J.J. Ephraums, eds.). Cambridge Univ. Press, 137–164.

Mock S.J., A.D. Hartwell and W.D. Hibler (1972). Spatial aspects of pressure ridge statistics. *J. Geophys. Res.*, *77*(30), 5945–5953.

Morison, J. (1986). Internal waves in the Arctic Ocean: a review. In *The Geophysics of Sea Ice* (ed. N. Untersteiner), Plenum, New York, 1163–1183.

Morison, J., M. Steele and R. Andreson (1998). Hydrography of the upper Arctic Ocean measured from the nuclear submarine USS *Pargo*. *Deep-Sea Res., 45* , 15–38.

Moritz, R.E. (1991). Sampling the temporal variability of sea ice draft distribution. *Eos* supplement, Fall AGU Meeting, 237–238 (abstract).

Moritz, R.E. and D.K. Perovich (eds.) (1996). Surface Heat Budget of the Arctic Ocean Science Plan. ARCSS/OAII Rept. no. 5, Univ. Washington, Seattle.

Muench, R.D. and R.L. Charnell (1977). Observations of medium-scale features along the seasonal ice edge in the Bering Sea. *J. Phys. Oceanogr., 7*(4), 602–606.

Muench, R.D., P.H. LeBlond and L.E. Hachmeister (1983). On some possible interactions between internal waves and sea ice in the marginal ice zone. *J. Geophys. Res., 88*(C5), 2819–2826.

Muench, R.D., M.G. McPhee, C.A. Paulson and J.H. Morison (1992). Winter oceanographic conditions in the Fram Strait - Yermak Plateau region. *J. Geophys. Res., 97*(C3), 3469–3483.

Multala, J., H. Hautaniemi, M. Oksama, M. Leppäranta, J. Haapala, A. Herlevi, K. Riska and M. Lensu (1995). Airborne electromagnetic surveying of Baltic sea ice. Univ. Helsinki, Dept. of Geophys., Rept. Srs. in Geophys. no. 31, 58pp.

Munk, W.H. and A.M.G. Forbes (1989). Global ocean warming: an acoustic measure? *J. Phys. Oceanogr., 19*, 1765–1778.

Munk, W.H. and C. Wunsch (1979). Ocean acoustic tomography: a scheme for large scale monitoring. *Deep-Sea Res., 26*, 123–161.

Murty, T.S. (1985). Modification of hydrographic characteristics, tides, and normal modes by ice cover. *Marine Geodesy, 9*(4), 451–468.

Nakawo, M. and N.K. Sinha (1981). Growth rate and salinity profile of first-year sea ice in the high Arctic. *J. Glaciol., 27*(96), 315–330.

Nansen, F. (1897). *Farthest North.* Constable, London, *1*, 299.

Nansen, F. (1902). The oceanography of the North Polar Basin. *Norwegian North Polar Expedition 1893–1896. Scientific Results.* Vol. III, 1–427.

Nansen, F. (1922). The strandflat and isostasy. Videnskapsselsk. Skr. Kl. I. Mat. Naturvidensk. Kl. no. 11.

Nazintsev, Y.L. (1964). Nekotorye dannye k raschetu teplovykh svoistv morskogo l'da (Some data on the calculation of thermal properties of sea ice). *Trudy Arkt. i Antarkt. Inst., 267*, 31–47.

Neshyba, S (1977). Upwelling by icebergs. *Nature, 267,* 507–508.

Nevel, D. (1970). Moving loads on a floating ice sheet. US Army Cold Regions Res. & Engng. Lab., Hanover N.H., Res. Rept. 261.

Newyear, K. and S. Martin (1997). A comparison of theory and laboratory measurements of wave propagation and attenuation in grease ice. *J. Geophys. Res., 102*(C11), 25091–25099.

Niebauer, H.J. (1998). Variability in Bering Sea ice cover as affected by a regime shift in the North Pacific in the period 1947–1996. *J. Geophys. Res., 103*(C12), 27717–27737.

Niedrauer, T.M. and S. Martin (1979). An experimental study of brine drainage and convection in young sea ice. *J. Geophys. res., 84*(C3), 1176–1186.

Nilsen, T. and N. Bøhmer (1994). Sources to radioactive contamination in Murmansk and Arkhangel'sk counties. Bellona Rept. no.1, 162pp, Bellona Foundation, Oslo.

Nilsen, T., I. Kudrik and A. Nikitin (1996). The Russian Northern Fleet. Sources of radioactive contamination. Bellona Rept. no. 2, 168pp, Bellona Foundation, Oslo.

NORCOR Engineering & Research Ltd. (1977). Probable behaviour and fate of a winter oil spill in the Beaufort Sea. Fisheries & Envt. Canada, Envtl. Impact Control Directorate, Technology Development Rept. EPS-4–EC-77-5, 111pp.

Norheim, G., J.U. Skaare and O. Wiig (1992). Some heavy metals, essential elements and chlorinated hydrocarbons in polar bear (*Ursus maritimus*) at Svalbard. *Envtl. Pollution, 77*, 51–57.

Nye, J.F. (1973). The physical meaning of the two dimensional stresses in a floating ice cover. *AIDJEX Bull., 21*, 1–8. Div. Marine Resources, Univ. Washington, Seattle.

Oceanography (1997). Special issue on High Frequency Radars for Coastal Oceanography. *Oceanography, 10*(2).

Ono, N. (1965). Thermal properties of sea ice. II. A method for determining the K/c value of a non-homogeneous ice sheet. *Low Temp. Sci. J., A23*, 177–183.

Ono, N. (1967). Specific heat and heat of fusion of sea ice. In *Physics of Snow and Ice* (H. Oura, ed.), Inst. Low Temp. Sci., Hokkaido Univ., Sapporo, *1*, 599–610.

Ono, N. (1968). Thermal properties of sea ice. IV. Thermal constants of sea ice. *Low Temp. Sci., A26*, 329–349.

Onstott, R.G., T.C. Grenfell, C. Maetzler, C.A. Luther and E.A. Svendsen (1987). Evolution of microwave sea ice signatures during early and mid summer in the marginal ice zone. *J. Geophys. Res., 92*(C7), 6825–6837.

Orheim, O. (1980). Physical characteristics and life expectancy of tabular Antarctic icebergs. *Ann. Glaciol., 1*, 11–18.

Orheim, O. (1984). Iceberg discharge and the mass balance of Antarctica. *Iceberg Res., 8*, 3–7. Scott Polar Res. Inst., Cambridge.

Ou, H.W. (1988). A time-dependent model of a coastal polynya. *J. Phys. Oceanogr., 18*(4), 584–590.

Overgaard, S., P. Wadhams and M. Leppäranta (1983). Ice properties in the Greenland and Barents Sea during summer. *J. Glaciol., 29*(101), 142–164.

Overland, J.E. (1985). Atmospheric boundary layer structure and drag coefficients over sea ice. *J. Geophys. Res., 90*(C5), 9029–9049.

Padman, L., A.J. Plueddemann, R.D. Muench and R. Pinkel (1992). Diurnal tides near the Yermak Plateau. *J. Geophys. Res., 97*(C8), 12639–12652.

Palmer, A.C. (1997). Geotechnical evidence of ice scour as a guide to pipeline burial depth. *Can. Geotech. J., 34*, 1002–1003.

Palmer, A.C., I. Konuk, G. Comfort and K. Been (1990). Ice gouging and the safety of marine pipelines. *Proc. 22nd Offshore Technology Conf., Houston, 3*, 235–244. OTC6371.

Panov, A.N. and A.O. Shpaikher (1964). Influence of Atlantic waters on some features of the hydrology of the Arctic Basin and adjacent seas. *Deep-Sea Res., 11*, 275–285.

Parkinson, C.L. (1991). Interannual variability of the spatial distribution of sea ice in the north polar region. *J. Geophys. Res., 96*(C3), 4791–4802.

Parkinson, C.L. (1995). Recent sea-ice advances in Baffin Bay / Davis Strait and retreats in the Bellingshausen Sea. *Ann. Glaciol., 21*, 348–352.

Parkinson, C.L. (1998). Length of the sea ice season in the Southern Ocean, 1988–1994. In *Antarctic Sea Ice: Physical Processes, Interactions and Variability*. Antarctic Res. Srs. 74, Am. Geophys. U., Washington, 173–186.

Parkinson, C.L. and W.M. Kellogg (1979). Arctic sea ice decay simulated for a CO_2-induced temperature rise. *Climatic Change*, 2, 149–162.

Parkinson, C.L. and W.M. Washington (1979). A large-scale numerical model of sea ice. *J. Geophys. Res.*, *84*(C1), 311–324.

Parkinson, C.L., J.C. Comiso, H.J. Zwally, D.J. Cavalieri, P. Gloeren and W.J. Campbell 91987). *Arctic Sea Ice, 1973–1976: Satellite Passive-Microwave Observations*. NASA Publn. SP-489, Washington. 296pp.

Parkinson, C.L., D.J. Cavalieri, P. Gloersen, H.J. Zwally and J.C. Comiso (1999). Arctic sea ice extents, areas and trends, 1978–1996. *J. Geophys. Res.*, *104*(C9), 20837–20856.

Parmerter R.R. and M. Coon (1972). Model of pressure ridge formation in sea ice. *J. Geophys. Res.*, *77*(33), 6565 - 6575.

Pease, C. H. (1987). The size of wind driven coastal polynyas. *J. Geophys. Res.*, *92*(C7), 7049–7059.

Pedlosky, J. (1979). *Geophysical Fluid Dynamics*, Springer Verlag, New York. 624pp

Perovich, D.K. (1998). The optical properties of sea ice. In *Physics of Ice-Covered Seas* (ed. M. Leppäranta), Univ. Helsinki Press, *1*, 195–230.

Peters, A.S. (1950). The effect of a floating mat on water waves. *Communs. Pure Appl. Math.*, *3*, 319–354.

Petrenko, V.F. and R.W. Whitworth (1999). *Physics of Ice*. Oxford Univ. Press, 373pp.

Pfirman, S.L., H. Eicken, D. Bauch and W.F. Weeks (1995). The potential transport of pollutants by Arctic sea ice. *Sci. of Total Envt.*, *159*, 129–146.

Pilkington, G.R. and R.W. Marcellus (1981). Methods of determining pipeline trench depths in the Canadian Beaufort Sea. *Proc. 6th Intl. Conf. Port & Ocean Engng. under Arctic Condns.*, *Quebec*, 2, 674–687.

Pilkington, G.R. and B.D. Wright (1991). Beaufort Sea ice thickness measurements from an acoustic, under ice, upward looking ice keel profiler. *Proc. 1st Intl. Offshore & Polar Engng Conf., Edinburgh, 11–16 Aug 1991*.

Pollard, J.H. (1977). *A Handbook of Numerical and Statistical Techniques.*, 232–233, Cambridge Univ. Press.

Poorooshasb, F., J.I. Clark and C.M. Woodworth-Lynas (1989). Small-scale modelling of iceberg scouring of the seabed. *Proc. 10th Intl. Conf. on Port & Ocean Engng. under Arctic Condns.*, *Lulea*, 1, 133–145.

Pritchard, R.S.(ed.) (1980). *Sea Ice Processes and Models*. Univ. Washington Press, Seattle, 474pp.

Proshutinsky, A. and M. Johnson (1997). Two circulation regimes of the wind-driven Arctic Ocean. *J. Geophys. Res., 102*(C6), 12493–12514.

Prowse, T.D. (1990). Heat and mass balance of an ablating ice jam. *Can. J. Civil Engng., 17*(4), 629–635.

Prowse, T.D. and N.C. Gridley (ed.) (1993). *Environmental Aspects of River Ice*. Nat. Hydrology Res. Inst., Saskatoon, Sci. Rept. no. 5, Environment Canada. 155pp.

Pullen, T.C. and C. Swithinbank (1991). Transits of the Northwest Passage 1906–90. *Polar Record, 27*(163), 365–367.

Rahmstorf, S. and A. Ganopolski (1999). Long-term global warming scenarios computed with an efficient coupled climate model. *Clim. Change, 43*(2), 353–367.

Reid, J.L., W.D. Nowlin Jr. and W.C. Patzert (1977). On the characteristics and circulation of the southwestern Atlantic Ocean. *J. Phys. Oceanogr.*, 7, 62–91.

Reimnitz, E. and K.F. Bruder (1972). River discharge into an ice-covered ocean and related sediment dispersal, Beaufort Sea coast of Alaska. *Geol. Soc. Am. Bull., 83*, 861–866.

Reimnitz, E. and E.W. Kempema (1984). Pack ice interaction with Stamukhi Shoal, Beaufort Sea, Alaska. In *The Alaskan Beaufort Sea: Ecosystems and Environments* (ed. P.W. Barnes, D.M. Schell, E. Reimnitz), Academic Press, Orlando, 159–183.

Reimnitz, E., C.A. Rodeick and S.C. Wolf (1974). Strudel scour: a unique Arctic marine geologic phenomenon. *J. Sed. Petrol.*, 44(2), 409–420.

Reimnitz, E., P.W Barnes, L.J. Toimil and J. Melchior (1977). Ice gouge recurrence and rates of sediment reworking, Beaufort Sea, Alaska. *Geology*, 5(7), 405–408.

Reimnitz, E., L. Toimil and P.W. Barnes (1978). Arctic continental shelf morphology related to sea ice zonation, Beaufort Sea, Alaska. *Marine Geol.*, 28, 179–210.

Reimnitz, E., E.W. Kempema and P.W. Barnes (1987). Anchor ice, seabed freezing and sediment dynamics in shallow Arctic seas. *J. Geophys. Res*, 92(C13), 14671–14678.

Reynolds, J.M. (1979). Icebergs are a frozen asset. *Geogr. Mag.*, 52, 177–185.

Reynolds, M., C.H. Pease and J.E. Overland (1985). Ice drift and regional meteorology in the southern Bering Sea: results from MIZEX West. *J. Geophys. Res.*, 90(C6), 11967–11981.

Rigby, F.A. (1976). Pressure ridge generated internal wave wakes at the base of the mixed layer in the Arctic Ocean. MSc thesis, Univ. Washington, Seattle.

Rigby F.A. and A. Hanson (1976). Evolution of a large Arctic pressure ridge. *AIDJEX Bull. 34*, 43–71. Div. Marine Resources, Univ. Washington, Seattle.

Rintoul, S.R. (1998). On the origin and influence of Adélie Land bottom water. In *Ocean, Ice and Atmosphere. Interactions at the Antarctic Continental Margin* (ed. S.S. Jacobs, R.F. Weiss), Antarctic Res. Srs. 75, Amer. Geophys. U., Washington, 151–171.

Roberts, D.M. (1978). Icebergs as a heat sink for power generation. In *Iceberg Utilization* (ed. A.A. Husseiny), Pergamon, New York, 674–689.

Robin, G. de Q. (1963). Wave propagation through fields of pack ice. *Phil. Trans. R. Soc.*, A255, 313–339.

Røed, L-P. and J.J. O'Brien (1983). A coupled ice-ocean model of upwelling in the marginal ice zone. *J. Geophys. Res.*, 88(C5), 2863–2872.

Rothrock, D.A. (1986). Ice thickness distribution - measurement and theory. In *The Geophysics of Sea Ice* (ed. N. Untersteiner), Plenum, New York, 551–575.

Rothrock, D.A. and A.S. Thorndike (1980). Geometric properties of the underside of sea ice. *J. Geophys. Res.*, 85(C7), 3955–3963.

Rothrock, D.A., Y. Yu and G.A. Maykut (1999). Thinning of the Arctic sea-ice cover. *Geophys. Res. Lett.*, 26(23), 3469–3472.

Rottier, P.J. (1991). Wave/ice interactions in the marginal ice zone and the generation of ocean noise., PhD thesis, Univ. Cambridge, Scott Polar Res. Inst.

Rudels, B. (1990). Haline convection in the Greenland Sea. *Deep-Sea Res.*, 37(9), 1491–1511.

Rudels, B. (1995). The thermohaline circulation of the Arctic Ocean and the Greenland Sea. *Phil. Trans. Roy. Soc.*, A352, 287–299.

Russell-Head, D.S. (1980). The melting of free-drifting icebergs. *Ann. Glaciol., 1*, 119–122.

Saito, T. and N. Ono (1980). Percolation of sea ice. II. Brine drainage channels in young sea ice. *Low Temp. Sci., Srs. A, 39*, 127–132.

Sanderson, T.J.O. (1988). *Ice Mechanics. Risks to Offshore Structures.* Graham and Trotman, London.

Sayed M. and R. Frederking (1986). On modelling of ice ridge formation. International Association for Hydraulic Research Ice Symposium, 603–614.

Schledermann, P. (1980). Polynyas and prehistoric settlement patterns. *Arctic, 33*(2), 292–302.

Schlosser, P., G. Bönisch, M. Rhein and R. Bayer (1991). Reduction of deepwater formation in the Greenland Sea during the 1980s: evidence from tracer data. *Science, 251*, 1054–1056.

Schneider, W and G. Budéus (1994). The north-east water polynya (Greenland Sea). 1. A physical concept of its generation. *Polar Biology, 14*(1), 1–9.

Schneider, W. and G. Budéus (1995). On the generation of the northeast water polynya. *J. Geophys. Res., 100*(C3), 4269–4286.

Scholander, P.F. and D.C. Nutt (1960). Bubble pressure in Greenland icebergs. *J. Glaciol., 3*(28), 671–678.

Schramm, J.L., G.M. Flato and J.A. Curry (1999). Towards the modeling of enhanced basal melting in ridge keels. *J. Geophys. Res.*, in press.

Schumaker, J.D., K. Aagaard, C.H. Pease and R.B. Tripp (1983). Effects of a shelf polynya on flow and water properties in the northern Bering Sea. *J. Geophys. Res., 88*(C5), 2723–2732.

Schwarzacher, W. (1959). Pack ice studies in the Arctic Ocean. *J. Geophys. Res., 64*(12), 2357–2367.

Schwertfeger, P. (1963). The thermal properties of sea ice. *J. Glaciol., 4, 789–807.*

Scoresby, W. Jr. (1815). On the Greenland or Polar Ice. *Memoirs of the Wernerian Soc., 2*, 328–336. (Reprinted 1980, Caedmon of Whitby Press).

Scoresby, W. Jr. (1820). *An Account of the Arctic Regions, With a History and Description of the Northern Whale-Fishery.* Constable, Edinburgh, 2 vols. (Reprinted 1969, David and Charles, Newton Abbot).

Semtner, A.J. (1976). A model for the thermodynamic growth of sea ice in numerical investigations of climate. *J. Phys. Oceanogr., 6*(3), 379–389.

Serson, H.V. (1972). Investigation of a plug of multi-year old sea ice in the mouth of Nansen Sound. Defence Res. Establ., Ottawa, Tech. Note 72–6, 4pp.

Serson, H.V. (1974). Sverdrup Channel. Defence Res. Establ., Ottawa, Tech. Note 74–10.

Serreze, M.C., R.G. Barry and A.S. McLaren (1989). Seasonal variations in sea ice motion and effects on ice concentration in the Canada Basin. *J. Geophys. Res., 94*(C8), 10955–10970.

Shapiro, A. and L.S. Simpson (1953). The effect of a broken icefield on water waves. *Trans. Am. Geophys. U., 34*(1), 36–42.

Shapiro, M.A., T. Hampel and L.S. Fedor (1989). Research aircraft observations of an arctic front over the Barents Sea. In *Polar and Arctic Lows* (ed. P.F. Twitchell, E.A. Rasmussen and K. L. Davidson), A. Deepak, Hampton, Va., 279–289.

Shearer, J.M., R.F. MacNab, B.R. Pelletier and T.B. Smith (1971). Submarine pingos in the Beaufort Sea. *Science, 174*(4011), 816–818.

Shen, H.H. and S.F. Ackley (1991). A one-dimensional model for wave-induced ice-floe collisions. *Ann. Glaciol., 15*, 87–95.

Shen, H.H., W.D. Hibler III and M. Leppäranta (1987). The role of ice floe collisions in sea ice rheology. *J. Geophys. Res., 92*(C7), 7085–7096.

Shumskiy, P.A. (1964). *Principles of Structural Glaciology.* Dover, New York. 497pp.

Shumskiy, P.A., A.N. Krenke and I.A. Zotikov (1964). Ice and its changes. In *Solid Earth and Interface Phenomena, Res. in Geophysics*, vol. 2, 425–460, MIT Press, Cambridge, Mass.

Skinner, B.J. and K.K. Turekian (1973). *Man and the Ocean.* Prentice-Hall, Englewood Cliffs, N.J.

Smith, D.C., J.H. Morison, J.A. Johannessen and N. Untersteiner (1984). Topographic generation of an eddy at the edge of the East Greenland Current. *J. Geophys. Res., 89*(C5), 8205–8208.

Smith, R.A. (1978). Iceberg cleaving and fracture mechanics — a preliminary survey. In *Iceberg Utilization* (ed. A.A. Husseiny), 176–190, Pergamon, New York.

Smith, S.D. (1988). Coefficients of sea surface wind stress, heat flux, and wind profiles as a function of wind speed and temperature. *J. Geophys. Res., 93*(C12) 15467–15472.

Smith, S.D., E.G. Banke and O.M. Johannessen (1970). Wind stress and turbulence over ice in the Gulf of St. Lawrence. *J. Geophys. Res., 75*(15), 2803–2812.

Smith, S.D., R.D. Muench and C.H. Pease (1990). Polynyas and leads: an overview of physical processes and environment. *J. Geophys. Res., 95*(C6), 9461–9479.

Sodhi, D.S. and M. El-Tahan (1980). Prediction of iceberg drift trajectory during a storm. *Ann. Glaciol., 1*, 77–82.

Squire, V.A. (1983). Numerical modelling of realistic ice floes in ocean waves. *Ann. Glaciol., 4*, 277–282.

Squire, V.A., J. Dugan, P. Wadhams, A. Liu and P. Rottier (1995). Of ocean waves and sea ice. *Ann. Rev. Fluid Mech., 27*, 115–168.

Squire, V.A., R.J. Hosking, A.D. Kerr and P.J. Langhorne (1996). *Moving Loads on Ice Plates.* Kluwer, Dordrecht. 230pp.

Steele, M. and T. Boyd (1998). Retreat of the cold halocline in the Arctic Ocean. *J. Geophys. Res., 103*(C5), 10419–10435.

Stefan, J. (1890). Über die Theorie der Eisbildung, insbesondere über die Eisbildung im Polarmeere. *Sitzber. Akad. Wiss. Wien.* 98pp.

Steffen, K. (1985). Warm water cells in the North Water, northern Baffin Bay, during winter. *J. Geophys. Res., 90* (C5), 9129–9136.

Stirling, I. (1980). The biological importance of polynyas in the Canadian Arctic. *Arctic, 33*(2), 303–315.

Stouffer, R.J., S. Manabe and K. Bryan (1989). Interhemispheric asymmetry in climate response to a gradual increase of atmospheric CO_2. *Nature, 342*, 660–662.

Strass, V.H. and E. Fahrbach (1998). Temporal and regional variation of sea ice draft and coverage in the Weddell Sea obtained from upward looking sonars. In *Antarctic Sea Ice. Physical Processes, Interactions and Variability* (ed. M.O. Jeffries). Antarctic Res. Srs., 74, 123–140, Amer. Geophys. U., Washington.

Stringer, W.J. (1974). Sea ice morphology of the Beaufort shorefast ice. In *The Coast and Shelf of the Beaufort Sea* (ed. J.C. Reed, J.A. Sater), Arctic Inst. N. Amer., Arlington, 165–172.

Sturm, M., K. Morris and R.A. Massom (1998). The winter snow cover of the West Antarctic pack ice: its spatial and temporal variability. In *Antarctic Sea Ice. Physical Processes, Interactions and Variability* (ed. M.O. Jeffries). Antarctic Res. Srs., 74, 1–18, Amer. Geophys. U., Washington.

Sullivan, C.W., A.C. Palmisano, S.T. Kottmeier, S. McGrath Grossi and R. Moe (1985). The influence of light on growth and development of the sea-ice microbial community of McMurdo Sound. In *Antarctic Nutrient Cycles and Food Webs* (ed. W.R. Siegfried, P.R. Condy and R.M. Laws), Springer, Berlin, 84–88.

Sverdrup, H.U. (1926). Dynamics of tides on the northern Siberian shelf. *Geophys. Publ., 4, 5.* Oslo, 76pp.

Swithinbank, C.W.M., E.P. McClain and P. Little (1977). Drift tracks of Antarctic icebergs. *Polar Rec., 18*(116), 495–501.

Tchernia, P. and P.F. Jeannin (1984). Circulation in Antarctic waters as revealed by iceberg tracks 1972–1983. *Polar Rec., 22(138),* 263–269.

Thomas, C.W. (1963). On the transfer of visible radiation through sea ice and snow. *J. Glaciol., 4*(34), 481–484.

Thompson, D.W.J. and J.M. Wallace (1998). The Arctic Oscillation signature in the wintertime geopotential height and temperature fields. *Geophys. Res. Lett., 25,* 1297–1300.

Thorndike, A.S. and R. Colony (1980). Large-scale ice motion in the Beaufort Sea during AIDJEX, April 1975 – April 1976. In In *Sea Ice Processes and Models* (ed. R.S. Pritchard), Univ. Washington Press, Seattle, 249–260.

Thorndike, A.S. and R. Colony (1982). Sea ice motion in response to geostrophic winds. *J. Geophys. Res., 87*(C8), 5845–5852.

Thorndike, A.S., D.A. Rothrock, G.A. Maykut and R. Colony (1975). The thickness distribution of sea ice. *J. Geophys. Res., 80*(33), 4501–4513.

Thorndike, A.S., C. Parkinson and D.A. Rothrock (eds) (1992). Report of the Sea Ice Thickness Workshop, 19–21 November 1991, New Carrollton, Maryland. Polar Science Center, Applied Physics Lab., Univ. Washington., Seattle.

Tonge, A.M. (1992). An incremental approach to autonomous underwater vehicles. Proc. Oceanology International '92, Brighton, 10–13 March 1992.

Trombetta-Panigadi, F. (1996). The exploitation of Antarctic icebergs in international law. In *International Law for Antarctica* (ed. F. Francioni, T. Scovazzi), Kluwer Law International, The Hague, 2nd ed., 225–260.

Tsang, G. (1982). Frazil and anchor ice - a monograph. Nat. Water Res. Inst., Canada Centre for Inland Waters, Burlington. Nat. Res. Counc., Ottawa. 90pp.

Tsatsoulis, C. and R. Kwok (eds.) (1998). *Analysis of SAR Data of the Polar Oceans. Recent Advances.* Springer, Berlin. 290pp.

Tucker W.B. III (1989). An overview of the physical properties of sea ice. Proceedings of Workshop on Ice Properties, Associate Committee on Geotechnical Research, National Research Council Canada (June 21–22, 1988, St. Johns, Newfoundland). Technical Memorandum No. 144, NRCC 30358.

Tucker, W.B. III and D.K. Perovich (1992). Stress measurements in drifting pack ice. *Cold Regions Sci. & Technol., 20,* 119–139.

Tucker, W.B. III, A.J. Gow and W.F. Weeks (1987). Physical properties of summer sea ice in the Fram Strait. *J. Geophys. Res., 92*(C7), 6787–6803.

Tucker, W.B. III, W.F. Weeks and M. Frank (1979). Sea ice ridging over the Alaskan continental shelf. *J. Geophys. Res., 84*(C8), 4885–4897.

Tucker, W.B. III, W.F. Weeks, A. Kovacs and A.J. Gow (1980). Nearshore ice motion at Prudhoe Bay, Alaska. In *Sea Ice Processes and Models* (ed. R.S. Pritchard), Univ. Washington Press, Seattle, 261–272.

Turner, D., N. Owens and J. Priddle (eds.) (1995). Southern Ocean JGOFS: the U.K. "Sterna" study in the Bellingshausen Sea. *Deep-Sea Res. II, 42*(4–5), 905–1335.

Untersteiner, N. (1961). On the mass and heat budget of arctic sea ice. *Arch. Met. Geophys. Bioklim., A*(12), 151–182.

Untersteiner, N. (1964). Calculations of temperature regime and heat budget of sea ice in the Central Arctic. *J. Geophys. Res., 69*(22), 4755–4766.

Untersteiner, N. (1968). Calculating the thermal regime and mass budget of sea ice. In *Proc. Symp. on the Arctic Heat Budget and Atmospheric Circulation*, ed. J.O. Fletcher, 203–214. RAND Corp., Santa Monica, Rept. RM-5233–NSF.

Untersteiner, N. (1984). The cryosphere. In *The Global Climate* (ed. J.T. Houghton), Cambridge Univ. Press, Cambridge, 121–140.

Ursell, F. (1947). The effect of a fixed vertical barrier on surface waves in deep water. *Proc. Camb. phil. Soc. math. phys. Sci., 43*, 374–382.

Ushio, S., and M. Wakatsuchi, (1993). A laboratory study on supercooling and frazil ice production processes in winter coastal polynyas. *J. Geophys. Res., 98*(C11), 20321–20328.

USNOO (1963–71). Project BIRDS EYE reports. Informal Repts. IR 010–63 to IR 71–10, US Naval Oceanogr. Office, Washington D.C.

Vinje, T.E. (1977a). Sea ice studies in the Spitzbergen-Greenland area. Landsat Rept. E77–10206, US Dept. Commerce, Natl. Tech. Info. Service, Springfield, Va.

Vinje, T.E. (1977b). Sea ice conditions in the European sector of the marginal seas of the Arctic, 1966–75. *Norsk Polarinstitutt Årbok 1975*, 163–174.

Vinje, T.E. (1979). On the drift ice conditions in the Atlantic sector of the Antarctic. In *Proc. 5th Int;l. Conf. Port & Ocean Engng. under Arctic Condns., 3*, 75–82. Norwegian Inst. Technol., Univ. Trondheim.

Vinje, T.E. (1982). A grounded iceberg in Fram Strait. *Polar Rec., 21*(131), 174–175.

Vinje, T.E. (1989). An upward looking sonar ice draft series. In *Proc. 10th Intl. Conf. Port & Ocean Engng. under Arctic Condns.* (ed. K.B.E. Axelsson, L.A. Fransson). Luleå Univ. Technology, *1*, 178–187.

Vinje, T.E., M. Nyborg and G. Kjaernli (1996). Sea-ice variation in the Greenland Sea during the nineteenth century. In *ESOP. European Subpolar Ocean Programme. Sea ice - ocean interactions* (ed. P. Wadhams, J.P. Wilkinson and S.C.S. Wells). Rept. on EU Programme MAS2–CT93–0057, Scott Polar Res. Inst., Cambridge, *1*, 101–103.

Vinje, T.E., N. Nordlund and A. Kvambekk (1998). Monitoring ice thickness in Fram Strait. *J. Geophys. Res., 103*(C5), 10437–10449.

Wadhams, P. (1971). "Hudson-70" Canadian Oceanographic Expedition 1969–70. *Polar Rec., 15*(97), 524–526.

Wadhams, P. (1972). Measurement of wave attenuation in pack ice by inverted echo sounding. In *Sea Ice* (T. Karlsson, ed.), Nat. Res. Counc. of Iceland, Reykjavik, 255–260.

Wadhams, P. (1973). The effect of a sea ice cover on ocean surface waves. PhD thesis, Univ. Cambridge.

Wadhams, P. (1975). Airborne laser profiling of swell in an open ice field. *J. Geophys. Res.*, *80*(33), 4520–4528.

Wadhams, P. (1976a). Sea ice topography in the Beaufort Sea and its effect on oil containment. *AIDJEX Bull.*, *33*, 1–52. Div. Marine Resources, Univ. Washington, Seattle.

Wadhams, P. (1976b). Oil and ice in the Beaufort Sea. *Polar Record*, *18*(114), 237–250.

Wadhams P. (1978a). Characteristics of deep pressure ridges in the Arctic Ocean. *POAC 77*, Proceedings of the Fourth International Conference on Port and Ocean Engineering under Arctic Conditions, St. John's, *1*, 544 - 555. Memorial Univ., St. John's., Nfld.

Wadhams, P. (1978b). Wave decay in the marginal ice zone measured from a submarine. *Deep-Sea Res.*, *25*, 23–40.

Wadhams, P. (1979). Field experiments on wave-ice interaction in the Labrador and East Greenland Currents, 1978. *Polar Rec.*, *19*(121), 373–376.

Wadhams, P. (1980). A comparison of sonar and laser profiles along corresponding tracks in the Arctic Ocean. In *Sea Ice Processes and Models* (ed. R.S. Pritchard), Univ. Washington Press, Seattle, 283–299.

Wadhams, P. (1981). Sea ice topography of the Arctic Ocean in the region 70°W to 25°E, *Phil. Trans. Roy. Soc., Lond.*, *A302*(1464), 45–85, 1981.

Wadhams, P. (1981b). Oil and ice in the Beaufort Sea - the physical effects of a hypothetical blowout. In *Petroleum and the Marine Environment*, Graham and Trotman, London, 299–318.

Wadhams, P. (1981c). The ice cover in the Greenland and Norwegian Seas. *Rev. Geophys.. Space Phys.*, *19*(3), 345–393.

Wadhams, P. (1982). Icebergs for sale. *Sci. Now*, *2*, 38–41.

Wadhams, P. (1983). Sea ice thickness distribution in Fram Strait. *Nature*, *305*, 108–111.

Wadhams, P. (1983b). The prediction of extreme keel depths from sea ice profiles. *Cold Regions Sci. Technol.*, *6*, 257–266.

Wadhams, P. (1983c). A mechanism for the formation of ice edge bands. *J. Geophys. Res.*, *88*(C5), 2813–2818.

Wadhams, P. (1986). The seasonal ice zone. In *The Geophysics of Sea Ice* (ed. N. Untersteiner), Plenum Press, New York, 825–991.

Wadhams, P. (1988). Winter observations of iceberg frequencies and sizes in the South Atlantic Ocean. *J. Geophys. Res.*, *93*(C4), 3583–3590.

Wadhams, P. (1989). Sea-ice thickness in the Trans Polar Drift Stream. *Rapp. P-v Reun Cons. Int. Explor. Mer*, *188*, 59–65.

Wadhams, P. (1990a). Evidence for thinning of the Arctic ice cover north of Greenland. *Nature, 345*, 795–797.

Wadhams, P. (1990b). Ice thickness distribution in the Arctic Ocean. In *Ice Technology for Polar Operations* (T.K.S. Murthy, J.G. Paren, W.M. Sackinger, P. Wadhams, eds.), Computational Mechanics Publns., Southampton, 3–20.

Wadhams, P. (1990c). Sea ice and economic development in the Arctic Ocean - a glaciologist's experience. In *Arctic Technology and Economy. Present situation and problems, future issues*. Bureau Veritas, Paris, 1–23.

Wadhams, P. (1991). Atmosphere-ice-ocean interaction in the Antarctic. In *Antarctica And Global Climatic Change* (C. Harris, B. Stonehouse, eds.), Belhaven Press, London, 65–81.

Wadhams P. (1992). Sea ice thickness distribution in the Greenland Sea and Eurasian Basin, May 1987. *J. Geophys. Res., 97*(C4), 5331–5348.

Wadhams, P. (1997a). Ice thickness in the Arctic Ocean: the statistical reliability of experimental data. *J. Geophys. Res., 102* (C13), 27951–27959.

Wadhams, P. (1997b). Variability of Arctic sea ice thickness - statistical significance and its relationship to heat flux. In *Operational Oceanography. The Challenge for European Co-operation* (ed. J.H. Stel, H.W.A. Behrens, J.C. Borst, L.J. Droppert, J.P. Van der Meulen). Elsevier Oceanogr. Srs. 62, Amsterdam, 368–384.

Wadhams, P. (1999). The Odden ice tongue and Greenland Sea convection. *Weather, 54*(3), 83–84, 91–97.

Wadhams, P. and J.C. Comiso (1992). The ice thickness distribution inferred using remote sensing techniques. In *Microwave Remote Sensing of Sea Ice* (ed. F. Carsey), Geophysical Monograph 68, Amer. Geophys. U., Washington, ch. 21, pp. 375–383.

Wadhams, P. and D.R. Crane (1991). SPRI participation in the Winter Weddell Gyre Study 1989. *Polar Record, 27*(160), 29–38.

Wadhams, P. and N.R. Davis (1994). The fractal properties of the underside of Arctic sea ice. In *Marine, Offshore and Ice Technology* (ed. T.K.S. Murthy, P.A. Wilson and P. Wadhams). Computational Mechanics Publns., Southampton, 353–363.

Wadhams, P. and T. Davy (1986). On the spacing and draft distributions for pressure ridge keels. *J. Geophys. Res., 91*(C9), 10697–10708.

Wadhams, P. and B. Holt (1991). Waves in frazil and pancake ice and their detection in Seasat synthetic aperture radar imagery. *J. Geophys. Res., 96*(C5), 8835–8852.

Wadhams, P. and R.J. Horne (1980). An analysis of ice profiles obtained by submarine sonar in the Beaufort Sea. *J. Glaciol., 25*(93), 401–424.

Wadhams, P. and R.T. Lowry (1977). A joint topside-bottomside remote sensing experiment on Arctic sea ice. *Proc. 4th Canadian Symp. on Remote Sensing, Quebec, 16–18 May 1977.* Canadian Remote Sensing Soc., 407–423.

Wadhams, P. and S. Martin (1990). Processes determining the bottom topography of multiyear Arctic sea ice. In *Sea Ice Properties and Processes* (ed. S.F. Ackley and W.F. Weeks), US Army Cold Regions Res. & Engng Lab., Hanover N.H., Monograph 90–1, 136–141.

Wadhams, P. and V.A. Squire (1983). An ice-water vortex at the edge of the East Greenland Current. *J. Geophys. Res., 88*(C5), 2770–2780.

Wadhams, P,. and T. Viehoff (1993). The Odden ice tongue in the Greenland Sea: SAR imagery and field observations of its development in 1993. *Proc. 2nd ERS-1 Symp. — Space at the Service of our Environment, Hamburg, 11–14 October 1993.* ESA, Paris, SP361, 2, 1285–1294.

Wadhams, P. and J.P. Wilkinson (1999).The physical properties of sea ice in the Odden ice tongue. *Deep-Sea Res. II, 46*(6–7), 1275–1300.

Wadhams, P., A.E. Gill and P.F. Linden (1979). Transect by submarine of the East Greenland Polar Front. *Deep-Sea Res., A26*(12), 1311–1328.

Wadhams, P., M. Kristensen and O. Orheim (1983). The response of Antarctic icebergs to ocean waves. *J. Geophys. Res., 88*(C10), 6053–6065.

Wadhams, P., A.S. McLaren and R. Weintraub (1985). Ice thickness distribution in Davis Strait in February from submarine sonar profiles. *J. Geophys. Res., 90*(C1), 1069–1077.

Wadhams, P., V.A. Squire, J.A. Ewing and R.W. Pascal (1986). The effect of the marginal ice zone on the directional wave spectrum of the ocean. *J. Phys. Oceanogr., 6*(2), 358–376.

Wadhams, P., M.A. Lange and S.F. Ackley (1987). The ice thickness distribution across the Atlantic sector of the Antarctic Ocean in midwinter. *J. Geophys. Res., 92*(C13), 14535–14552.

Wadhams, P., V.A. Squire, D.J. Goodman, A.M. Cowan and S.C. Moore (1988). The attenuation rates of ocean waves in the marginal ice zone. *J. Geophys. Res., 93*(C6), 6799–6818.

Wadhams, P., J.C. Comiso, J. Crawford, G. Jackson, W. Krabill, R. Kutz, C.B. Sear, R. Swift, W.B. Tucker and N. R. Davis (1991). Concurrent remote sensing of Arctic sea ice from submarine and aircraft. *Int. J. Remote Sensing, 12*(9), 1829–1840.

Wadhams, P., W.B. Tucker III, W.B. Krabill, R.N. Swift, J.C. Comiso and N.R. Davis (1992). Relationship between sea ice freeboard and draft in the Arctic Basin, and implications for ice thickness monitoring. *J. Geophys. Res., 97*(C12), 20325–20334.

Wadhams, P., J.C. Comiso, E. Prussen, S. Wells, M. Brandon, E. Aldworth, T. Viehoff, R. Allegrino and D.R. Crane (1996). The development of the Odden ice tongue in the Greenland Sea during winter 1993 from remote sensing and field observations. *J. Geophys. Res., 101*(C8), 18213–18235.

Wadhams, P., J-C. Gascard and L. Miller (eds) (1999). The European Subpolar Ocean Programme: ESOP. *Deep-Sea Res. II, 46*(6–7), 1011–1530.

Wakasutchi, M. and K.I. Ohshima (1990). Observations of ice-ocean eddy streets in the Sea of Okhotsk off the Hokkaido coast using radar images. *J. Phys. Oceanogr., 20*(4), 585–599.

Walker, E.R. and P. Wadhams (1979). Thick sea-ice floes. *Arctic, 32*(2), 140–147.

Walsh, J.E. and W.L. Chapman (1991). Model simulation of changes in Arctic sea ice thickness, 1960–1989. Proc. 5th AMS Conf. on Climate Variations, Denver, October 1991.

Weber, J.E. (1987). Wave attenuation and wave drift in the marginal ice zone. *J. Phys. Oceanogr., 17*(12), 2351–2361.

Weber, J.R. and M. Erdelyi (1976). Ice and ocean tilt measurements in the Beaufort Sea. *J. Glaciol., 17(75),* 61–71.

Weeks, W.F. and S.F. Ackley (1986). The growth, structure, and properties of sea ice. In *The Geophysics of Sea Ice* (ed. N. Untersteiner), Plenum, New York, 9–164.

Weeks, W.F. and W.J. Campbell (1973). Icebergs as a fresh water source: an appraisal. *J. Glaciol., 12*(65), 207–232.

Weeks, W.F. and A.J. Gow (1978). Preferred crystal orientations in the fast ice along the margins of the Arctic Ocean. *J. Geophys. Res., 83*(C10), 5105–5121.

Weeks, W.F. and M. Mellor (1978). Some elements of iceberg technology. In *Iceberg Utilization* (ed. A.A. Husseiny), Pergamon, New York, 45–98.

Weeks, W.F., A. Kovacs and W.D. Hibler III (1971). Pressure ridge characteristics in the Arctic coastal environment. *Proc. 1st Intl. Conf. Port & Ocean Engng. under Arctic Condns* (ed. S.S. Wetteland, P. Bruun), Tech. Univ. Norway, Trondheim, 152–183.

Weeks, W.F., P.W. Barnes, D.M. Rearic and E. Reimnitz (1984). Some probabilistic aspects of ice gouging on the Alaskan Shelf of the Beaufort Sea. In *The Alaskan Beaufort Sea — Ecosystems and Environments* (ed. P.W. Barnes, D. Schell and E. Reimnitz), Academic Press, London, 213–236.

Weeks, W.F., S.F. Ackley and J. Govoni (1989). Sea ice ridging in the Ross Sea, Antarctica, as compared with sites in the Arctic. *J. Geophys. Res., 94*(C4), 4984–4988.

Weitz, M. and J.B. Keller (1950). Reflection of water waves from floating ice in water of finite depth. *Communs. Pure Appl. Math., 3*(3), 305–318.

Weller, G.E. (1968). The heat budget and heat transfer processes in Antarctic plateau ice and sea ice. ANARE Sci. Repts., Srs. A(IV), Glaciology, Publn. 102, 155pp.

Weller, G.E. and P. Schwerdtfeger (1967). Radiation penetration in antarctic plateau and sea ice. In *Polar Meteorol., World Meteorol. Org. Tech. Note 87*, 120–141.

Wheeler, P.A. (ed.) (1997). 1994 Arctic Ocean section. *Deep-Sea Res. II, 44*(8), 1487–1757.

White, W.B. and R.G. Peterson (1996). An Antarctic circumpolar wave in surface pressure, wind, temperature and sea-ice extent. *Nature, 380*, 699–702.

Williams, C.B. (1940). A note on the statistical analysis of sentence-length as a criterion of literary style. *Biometrika, 31*, 356.

Williams, P.D.L. (1975). Limitations of radar techniques for the detection of small surface agents in clutter. *Radio & Electron. Eng., 45*(8), 379–389.

Williams, P.D.L. (1979). The detection of ice at sea by radar. *Radio & Electron. Eng., 49*(6), 275–287.

Wingham, D.J. (1999). The first of ESA's first Opportunity Missions: Cryosat. *Earth Obsn. Quarterly*, ESA, Noordwijk, no. 63, 21–24.

Winsor, P. (1997). A wind-driven coastal polynya model. MSc thesis, Institutionen for geovetenskaper, Gøteborgs Universitet.

WMO (1970). WMO sea ice nomenclature. WMO, Geneva, Rep. 259., T.P. 145, 147 pp + 8 suppl.

Wood, R.A., A.B. Keen, J.F.B. Mitchell and J.M. Gregory (1999). Changing spatial structure of the thermohaline circulation in response to atmospheric CO_2 forcing in a climate model. *Nature, 399*, 572–575.

Woodworth-Lynas, C.M.L. and J.Y. Guigné (1990). Iceberg scours in the geological record: examples from glacial Lake Agassiz. *Glaciomarine Environments: Processes and Sediments*. Geological Soc., Special Publn. *53*, 217–233.

Woodworth-Lynas, C.M.L., A. Simms and C.M. Rendell (1984). Grounding and scouring icebergs on the Labrador Shelf. *Iceberg Res., 7*, 13–20. Scott Polar Res. Inst., Cambridge.

Woodworth-Lynas, C.M.L., J.D. Nixon, R. Phillips and A.C. Palmer (1996). Subgouge deformation and the safety of Arctic marine pipelines. *Proc. 28th Annual Offshore Technol. Conf., Houston, 3*, 235–244.

Wordie, J.M. and S. Kemp (1933). Observations on certain Antarctic icebergs. *Geogr. J., 81*, 426–434.

Yen, Y.C. (1981). Review of thermal properties of snow, ice and sea ice. US Army Cold Regions Res. & Engng. Lab., Hanover N.H., Res. Rept. 81–10, 27pp.

Zhang, J. and W.D. Hibler III (1997). On an efficient numerical method for modeling sea ice dynamics. *J. Geophys. Res., 102*(C4), 8691–8702.

Zubov N.N. (1945). *Arctic Ice.* Izdatel'stvo Glavsermorputi, Moscow. (Translation AD426972, National Technical Information Service, Springfield, Va.)

Zwally, H.J. and P. Gloersen (1977). Passive-microwave images of the polar regions and research applications. *Polar Rec., 18*(116), 431–450.

Zwally, H.J., T.T. Wilheit, P. Gloersen and J.L. Mueller (1976). Characteristics of Antarctic sea ice as determined by satellite-borne microwave imagers. In *Proc. Symp. on Met. Observations from Space: Their Contribution to the first GARP Global Expt.* Committee on Space Res., Int. Counc. Sci. Unions, Philadelphia, 94–97.

Zwally, H.J., J.C. Comiso, C.L. Parkinson, W.J. Campbell, F.D. Carsey and P. Gloersen (1983). *Antarctic Sea Ice 1973–1976: Satellite Passive Microwave Observations.* NASA, Washington D.C., Rept. SP-459.

Zwally, H.J., J.C. Comiso and A.L. Gordon (1985). Antarctic offshore leads and polynyas and oceanographic effects. In *Oceanology of the Antarctic Continental Shelf* (S.S. Jacobs, ed.), Antarctic Res. Srs. 43, American Geophysical Union, Washington, 203–226.

INDEX